PROBLEMS AND SOLUTIONS IN BIOLOGICAL SEQUENCE ANALYSIS

MARK BORODOVSKY AND SVETLANA EKISHEVA

CAMBRIDGE
UNIVERSITY PRESS

CAMBRIDGE UNIVERSITY PRESS
Cambridge, New York, Melbourne, Madrid, Cape Town, Singapore, São Paulo

Cambridge University Press
The Edinburgh Building, Cambridge CB2 2RU, UK

Published in the United States of America by Cambridge University Press, New York

www.cambridge.org
Information on this title: www.cambridge.org/9780521847544

First published 2006

Printed in the United Kingdom at the University Press, Cambridge

A catalogue record for this publication is available from the British Library

ISBN-13 978-0-521-84754-4 hardback
ISBN-10 0-521-84754-0 hardback

ISBN-13 978-0-521-61230-2 paperback
ISBN-10 0-521-61230-6 paperback

M. B.:
To Richard and Judy Lincoff

S. E.:
To Sergey and Natasha

Contents

Preface

Bioinformatics, an integral part of post-genomic biology, creates principles and ideas for computational analysis of biological sequences. These ideas facilitate the conversion of the flood of sequence data unleashed by the recent information explosion in biology into a continuous stream of discoveries. Not surprisingly, the new biology of the twenty-first century has attracted the interest of many talented university graduates with various backgrounds. Teaching bioinformatics to such a diverse audience presents a well-known challenge. The approach requiring students to advance their knowledge of computer programming and statistics prior to taking a comprehensive core course in bioinformatics has been accepted by many universities, including the Georgia Institute of Technology, Atlanta, USA.

In 1998, at the start of our graduate program, we selected the then recently published book *Biological Sequence Analysis (BSA)* by Richard Durbin, Anders Krogh, Sean R. Eddy, and Graeme Mitchison as a text for the core course in bioinformatics. Through the years, *BSA*, which describes the ideas of the major bioinformatic algorithms in a remarkably concise and consistent manner, has been widely adopted as a required text for bioinformatics courses at leading universities around the globe. Many problems included in *BSA* as exercises for its readers have been repeatedly used for homeworks and tests. However, the detailed solutions to these problems have not been available. The absence of such a resource was noticed by students and teachers alike.

The goal of this book, *Problems and Solutions in Biological Sequence Analysis* is to close this gap, extend the set of workable problems, and help its readers develop problem-solving skills that are vitally important for conducting successful research in the growing field of bioinformatics. We hope that this book will facilitate understanding of the content of the *BSA* chapters and also will provide an additional perspective for in-depth *BSA* reading by those who might not be able to take a formal bioinformatics course. We have augmented the set of original *BSA* problems with many new problems, primarily those that were offered to the Georgia Tech graduate students.

Probabilistic modeling and statistical analysis are frequently used in bioinformatics research. The mainstream bioinformatics algorithms, those for pairwise and multiple sequence alignment, gene finding, detecting orthologs, and building phylogenetic trees, would not work without rational model selection, parameter estimation, properly justified scoring systems, and assessment of statistical significance. These and many other elements of efficient bioinformatic tools require one to take into account the random nature of DNA and protein sequences.

As it has been illustrated by the *BSA* authors, probabilistic modeling laid the foundation for the development of powerful methods and algorithms for biological sequence interpretation and the revelation of its functional meaning and evolutionary connections. Notably, probabilistic modeling is a generalization of strictly deterministic modeling, which has a remarkable tradition in natural science. This tradition could be traced back to the explanation of astronomic observations on the motion of solar system planets by Isaac Newton, who suggested a concise model combining the newly discovered law of gravity and the laws of dynamics.

The maximum likelihood principle of statistics, notwithstanding the fashion of its traditional application, also has its roots in "deterministic" science that suggests that the chosen structure and parameters of a theoretical model should provide the best match of predictions to experimental observations. For instance, one could recognize the maximum likelihood approach in Francis Crick and James Watson's inference of the DNA double helix model, chosen from the combinatorial number of biochemically viable alternatives as the best fit to the X-ray data on DNA three-dimensional structure and other experimental data available.

In studying the processes of inheritance and molecular evolution, where random factors play important roles, fully fledged probabilistic models enter the picture. A classic cycle of experiments, data analysis, and modeling with search for a best fit of the models to data was designed and implemented by Gregor Mendel. His remarkable long term research endeavor provided proof of the existence of discrete units of inheritance, the genes.

When we deal with data coming from a less controllable environment, such as data on natural biological evolution spanning time periods on a scale of millions of years, the problem is even more challenging. Still, the situation is hopeful. The models of molecular evolution proposed by Dayhoff and co-authors, Jukes and Cantor, and Kimura, are classical examples of fundamental advances in modeling of the complex processes of DNA and protein evolution. Notably these models focus on only a single site of a molecular sequence and require the further simplifying assumption that evolution of sequence sites occurs independently from each other. Nevertheless, such models are useful starting points for understanding the function and evolution of biological sequences as well as for designing algorithms elucidating these functional and evolutionary connections.

For instance, amino acid substitution scores are critically important parameters of the optimal global (Needleman and Wunsch) and local (Smith and Waterman) sequence alignment algorithms. Biologically sensible derivation of the substitution scores is impossible without models of protein evolution.

In the mid 1990s the notion of the hidden Markov model (HMM), having been of great practical use in speech recognition, was introduced to bioinformatics and quickly entered the mainstream of the modeling techniques in biological sequence analysis.

Theoretical advances that have occurred since the mid 1990s have shown that the sequence alignment problem has a natural probabilistic interpretation in terms of hidden Markov models. In particular, the dynamic programming (DP) algorithm for pairwise and multiple sequence alignment has the HMM-based algorithmic equivalent, the Viterbi algorithm. If the type of probabilistic model for a biological sequence has been chosen, parameters of the model could be inferred by statistical (machine learning) methods. Two competitive models could be compared to identify the one with the best fit.

The events and selective forces of the past, moving the evolution of biological species, have to be reconstructed from the current biological sequence data containing significant noise caused by all the changes that have occurred in the lifetime of disappeared generations. This difficulty can be overcome to some extent by the use of the general concept of self-consistent models with parameters adjusted iteratively to fit the growing collection of sequence data. Subsequently, implementation of this concept requires the expectation–maximization type algorithms able to estimate the model parameters simultaneously with rearranging data to produce the data structure (such as a multiple alignment) that fits the model better. *BSA* describes several algorithms of expectation–maximization type, including the self-training algorithm for a profile HMM and the self-training algorithm for a phylogenetic HMM. Given that the practice with many algorithms described in *BSA* requires significant computer programming, one may expect that describing the solutions would lead us into heavy computer codes, thus moving far away from the initial concepts and ideas. However, the majority of the *BSA* exercises have analytical solutions. On several occasions we have illustrated the implementations of the algorithms by "toy" examples. The computer codes written in C++ and Perl languages for such examples are available at opal.biology.gatech.edu/PSBSA. Note, that in the "Further reading" sections we include mostly papers that were published later than 1998, the year of *BSA* publication. Finally, we should mention that the references in the text to the pages in the BSA book cite the 2006 edition.

Acknowledgements

We thank Sergey Latkin, Svetlana's husband, for the remarkable help with preparation of LaTex figures and tables. We are grateful to Alexandre Lomsadze, Ryan Mills, Yuan Tian, Burcu Bakir, Jittima Piriyapongsa, Vardges Ter-Hovhannisyan, Wenhan Zhu, Jeffrey Yunes, and Matthew Berginski for invaluable technical assistance in preparation of the book materials; to Soojin Yi, and Galina Glazko for useful references on molecular evolution; to Michael Roytberg for helpful discussions on transformational grammars and finite automata. We cordially thank our editor Katrina Halliday for tremendous patience and constant support, without which this book would never have come to fruition. We are especially grateful to Richard Durbin, Anders Krogh, Sean R. Eddy, and Graeme Mitchison, for encouragement, helpful criticism and suggestions. Further, it is our pleasure to acknowledge firm support from the Georgia Tech School of Biology and the Wallace H. Coulter Department of Biomedical Engineering at Georgia Tech and Emory University. Finally, we wish to express our particular gratitude to our families for great patience and constant understanding.

M.B. and S.E.

1

Introduction

The reader will quickly discover that the organization of this book was chosen to be parallel to the organization of *Biological Sequence Analysis* by Durbin *et al.* (1998). The first chapter of *BSA* contains an introduction to the fundamental notions of biological sequence analysis: sequence similarity, homology, sequence alignment, and the basic concepts of probabilistic modeling.

Finding these distinct concepts described back-to-back is surprising at first glance. However, let us recall several important bioinformatics questions. How could we construct a pairwise sequence alignment? How could we build an alignment of multiple sequences? How could we create a phylogenetic tree for several biological sequences? How could we predict an RNA secondary structure? None of these questions can be consistently addressed without use of probabilistic methods. The mathematical complexity of these methods ranges from basic theorems and formulas to sophisticated architectures of hidden Markov models and stochastic grammars able to grasp fine compositional characteristics of empirical biological sequences.

The explosive growth of biological sequence data created an excellent opportunity for the meaningful application of discrete probabilistic models. Perhaps, without much exaggeration, the implications of this new development could be compared with implications of the revolutionary use of calculus and differential equations for solving problems of classic mechanics in the eighteenth century.

The problems considered in this introductory chapter are concerned with the fundamental concepts that play an important role in biological sequence analysis: the maximum likelihood and the maximum *a posteriori* (Bayesian) estimation of the model parameters. These concepts are crucial for understanding statistical inference from experimental data and are impossible to introduce without notions of conditional, joint, and marginal probabilities.

The frequently arising problem of model parameterization is inherently difficult if only a small training set is available. One may still attempt to use methods suitable for large training sets. But this move may result in overfitting and the generation of biased parameter estimates. Fortunately, this bias can be eliminated to some degree; the model can be generalized as the training set is augmented by artificially introduced observations, pseudocounts.

Problems included in this chapter are intended to provide practice with utilizing the notions of marginal and conditional probabilities, Bayes' theorem, maximum likelihood, and Bayesian parameter estimation. Necessary definitions of these notions and concepts frequently used in *BSA* can be found in undergraduate textbooks on probability and statistics (for example, Meyer (1970), Larson (1982), Hogg and Craig (1994), Casella and Berger (2001), and Hogg and Tanis (2005)).

1.1 Original problems

Problem 1.1 Consider an occasionally dishonest casino that uses two kinds of dice. Of the dice 99% are fair but 1% are loaded so that a six comes up 50% of the time. We pick up a die from a table at random. What are $P(\text{six}|D_{\text{loaded}})$ and $P(\text{six}|D_{\text{fair}})$? What are $P(\text{six}, D_{\text{loaded}})$ and $P(\text{six}, D_{\text{fair}})$? What is the probability of rolling a six from the die we picked up?

Solution All possible outcomes of a fair die roll are equally likely, i.e. $P(\text{six}|D_{\text{fair}}) = 1/6$. On the other hand, the probability of rolling a six from the loaded die, $P(\text{six}|D_{\text{loaded}})$, is equal to $1/2$. To compute the probability of the combined event $(\text{six}, D_{\text{loaded}})$, rolling a six and picking up a loaded die, we use the definition of conditional probability:

$$P(\text{six}, D_{\text{loaded}}) = P(D_{\text{loaded}})P(\text{six}|D_{\text{loaded}}). \qquad (1.1)$$

As the probability of picking up a loaded die is $1/100$, Equality (1.1) yields

$$P(\text{six}, D_{\text{loaded}}) = \frac{1}{100} \times \frac{1}{2} = \frac{1}{200}.$$

By a similar argument,

$$P(\text{six}, D_{\text{fair}}) = P(\text{six}|D_{\text{fair}})P(D_{\text{fair}}) = \frac{1}{6} \times \frac{99}{100} = \frac{33}{200}.$$

The probability of rolling a six from the die picked up at random is computed as the total probability of event "six" occurring in combination either with event D_{loaded} or with event D_{fair}:

$$P(\text{six}) = P(\text{six}, D_{\text{loaded}}) + P(\text{six}, D_{\text{fair}}) = \frac{34}{200} = \frac{17}{100}. \qquad \square$$

Problem 1.2 How many sixes in a row would we need to see in Problem 1.1 before it is more likely that we had picked a loaded die?

Solution Bayes' theorem is all we need to determine the conditional probability of picking up a loaded die, $P(D_{\text{loaded}}|n \text{ sixes})$, given that n sixes in a row have been rolled:

$$P(D_{\text{loaded}}|n \text{ sixes}) = \frac{P(n \text{ sixes}|D_{\text{loaded}})P(D_{\text{loaded}})}{P(n \text{ sixes})}$$

$$= \frac{P(n \text{ sixes}|D_{\text{loaded}})P(D_{\text{loaded}})}{P(n \text{ sixes}|D_{\text{loaded}})P(D_{\text{loaded}}) + P(n \text{ sixes}|D_{\text{fair}})P(D_{\text{fair}})}.$$

Rolls of both fair or loaded dice are independent, therefore

$$P(D_{\text{loaded}}|n \text{ sixes}) = \frac{(1/100) \times (1/2)^n}{(99/100) \times (1/6)^n + (1/100) \times (1/2)^n} = \frac{1}{11 \times (1/3)^{n-2} + 1}.$$

This result indicates that $P(D_{\text{loaded}}|n \text{ sixes})$ approaches one as n, the length of the observed run of sixes, increases. The inequality

$$P(D_{\text{loaded}}|n \text{ sixes}) > 1/2$$

tells us that it is more likely that a loaded die was picked up. This inequality holds if

$$\left(\frac{1}{3}\right)^{n-2} < \frac{1}{11}, \quad n \geq 5.$$

Therefore, seeing five or more sixes in a row indicates that it is more likely that the loaded die was picked up. □

Problem 1.3 Use the definition of conditional probability to prove Bayes' theorem,

$$P(X|Y) = \frac{P(X)P(Y|X)}{P(Y)}.$$

Solution For any two events X and Y such that $P(Y) > 0$ the conditional probability of X given Y is defined as

$$P(X|Y) = \frac{P(X \cap Y)}{P(Y)}.$$

Applying this definition once again to substitute $P(X \cap Y)$ by $P(X)P(Y|X)$, we arrive at the equation which is equivalent to Bayes' theorem:

$$P(X|Y) = \frac{P(X)P(Y|X)}{P(Y)}.$$ □

Problem 1.4 A rare genetic disease is discovered. Although only one in a million people carry it, you consider getting screened. You are told that the genetic test is extremely good; it is 100% sensitive (it is always correct if you have the disease) and 99.99% specific (it gives a false positive result only 0.01% of the time). Using Bayes' theorem, explain why you might decide not to take the test.

Solution Before taking the test, the probability $P(D)$ that you have the genetic disease is 10^{-6} and the probability $P(H)$ that you do not is $1 - 10^{-6}$. By how much will the test change this uncertainty? Let us consider two possible outcomes.

If the test is positive, then the Bayesian posterior probabilities of having and not having the disease are as follows:

$$P(D|\text{positive}) = \frac{P(\text{positive}|D)P(D)}{P(\text{positive})}$$

$$= \frac{P(\text{positive}|D)P(D)}{P(\text{positive}|D)P(D) + P(\text{positive}|H)P(H)}$$

$$= \frac{10^{-6}}{10^{-6} + 0.999999 \times 10^{-4}} = 0.0099,$$

$$P(H|\text{positive}) = \frac{P(\text{positive}|H)P(H)}{P(\text{positive})} = 0.9901.$$

If the test is negative, the Bayesian posterior probabilities become

$$P(D|\text{negative}) = \frac{P(\text{negative}|D)P(D)}{P(\text{negative})}$$

$$= \frac{P(\text{negative}|D)P(D)}{P(\text{negative}|D)P(D) + P(\text{negative}|H)P(H)}$$

$$= \frac{0}{0 + 0.9999 \times (1 - 10^{-6})} = 0,$$

$$P(H|\text{negative}) = \frac{P(\text{negative}|H)P(H)}{P(\text{negative})} = 1.$$

Thus, the changes of prior probabilities $P(D)$, $P(H)$ are very small:

$$|P(D) - P(D|\text{positive})| = 0.0099, \quad |P(D) - P(D|\text{negative})| = 10^{-6},$$

$$|P(H) - P(H|\text{positive})| = 0.0099, \quad |P(H) - P(H|\text{negative})| = 10^{-6}.$$

We see that even if the test is positive the probability of having the disease changes from 10^{-6} to 10^{-2}. Thus, taking the test is not worthwhile for practical reasons. □

Problem 1.5 We have to examine a die which is expected to be loaded in some way. We roll a die ten times and observe outcomes of 1, 3, 4, 2, 4, 6, 2, 1, 2, and 2. What is our maximum likelihood estimate for p_2, the probability of rolling a two? What is the Bayesian estimate if we add one pseudocount per category? What if we add five pseudocounts per category?

Solution The maximum likelihood estimate for p_2 is the (relative) frequency of outcome "two," thus $\hat{p}_2 = 4/10 = 2/5$. If one pseudocount per category is added, the Bayesian estimate is $\hat{p}_2 = 5/16$. If we add five pseudocounts per category, then $\hat{p}_2 = 9/40$. In the last case the Bayesian estimate $\hat{p}_2$ is closer to the probability of the event "two" upon rolling a fair die, $p_2 = 1/6$.

In any case, it is difficult to assess the validity of these alternative approaches without additional information. The best way to improve the estimate is to collect more data. $\quad\square$

1.2 Additional problems

The following problems motivated by questions arising in biological sequence analysis require the ability to apply formulas from combinatorics (Problems 1.6, 1.7, 1.9, and 1.10), elementary calculation of probabilities (Problems 1.8 and 1.16), as well as a knowledge of properties of random variables (Problems 1.13 and 1.18). Our goal here is to help the reader recognize the probabilistic nature of these (and similar) problems about biological sequences.

Basic probability distributions are used in this section to describe the properties of DNA sequences: a geometric distribution to describe the length distribution of restriction fragments (Problem 1.12) and open reading frames (Problem 1.14); a Poisson distribution as a good approximation for the number of occurrences of oligonucleotides in DNA sequences (Problems 1.11, 1.17, 1.19, and 1.22). We will use the notion of an "independence model" for a sequence of independent identically distributed (i.i.d.) random variables with values from a finite alphabet $\mathcal{A}$ (i.e. the alphabet of nucleotides or amino acids) such that the probability of occurrence of symbol a at any sequence site is equal to q_a, $\sum_{a \in \mathcal{A}} q_a = 1$. Thus, a DNA or protein sequence fragment $x_1, \ldots, x_n$ generated by the independence model has probability $\prod_{i=1}^{n} q_{x_i}$. Note that the same model is called the random sequence model in the BSA text (Durbin *et al.*, 1998). The independence model is used to describe DNA sequences in Problems 1.12, 1.14, 1.16, and 1.17.

The introductory level of Chapter 1 still allows us to deal with the notion of hypotheses testing. In Problem 1.20 such a test helps to identify *CpG*-islands in a DNA sequence, while in Problem 1.21 we consider the test for discrimination between DNA sequence regions with higher and lower $G + C$ content.

Finally, issues of the probabilistic model comparison are considered in Problems 1.16, 1.18, and 1.19.

Problem 1.6 In the herpesvirus genome, nucleotides C, G, A, and T occur with frequencies 35/100, 35/100, 15/100, and 15/100, respectively. Assuming the independence model for the genome, what is the probability that a randomly selected 15 nt long DNA fragment contains eight C's or G's and seven A's or T's?

Solution The probability of there being eight C's or G's and seven A's or T's in a 15 nt fragment, given the frequencies 0.7 and 0.3 for each group C & G and A & T, respectively, is $0.7^8 \times 0.3^7 = 0.0000126$. This number must be multiplied by $\binom{15}{8} = 15!/8!7!$, the number of possible arrangements of representatives of these nucleotide groups among fifteen nucleotide positions. Thus, we get the probability 0.08. □

Problem 1.7 A DNA primer used in the polymerase chain reaction is a one-strand DNA fragment designed to bind (to hybridize) to one of the strands of a target DNA molecule. It was observed that primers can hybridize not only to their perfect complements, but also to DNA fragments of the same length having one or two mismatching nucleotides. If the genomic DNA is "sufficiently long," how many different DNA sequences may bind to an eight nucleotide long primer? The notion of "sufficient length" implies that all possible oligonucleotides of length 8 are present in the target genomic DNA.

Solution We consider a more general situation with the length of primer equal to n. There are three possible cases of hybridization between the primer and the DNA: with no mismatch, with one mismatch, and with two mismatches. The first case obviously identifies only one DNA sequence exactly complementary to the primer. The second case, one mismatch, with the freedom to choose one of three mismatching types of nucleotides in one position of the complementary sequence, gives $3n$ possible sequences. Finally, two positions carrying mismatching nucleotides can occur in $n(n-1)/2$ ways. Each choice of these two positions generates nine possibilities to choose two nucleotides different from the matching types. This gives a total of $9n(n-1)/2$ possible sequences with two mismatches. Hence, for $n = 8$, there are

$$1 + 3 \times 8 + \frac{9 \times 8 \times 7}{2} = 277$$

different sequences able to hybridize to the given primer. □

Problem 1.8 A DNA sequencing reaction is performed with an error rate of 10%, thus a given nucleotide is wrongly identified with probability 0.1. To minimize the error rate, DNA is sequenced by $n = 3$ independent reactions, the newly sequenced fragments are aligned, and the nucleotides are identified by the following majority rule. The type of nucleotide at a particular position is identified as α, $\alpha \in \{T, C, A, G\}$, if more nucleotides of type α are aligned in this position than all other types combined. If at an alignment position no nucleotide type appears more than $n/2$ times, the type of nucleotide is not identified (type N).

What is the expected percentage of (a) correctly and (b) incorrectly identified nucleotides? (c) What is the probability that at a particular site identification is impossible? (d) How does the result of (a) change if $n = 5$; what about for $n = 7$? Assume that there are only substitution type errors (no insertions or deletions) with no bias to a particular nucleotide type.

Solution (a) In a given position, we consider the three sequencing reaction calls as outcomes of the three Bernoulli trials with "success" taking place if the nucleotide is identified correctly (with probability $p = 0.9$) and "failure" otherwise (with probability $q = 0.1$). Then the probabilities of the following events are described by the binomial distribution and can be determined immediately:

$$P_3 = P(\text{"success" is observed three times}) = p^3 = 0.9^3 = 0.729,$$

$$P_2 = P(\text{"success" is observed twice}) = \binom{3}{2} p^2 q$$

$$= 3 \times 0.9^2 \times 0.1 = 0.243.$$

Under the majority rule, the expected percentage $\mathbf{E}$ of correctly identified nucleotides is given by

$$\mathbf{E}^c_{n=3} = P(\text{"success" is observed at least twice}) \times 100\%$$

$$= (P_3 + P_2) \times 100\% = 97.2\%.$$

(b) To determine the probability of identifying a nucleotide at a given site incorrectly, we have to be able to classify the "failure" outcomes; thus, we need to generalize the binomial distribution to a multinomial one. Specifically, in each independent trial (carried out at a given sequence site) we can have "success" (with probability $p = 0.9$) and three other outcomes: "failure 1," "failure 2," and "failure 3" (with equal probabilities $q_1 = q_2 = q_3 = 1/30$). To identify a nucleotide incorrectly would mean to observe at least two "failure i" outcomes, $i = 1, 2, 3$, among $n = 3$ trials. Therefore,

$$P'_3 = (\text{"failure } i\text{" is observed three times}) = q_i^3 = (1/30)^3 = 0.000037,$$

$$P'_2 = P(\text{"failure } i\text{" is observed twice}) = 2\binom{3}{2}q_i^2 q_j + \binom{3}{2}q_i^2 p$$

$$= 6 \times (1/30)^3 + 3 \times (1/30)^2 \times 0.9 = 0.00356.$$

Finally, for the expected percentage of wrongly identified nucleotides we have

$$\mathbf{E}^w_{n=3} = \left(\sum_{i=1,2,3} (P'_3 + P'_2) \right) \times 100\%$$

$$= 3(P'_3 + P'_2) \times 100\% = 1.1\%.$$

(c) At a particular site, the base calling results in three mutually exclusive events: "correct identification," "incorrect identification," or "identification impossible." Then, the probability of the last outcome is given by

$$P(\text{nucleotide cannot be identified}) = 1 - (P_3 + P_2) - 3(P'_3 + P'_2) = 0.0172.$$

(d) To calculate the expected percentage $\mathbf{E}^c_n$ of correctly identified nucleotides for $n = 5$ and $n = 7$, we apply the same arguments as in section (a), only instead of three Bernoulli trials we consider five and seven, respectively. We find:

$$\mathbf{E}^c_{n=5} = P(\text{at least three "successes" among five trials}) \times 100\%$$

$$= p^5 + 5 \times 0.9^4 \times 0.1 + 10 \times 0.9^3 0.1^2 = 99.14\%.$$

Similarly,

$$\mathbf{E}^c_{n=7} = P(\text{at least four "successes" among seven trials}) \times 100\% = 99.73\%.$$

As expected, the increase in the number of independent reactions improves the quality of sequencing. $\qquad\square$

Problem 1.9 Due to redundancy of genetic code, a sequence of amino acids could be encoded by several DNA sequences. For a given ten amino acid long protein fragment, what are the lower and upper bounds for the number of possible DNA sequences that could carry code for this protein fragment?

Solution The lower bound of one would be reached if all ten amino acids are methionine or tryptophan, the amino acids encoded by a single codon. In this case the amino acid sequence uniquely defines the underlying nucleotide sequence. The

Table 1.1. *The maximum number I_α of nucleotides C and G that appear in one of the synonomous codons for given amino acid α*

I_α	Amino acid α
1	Asn, Ile, Lys, **Met**, Phe, Tyr
2	Asp, Cys, Gln, Glu, His, Leu, Ser, Thr, **Trp**, Val
3	Ala, Arg, Gly, Pro

upper bound would be reached if the amino acid sequence consists of leucine, arginine, or serine, the amino acids encoded by six codons each. A ten amino acid long sequence consisting of any arrangement of *Leu*, *Ser*, or *Arg* can be encoded by as many as $6^{10} = 60\,466\,176$ different nucleotide sequences. □

Problem 1.10 Life forms from planet XYZ were discovered to have a DNA and protein basis with proteins consisting of twenty amino acids. By analysis of the protein composition, it was determined that the average frequencies of all amino acids excluding *Met* and *Trp* were equal to 1/19, while the frequencies of *Met* and *Trp* were equal to 1/38. Given the high temperature on the XYZ surface, it was speculated that the DNA has an extremely high $G + C$ content. What could be the highest average $G + C$ content of protein-coding regions (given the average amino acid composition as stated above) if the standard (the same as on planet Earth) genetic code is used to encode *XYZ* proteins?

Solution To make the highest possible $G + C$ content of protein-coding region that would satisfy the restrictions on amino acid composition, synonymous codons with highest $G + C$ content should be used on all occasions. The distribution of the high $G + C$ content codons according to the standard genetic code is as shown in Table 1.1 (where I_α designates the highest number of C and G nucleotides in a codon encoding amino acid α).

Therefore, the average value of the $G + C$ content of a protein-coding region is given by

$$\langle G + C \rangle = \sum_\alpha \frac{I_\alpha}{3} f_\alpha$$

$$= \frac{1}{3}\left(\frac{1}{19}(5 \times 1 + 9 \times 2 + 4 \times 3) + \frac{1}{38}(1 + 2) \right) = 0.64.$$

Here f_α is the frequency of amino acid α.

Remark Similar considerations can provide estimates of upper and lower bounds of $G + C$ content for prokaryotic genomes (planet Earth), where protein-coding regions typically occupy about 90% of total DNA length.   □

Problem 1.11 A restriction enzyme is cutting DNA at a palindromic site 6 nt long. Determine the probability that a circular chromosome, a double-stranded DNA molecule of length $L = 84\,000$ nt, will be cut by the restriction enzyme into exactly twenty fragments. It is assumed that the DNA sequence is described by the independence model with equal probabilities of nucleotides T, C, A, and G. Hint: use the Poisson distribution.

Solution The probability that a restriction site starts in any given position of the DNA sequence is $p = (1/4)^6 = 0.0002441$. If we do not take into account the mutual dependence of occurrences of restriction sites in positions i and j, $|i-j| \le 6$, the number X of the restriction sites in the DNA sequence can be considered as the number of successes (with probability p) in a sequence of L Bernoulli trials; therefore, X has a binomial distribution with parameters p and L. Since L is large and p is small, we can use the Poisson distribution with parameter $\lambda = pL = 20.5$ as an approximation of the binomial distribution. Then

$$P(X = 20) = e^{-\lambda}\frac{\lambda^{20}}{20!} = 0.088.$$

Notably, the probability of cutting this DNA sequence into any other particular number of fragments will be lower than $P(X = 20)$. Indeed, the ratio R_k of probabilities of two consecutive values of X,

$$R_k = \frac{P(X = k + 1)}{P(X = k)} = \frac{\lambda}{k + 1},$$

shows that $P(X = k)$ increases as k grows from 0 to λ, and decreases as k grows from λ to L, thus attaining its maximum value at point $k = \lambda$. In other words, if λ is not an integer, the most probable value of the Poisson distributed random variable is equal to $[\lambda]$, where $[\lambda]$ stands for the largest integer not greater than λ. Otherwise, the most probable values are both $\lambda - 1$ and λ.   □

Problem 1.12 Determine the average length of the restriction fragments produced by the six-cutter restriction enzyme *SmaI* with the restriction site *CCCGGG*. Consider (a) a genome with a $G + C$ content of 70% and (b) a genome with a $G + C$ content of 30%. It is assumed that the genomic sequence can be represented by the independence model with probabilities of nucleotides such that $q_G = q_C$, $q_A = q_T$. Note that enzyme *SmaI* cuts the double strand of DNA in the middle of site *CCCGGG*.

Solution We denote the probability that the restriction site starts in a particular sequence position as P and the length of a restriction fragment as L. We associate the number 1 with a sequence position where the restriction site starts and the number 0 otherwise. Then in the generated sequence of ones and zeros the lengths of runs of zeros (equal to the lengths of restriction fragments) can be considered as values of random variable L. If we do not take into account the mutual dependence of occurrences of restriction sites at positions i and j, $|i-j| \leq 6$, the random variable L has the geometric distribution: $P(L = n) = (1 - P)^{n-1}P$. The expected value of L is defined by

$$\mathbf{EL} = \sum_{n=1}^{+\infty} n(1 - P)^{n-1}P = P\sum_{n=1}^{+\infty} -\frac{d(1 - P)^n}{dP}$$

$$= -P\frac{d\left(\sum_{n=1}^{+\infty}(1 - P)^n\right)}{dP} = \frac{P}{P^2} = \frac{1}{P}.$$

For (a) we have

$$P_a = P(CCCGGG) = (0.35)^6 = 1.8 \times 10^{-3},$$

and the average length of restriction fragment is $\mathbf{EL}_a = 1/P_a = 544$ nt.
  Similarly, for (b),

$$P_b = P(CCCGGG) = (0.15)^6 = 1.14 \times 10^{-5},$$

and the average length of the restriction fragment is $\mathbf{EL}_b = 87\,788$ nt. The longer average length of restriction fragments in (b) could be expected as the $G + C$-rich restriction site $CCCGGG$ would appear less frequently in the $A + T$-rich genomic DNA. $\qquad\square$

Problem 1.13 Consider a DNA sequence of length n described by the independence model with equal probabilities of nucleotides. Let X be the number of occurrences of dinucleotide AA and Y be the number of occurrences of dinucleotide AT in this sequence. What are the expected values and variances of random variables X and Y? For simplicity consider a circular DNA of length n.

Solution Let us define random variables x_i and y_i, $i = 1, \ldots, n$, as follows:

$$x_i = \begin{cases} 1, & \text{if dinucleotide } AA \text{ starts in } i\text{th position in the sequence}, \\ 0, & \text{otherwise}; \end{cases}$$

$$y_i = \begin{cases} 1, & \text{if dinucleotide } AT \text{ starts in } i\text{th position in the sequence}, \\ 0, & \text{otherwise}. \end{cases}$$

Obviously, $X = \sum_{i=1}^{n} x_i$, $Y = \sum_{i=1}^{n} y_i$. The expected values of x_i and y_i, $i = 1, \ldots, n$, under the uniform independence model are given by

$$\mathbf{E}x_i = \mathbf{E}y_i = P(x_i = 1) = P(y_i = 1) = \left(\frac{1}{4}\right)^2 = \frac{1}{16}.$$

Thus, the mean value of X, Y is $\mathbf{E}X = \mathbf{E}Y = n/16$. Similarly, we can state that the expected number of occurrences of any other dinucleotide in the sequence is also $n/16$.

We denote the shortest distance between positions i and j in the circular DNA as $r(i, j)$ and find the second moment of X:

$$\mathbf{E}X^2 = \mathbf{E}\left(\sum_{i=1}^{n} x_i\right)^2 = \sum_{i=1}^{n} \mathbf{E}x_i^2 + \sum_{i,j:r(i,j)\geq 2} \mathbf{E}x_i x_j + \sum_{i,j:r(i,j)=1} \mathbf{E}x_i x_j. \qquad (1.2)$$

As $x_i^2 = x_i$, the first sum in (1.2) is $\sum_{i=1}^{n} \mathbf{E}x_i^2 = n\mathbf{E}x_i = n/16$.

If the distance $r(i, j) \geq 2$, the random variables x_i and x_j are independent and

$$\sum_{i,j:r(i,j)\geq 2} \mathbf{E}x_i x_j = \sum_{i,j:r(i,j)\geq 2} \mathbf{E}x_i \mathbf{E}x_j = \frac{n(n-3)}{256}.$$

If $r(i, j) = 1$, then positions i and j are adjacent and, for certainty, we assume that position i precedes j. Then product $x_i x_j$ takes the following values:

$$x_i x_j = \begin{cases} 1, & \text{if triplet } AAA \text{ starts in position } i, \\ 0, & \text{otherwise,} \end{cases}$$

and $\mathbf{E}x_i x_j = P(x_i x_j = 1) = (1/4)^3 = 1/64$. Therefore, the second moment of X becomes

$$\mathbf{E}X^2 = \frac{n}{16} + \frac{n(n-3)}{256} + \frac{2n}{64} = \frac{n(n+21)}{256},$$

and the variance of X is given by

$$\mathbf{Var}X = \mathbf{E}X^2 - (\mathbf{E}X)^2 = \frac{n(n+21)}{256} - \frac{n^2}{256} = \frac{21n}{256}.$$

Similarly, for the second moment of Y we have:

$$\mathbf{E}Y^2 = \mathbf{E}\left(\sum_{i=1}^{n} y_i\right)^2 = \sum_{i=1}^{n} \mathbf{E}y_i^2 + \sum_{i,j:r(i,j)\geq 2} \mathbf{E}y_i y_j + \sum_{i,j:r(i,j)=1} \mathbf{E}y_i y_j. \qquad (1.3)$$

The first two sums in (1.3) are the same as in Equation (1.2). However, if $r(i, j) = 1$, the product $y_i y_j$ is always zero, because dinucleotide AT cannot start in two adjacent positions i and j of the sequence. Therefore,

$$\mathbf{E}Y^2 = \frac{n}{16} + \frac{n(n-3)}{256} = \frac{n(n+13)}{256}$$

and

$$\mathbf{Var}Y = \mathbf{E}Y^2 - (\mathbf{E}Y)^2 = \frac{n(n+13)}{256} - \frac{n^2}{256} = \frac{13n}{256}.$$

We see that the variance of the number of occurrences of a dinucleotide depends on its structure: if it consists of different letters (thus, the dinucleotide cannot overlap with the neighbor of the same type), the variance is $13n/256$; if dinucleotide consists of the same letter repeated twice (and can overlap with the neighbor of the same type), the variance increases to $21n/256$.

Remark For an extended discussion of the first and second moments of frequencies of words in biological sequences, see Pevzner, Borodovsky, and Mironov (1989).

□

Problem 1.14 A prokaryotic protein-coding gene normally consists of an uninterrupted sequence of nucleotide triplets, codons. This sequence starts with a specific start codon (*ATG* is most frequent) and ends with one of the three stop codons: *TAA, TAG, TGA*. A sequence with such a structure is called an "open reading frame' (ORF). However, not every ORF found in prokaryotic genomic DNA is a functional gene. Assuming that *ATG* is the only possible start codon, what is the length distribution of ORFs that occur by chance? Consider an independence model with equal probabilities of four nucleotide types.

Solution There are $4^3 = 64$ triplets (codons) that will appear in the sequence with equal probabilities. Three out of the sixty four are stop codons. Therefore, the probability of encountering a stop codon upon scanning a sequence, triplet by triplet, is $3/64 = 0.047$. For the probability of occurrence of ORF of length L (in codons) we have

$$P(\text{ORF of length } L \text{ starts in a given position})$$

$$= P(ATG) \times P(\text{non-stop codon})^{L-2} \times P(\text{stop codon})$$

$$= \frac{1}{64} \times \left(1 - \frac{3}{64}\right)^{L-2} \times \frac{3}{64} = \frac{3}{4096}\left(\frac{61}{64}\right)^{L-2}.$$

To derive the ORF length distribution, we use the definition of conditional probability:

$$P(\text{length of ORF is equal to } L)$$

$$= \frac{P(\text{ORF of length } L \text{ starts in a given position})}{P(\text{any ORF starts in a given position})}$$

$$= \frac{P(\text{ORF of length } L \text{ starts in a given position})}{\sum_{L=2}^{+\infty} P(\text{ORF of length } L \text{ starts in a given position})}$$

$$= \frac{(3/4096)(61/64)^{L-2}}{1/64} = \frac{3}{64}\left(\frac{61}{64}\right)^{L-2}.$$

Thus, we have derived the geometric distribution of the lengths of random ORFs along with the parameters of the distribution. □

Problem 1.15 Assuming that non-coding DNA is described by the independence model with probabilities of nucleotides equal to 1/4, show that a gene start (under the assumption that the only start codon is the *ATG* codon) in 75% of cases is expected to coincide with the "longest ORF" start.

Solution Let us assume that a particular *ATG* codon is a *real* start of a gene, not overlapped by an adjacent gene. Then the DNA sequence located upstream to the *ATG* is non-coding DNA described by the independence model. Each possible triplet appears in sequence described by this model with probability 1/64. To find the probability that a given *ATG* situated at the *real* gene start is the 5′-most *ATG* in the ORF, we consider the complementary event that there is yet another *ATG* upstream to *real* start that would make an even longer ORF. By examining non-overlapping triplets upstream to the given *ATG* one at a time, starting with the one immediately adjacent to *ATG*, we observe one of the following possible outcomes. (i) The picked up triplet is one of sixty that are not *ATG*, *TAA*, *TGA*, *TAG*. In this case, we continue the process of triplet examining. (ii) This triplet is one of the three stop codons (*TAA*, *TGA*, *TAG*). We stop and infer that the initially considered *ATG* is the leftmost *ATG* in the ORF. (iii) The triplet under examination is *ATG*. We stop and infer that the initially considered *real* gene start is not the leftmost *ATG* in the ORF. Obviously, the termination of the scanning procedure by reaching one of the stop codons will occur three times more frequently than the termination by reaching the *ATG* codon. Therefore, the *ATG* start of a real gene in 75% of cases coincides with the leftmost *ATG* of the ORF, which defines the longest ORF for the fixed stop codon on the 3′ end. □

Problem 1.16 Suppose we consider two independence models of nucleotide sequence. The first model, M_1, has the same probabilities of nucleotides as defined in Problem 1.6. The second model, M_2, assigns to each nucleotide type the probability 1/4 to appear in any given position. Given the observed sequence $x = ACTGACGACTGAC$, compare the likelihoods of these models.

Solution The likelihood of a model is defined as the conditional probability of data (sequence x) given the model (Durbin *et al.* (1998), p. 6). Thus, we have to compare the probabilities of sequence x under each model. The likelihood of model M_1 is given by

$$P(x|M_1) = \left(\frac{3}{20}\right)^6 \left(\frac{7}{20}\right)^7.$$

Similarly, for the likelihood of model M_2 we have $P(x|M_2) = (1/4)^{13}$. The likelihood ratio is given by

$$\frac{P(x|M_2)}{P(x|M_1)} = \frac{20^{13}}{4^{13} \times 3^6 \times 7^7} = 2.0333 > 1.$$

Therefore, for the observed sequence x model M_2 has a greater likelihood than model M_1. □

Problem 1.17 A circular double-stranded DNA of $L = 3\,400$ nt long was cut by a restriction enzyme. A subsequent gel electrophoresis separation indicated the presence of five DNA pieces. It turned out that the absent-minded researcher could not recall the exact type of restriction enzyme that was used. Still, he knew that the chemical was picked up from a box containing equal number of 4-base cutters and 6-base cutters (restriction enzymes that cut specific 4 nt long sites and specific 6 nt long sites, respectively). What is the posterior probability that the 4-nucleotide cutter was used if the DNA sequence can be represented by the independence model with equal probabilities of nucleotides T, C, A, G.

Solution The probability of appearance of a restriction site in a particular position of DNA sequence is $p_1 = (1/4)^4 = 0.003906$ for the 4-base cutter and $p_2 = (1/4)^6 = 0.000244$ for the 6-base cutter.

We assume that in both cases the number X of restriction sites in the sequence can be approximated by the Poisson distribution with parameter $\lambda_1 = p_1 L = 13.28$ for the 4-base cutter and $\lambda_2 = p_2 L = 0.83$ for the 6-base cutter (see solution to Problem 1.11). Then we obtain

$$P(X = 5|4\text{-cutters}) = e^{-\lambda_1} \frac{(\lambda_1)^5}{5!} = 0.00588,$$

$$P(X = 5|6\text{-cutters}) = e^{-\lambda_2} \frac{(\lambda_2)^5}{5!} = 0.00143.$$

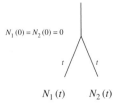

Figure 1.1. The simplest hylogenetic tree T with a pair of the homologous genes x^1 and x^2 being its leaves (see Problem 1.18).

We use Bayes' theorem to calculate the posterior probability that the 4-base cutter produced the restriction fragments:

$P(\text{4-cutters}|X = 5)$

$$= \frac{P(X = 5|\text{4-cutters})P(\text{4-cutters})}{P(X = 5|\text{4-cutters})P(\text{4-cutters}) + P(X = 5|\text{6-cutters})P(\text{6-cutters})}$$

$$= \frac{0.00588 \times 0.5}{0.00588 \times 0.5 + 0.00143 \times 0.5} = 0.804.$$

With 80.4% chance that the 4-base cutter was used, the initial uncertainty seems to be resolved.   □

Problem 1.18 One theory states that the latest common ancestor of birds and crocodiles lived 120 million years ago (MYA), while another theory suggests that this time is twice as long. Comparison of homologous genes x^1 and x^2 of two species, the Nile crocodile and the Mediterranean seagull, revealed on average 365 differences in 1000 nt long fragments. It is assumed that mutations at different DNA sites occur independently, and at each site the number of mutations fixed in evolution is approximated by the Poisson process. The rate of mutation fixation, p, per nucleotide site per year, is equal to 10^{-9}. Given the observed number of differences, (a) compare the likelihoods of the two theories, and (b) determine the maximum likelihood estimate of the divergence time. For simplicity, assume that no more than one mutation could occur at any given nucleotide site of the whole lineage.

Solution (a) Assuming that the divergence of the two species occurred t years ago, we consider the simplest phylogenetic tree T with leaves x^1 and x^2. The occurrence of substitutions along branches of the tree can be described by two independent Poisson processes $N_1(\tau)$ and $N_2(\tau)$ both with parameter p. The moment of divergence corresponds to $\tau = 0$ and the present time to $\tau = t$ (see Figure 1.1).

We will compare the likelihoods of tree T for two values of the elapsed time, $t = t_1 = 120\,\text{MYA}$ and $t = t_2 = 240\,\text{MYA}$, associated with the competing theories. The likelihood of a two-leaves tree with a molecular clock property depends on t only. Then the (conditional) likelihood at site u carrying matching nucleotides in DNA sequences is given by

$$
\begin{aligned}
L_u(t) &= P(x_u^1 = x_u^2 | t, \text{ no more than one mutation at site } i) \\
&= P(N_1(t) = 0, N_2(t) = 0 | N_1(t) + N_2(t) \le 1) \\
&= \frac{P(N_1(t) = 0, N_2(t) = 0, N_1(t) + N_2(t) \le 1)}{P(N_1(t) + N_2(t) \le 1)} \\
&= \frac{P(N_1(t) = 0, N_2(t) = 0)}{P(N_1(t) + N_2(t) \le 1)}.
\end{aligned}
\tag{1.4}
$$

The numerator of the last expression in Equation (1.4) is equal to

$$
P(N_1(t) = 0)P(N_2(t) = 0) = e^{-2pt}
$$

due to the independence of processes $N_1(\tau)$ and $N_2(\tau)$, while $N_1(t) + N_2(t)$ is again the Poisson random variable (say $N_3(t)$) with parameter $2p$ due to the known property of the Poisson distribution. Thus, we have

$$
\begin{aligned}
P(N_1(t) + N_2(t) \le 1) &= P(N_3(t) \le 1) = P(N_3(t) = 0) + P(N_3(t) = 1) \\
&= e^{-2pt} + 2pte^{-2pt},
\end{aligned}
$$

and the likelihood $L_u(t)$ from Equation (1.4) becomes

$$
L_u(t) = \frac{e^{-2pt}}{e^{-2pt} + 2pte^{-2pt}} = (1 + 2pt)^{-1}.
\tag{1.5}
$$

Similarly, at site u with mismatching nucleotides the likelihood is given by

$$
\begin{aligned}
L_u(t) &= P(x_u^1 \ne x_u^2 | t, \text{ no more than one mutation at site } i) \\
&= P(N_1(t) = 0, N_2(t) = 1 | N_1(t) + N_2(t) \le 1) \\
&\quad + P(N_1(t) = 1, N_2(t) = 0 | N_1(t) + N_2(t) \le 1) \\
&= \frac{2P(N_1(t) = 0)P(N_2(t) = 1)}{P(N_1(t) + N_2(t) \le 1)} = \frac{2e^{-pt}pte^{-pt}}{e^{-2pt} + 2pte^{-2pt}} = 2pt(1 + 2pt)^{-1}.
\end{aligned}
\tag{1.6}
$$

From Equations (1.5) and (1.6) we derive the likelihood of tree T with two leaves which are genomic sequences of length N aligned with M mismatches:

$$
L(t) = \prod_{u=1}^{N} L_u(t) = (1 + 2pt)^{-N}(2pt)^M.
\tag{1.7}
$$

To test the two theories, we calculate the log-odds ratio for $t_1 = 120$ MYA and $t_2 = 240$ MYA:

$$\ln \frac{L(t_1)}{L(t_2)} = N \ln(1 + 2pt_2) - N \ln(1 + 2pt_1) + M \ln(2pt_1) - M \ln(2pt_2)$$

$$= -252.99 < 0.$$

Therefore, the available data support the theory that birds and crocodiles diverged 240 MYA, since this theory has a greater (conditional) likelihood than the competing one.

(b) We determine the maximum likelihood estimate t^* of the time of divergence of the two species as a maximum point of the logarithm of likelihood $L(t)$, formula (1.7):

$$\frac{d \ln L(t)}{dt} = -\frac{2Np}{1 + 2pt^*} + \frac{2pM}{2pt^*} = 0,$$

$$t^* = \frac{M}{2p(N - M)} = 2.874 \times 10^8.$$

Thus, $t^* = 287.4$ MYA is the maximum likelihood divergence time, while the maximum likelihood value *per se* is given by

$$L_{max} = L(t^*) = (1 + 2pt^*)^{-N} (2pt^*)^M = 10^{-285}.$$ □

Problem 1.19 It is known that *CpG*-islands in high eukaryotes are relatively rich with *CpG* dinucleotides, while these dinucleotides are discriminated in the rest of a chromosome. It is assumed that the frequency of occurrences of *CpG* dinucleotides in a *CpG*-island can be approximated by the Poisson distribution with twenty-five *CpG* dinucleotides per 250 nt long fragment on average, while in the rest of the DNA this average is ten *CpG* per 250 nt. Suggest the Bayesian type algorithm for *CpG*-island identification. How will this algorithm characterize a 250 nt long DNA fragment containing nineteen *CpG* dinucleotides?

Solution We assume that that the numbers of occurrences of *CpG* dinucleotides in *CpG*-islands and non-*CpG*-islands are both described by the Poisson distribution with parameter $\lambda_1 = 25$ and $\lambda_2 = 10$, respectively.

If a given 250 nt long DNA fragment contains n dinucleotides *CpG*, how likely is it that the DNA fragment belongs to a *CpG*-island? We have to compare two *a posterior* probabilities: $P_1 = P$(being a *CpG*-island given n observed *CpG* dinucleotides) and $P_2 = P$(being a non-*CpG*-island given n observed *CpG* dinucleotides). Assuming that both alternatives, being a *CpG*-island and being a non-*CpG*-island, are *a priori* equally likely, we use Bayes' theorem to calculate P_1 and P_2:

$P_1 = P(\text{DNA fragment with } n \ CpC \text{ has Poisson distribution with } \lambda_1 = 25)$

$$= \frac{P(n \ CpG | \lambda_1 = 25)\frac{1}{2}}{P(n \ CpG | \lambda_1 = 25)\frac{1}{2} + P(n \ CpG | \lambda_2 = 10)\frac{1}{2}}$$

$$= \frac{25^n e^{-25}}{10^n e^{-10} + 25^n e^{-25}}.$$

In the above we applied the formula for Poisson distribution and canceled the common factor $n!$. Similarly,

$P_2 = P(\text{DNA fragment with } n \ CpC \text{ has Poisson distribution with } \lambda_2 = 10)$

$$= \frac{P(n \ CpG | \lambda_2 = 10)\frac{1}{2}}{P(n \ CpG | \lambda_1 = 25)\frac{1}{2} + P(n \ CpG | \lambda_2 = 10)\frac{1}{2}}$$

$$= \frac{10^n e^{-10}}{10^n e^{-10} + 25^n e^{-25}},$$

or $P_2 = 1 - P_1$. The simple identification algorithm for a *CpG*-island works as follows. For a given 250 nt long DNA fragment with n observed *CpG* dinucleotides value P_1 is computed. If $P_1 > 0.5$ ($P_1 > P_2$), the DNA fragment is identified as a part of a *CpG*-island. Otherwise, the fragment is identified as a part of a non-*CpG*-island.

For $n = 19$ we have

$$P_1 = \frac{(25)^{19} e^{-25}}{(25)^{19} e^{-25} + (10)^{19} e^{-10}} = 0.92,$$

$$P_2 = 1 - P_1 = 0.08,$$

and we conclude that the DNA fragment belongs to a *CpG*-island. $\square$

Problem 1.20   Given the conditions stated in Problem 1.19, the following decision-making rule is accepted: if more than eighteen *CpG* dinucleotides are observed in a 250 nt long DNA fragment, it is identified as a *CpG*-island. Determine false positive and false negative rates of this method.

Solution The false positive rate (*FPR*) is defined as the probability that the rule would identify a non-*CpG*-island as a *CpG*-island. Since the number X of *CpG* dinucleotides in a non-*CpG*-island is described by the Poisson distribution with parameter $\lambda = 10$, we have

$$FPR = P(\text{more than eighteen } CpG \text{ out of } 250 | \text{non-}CpG\text{-island})$$

$$= P(X > 18 | \lambda = 10) = \sum_{n=19}^{+\infty} P(X = n) = 1 - \sum_{n=0}^{18} P(X = n)$$

$$= 1 - \sum_{n=0}^{18} e^{-10} \frac{10^n}{n!} \approx 0.007.$$

The false negative rate (*FNR*) is defined as the probability that a *CpG*-island is identified as a non-*CpG*-island. Since the number Y of *CpG* dinucleotides in a *CpG*-island region has the Poisson distribution with parameter $\lambda = 25$, the false negative rate is given by

$$FNR = P(\text{less or equal to eighteen } CpG \text{ out of } 250 | CpG\text{-island})$$

$$= P(Y \leq 18 | \lambda = 25) = \sum_{n=0}^{18} P(Y = n)$$

$$= \sum_{n=0}^{18} e^{-25} \frac{25^n}{n!} \approx 0.09.$$

Note that $FPR < FNR$. This means that the classification rule is more likely to decide that *CpG*-island DNA is non-*CpG*-island DNA than vice versa. $\square$

Problem 1.21 An inhomogeneous DNA sequence is known to contain both $C + G$-rich composition regions and regions with unbiased (uniform) nucleotide composition. We assume that the independence model (P model) with parameters $p_T = 1/8$, $p_C = 3/8$, $p_A = 1/8$, $p_G = 3/8$, describes the regions with high $C + G$ content. Regions with uniform nucleotide composition are described by the independence model (Q model) with parameters $q_T = 1/4$, $q_C = 1/4$, $q_A = 1/4$, $q_G = 1/4$. For a given DNA fragment X, the log-odds ratio, $L = \log_2[P(X|P)/P(X|Q)]$ is determined, and, if $L \geq 0$, X is classified as a high $C+G$ composition fragment; if $L < 0$, X is classified as compositionally unbiased. Determine the probabilities of type-one error (false negative rate) and type-two error (false positive rate) of the classification of a DNA fragment of length n. Consider $n = 10, 20$ and 100.

Solution For a DNA sequence X of length n we test the null hypothesis,

$$H_0 = \{X \text{ belongs to a } C + G\text{-rich region}\} = \{X \in P\},$$

versus the alternative hypothesis,

$$H_a = \{X \text{ belongs to a region with uniform composition}\} = \{X \in Q\}.$$

The log-odds ratio is given by

$$L = \log_2[P(X|P)/P(X|Q)] = \log_2\left(\left(\frac{p_A}{q_A}\right)^{n_1}\left(\frac{p_C}{q_C}\right)^{n_2}\left(\frac{p_T}{q_T}\right)^{n_3}\left(\frac{p_G}{q_G}\right)^{n_4}\right)$$

$$= n_1\log_2\frac{1}{2} + n_2\log_2\frac{3}{2} + n_3\log_2\frac{1}{2} + n_4\log_2\frac{3}{2} \approx 0.585(n_2 + n_4) - (n_1 + n_3),$$

where n_1, n_2, n_3, and n_4 are numbers of nucleotides A, C, T, and G, respectively, observed in fragment X. We accept hypothesis H_0 (and reject H_a) if $L \geq 0$, i.e. if $P(X|P) \geq P(X|Q)$; and we reject H_0 (and accept H_a) otherwise. For the type-one error α (significance level of the test) we have

$$\alpha = P(\text{type-one error}) = P(H_0 \text{ is rejected}|H_0 \text{ is true}) = P(L < 0|X \in P)$$

$$= P(0.585(n_2 + n_4) - (n_1 + n_3) < 0|X \in P).$$

Next, we define the Bernoulli trial outcomes by interpreting an occurrence of A or T at a given site of sequence X as a "success" and an occurrence of C or G as a "failure." If p is the probability of "success," then the number of "successes" in n Bernoulli trials, $S = n_1 + n_3$, has a binomial distribution with parameters n and p. The type-one error, α, becomes

$$\alpha = P(0.585(n_2 + n_4) - (n_1 + n_3) < 0|X \in P)$$

$$= P(0.585(n - S) - S < 0|S \in B(n, p = 1/4))$$

$$= P(S > 0.369n|S \in B(n, p = 1/4))$$

$$= \sum_{k:0.369n<k\leq n}\binom{n}{k}\left(\frac{1}{4}\right)^k\left(\frac{3}{4}\right)^{n-k} = \left(\frac{1}{4}\right)^n\sum_{k:0.369n<k\leq n}\binom{n}{k}3^{n-k}.$$

It follows from the central limit theorem that as $n \to \infty$ the sequence of random variables $(S - \mathrm{E}S)/\sqrt{\mathrm{Var}S}$ weakly converges to the standard normal distribution . Therefore, for large n,

$$\alpha = P(S > 0.369n|S \in B(n, p = 1/4)) = P\left(\frac{S - \mathrm{E}S}{\sqrt{\mathrm{Var}S}} > \frac{0.369n - \mathrm{E}S}{\sqrt{\mathrm{Var}S}}\right)$$

$$= P\left(\frac{S - 0.25n}{0.25\sqrt{3n}} > 0.2748\sqrt{n}\right) \approx 1 - \Phi(0.2748\sqrt{n}).$$

Here

$$\Phi(x) = \frac{1}{\sqrt{2\pi}}\int_{-\infty}^{x}\exp\left(-\frac{t^2}{2}\right)dt$$

is the cumulative distribution function of a standard normal distribution.

Similarly, for the probability of the type-two error we have

$$\beta = P(\text{type-two error}) = P(H_0 \text{ is accepted}|H_0 \text{ is false}) = P(L \geq 0|X \in Q)$$

$$= P(0.585(n_2 + n_4) - (n_1 + n_3) \geq 0|X \in Q)$$

$$= P(0.585(n - S) - S \geq 0|S \in B(n, p = 1/2))$$

$$= P(S \leq 0.369n|S \in B(n, p = 1/2))$$

$$= \sum_{k:0 \leq k \leq 0.369n} \binom{n}{k} \left(\frac{1}{2}\right)^k \left(\frac{1}{2}\right)^{n-k} = \left(\frac{1}{2}\right)^n \sum_{k:0 \leq k \leq 0.369n} \binom{n}{k}.$$

Again, for large n the application of the central limit theorem leads to equality:

$$\beta = P(S \leq 0.369n|S \in B(n, p = 1/2)) = P\left(\frac{S - \mathbf{ES}}{\sqrt{\mathbf{Var}S}} \leq \frac{0.369n - \mathbf{ES}}{\sqrt{\mathbf{Var}S}}\right)$$

$$= P\left(\frac{S - 0.5n}{0.5\sqrt{n}} \leq -0.262\sqrt{n}\right) \approx \Phi(-0.262\sqrt{n}).$$

For $n = 10, 20, 100$ we calculate

$$\alpha_{10} = P(S \geq 4|S \in B(10, p = 1/4))$$

$$= 0.1460 + 0.0584 + 0.0162 + 0.0031 + 0.0004 = 0.2241,$$

$$\beta_{10} = P(S \leq 3|S \in B(10, p = 1/2))$$

$$= 0.001 + 0.0098 + 0.0439 + 0.1172 = 0.171;$$

$$\alpha_{20} = P(S \geq 8|S \in B(20, p = 1/4))$$

$$= 0.069 + 0.0271 + 0.0099 + 0.003 + 0.0008 + 0.0002 = 0.1019,$$

$$\beta_{20} = P(S \leq 7|S \in B(20, p = 1/2))$$

$$= 0.0002 + 0.0011 + 0.0046 + 0.0148 + 0.037 + 0.0739 = 0.1316;$$

$$\alpha_{100} = 1 - \Phi(2.748) \approx 0.0028,$$

$$\beta_{100} = \Phi(-2.62) \approx 0.0044.$$

As we expect, α and β decrease when the length of the nucleotide sequence increases and more statistical data become available. $\square$

Problem 1.22 Oligonucleotide *TTTTAAAA* was observed in protein-coding regions of a partially sequenced bacterial genome with frequency 0.008. The same oligonucleotide was observed in non-coding regions of the same genome with frequency 0.003. In a newly sequenced 400 nt long DNA fragment oligonucleotide *TTTTAAAA* was found four times. Find the posterior probability that this fragment is a part of a protein-coding gene. A prior probability that a randomly selected 400 nt long DNA fragment is located in a protein-coding region is assumed to be 0.8.

Solution We define the following events:

$B_1 = \{$a 400 nt fragment is located in a protein-coding region$\}$,

$B_2 = \{$a 400 nt fragment is located in a non-coding region$\}$,

$C = \{TTTTAAAA$ is found four times in a 400 nt fragment$\}$.

It is given that $P(B_1) = 0.8$, $P(B_2) = 1 - P(B_1) = 0.2$. The posterior probability $P(B_1|C)$ can be calculated by using Bayes' theorem:

$$P(B_1|C) = \frac{P(C|B_1)P(B_1)}{P(C|B_1)P(B_1) + P(C|B_2)P(B_2)}.$$

We use the Poisson distribution with parameter $\lambda = pN$ as an approximation of a binomial distribution of the number of occurrences of oligonucleotide $TTTTAAAA$ ("successes") since probability p of "success" is small and the number of trials N is large. Thus, we have $\lambda_1 = 3.2$, $\lambda_2 = 1.2$, for the coding and non-coding sequences, respectively. The probabilities of finding four oligonucleotides $TTTTAAAA$ are given by

$$P(C|B_1) = P(\text{four } TTTTAAAA|B_1) = e^{-3.2}\frac{3.2^4}{4!} = 0.1781,$$

$$P(C|B_2) = P(\text{four } TTTTAAAA|B_2) = e^{-1.2}\frac{1.2^4}{4!} = 0.0260.$$

Then the posterior probability that the 400 nt long DNA fragment in question is a part of a protein-coding gene becomes

$$P(B_1|C) = \frac{0.1781 \times 0.7}{0.1781 \times 0.7 + 0.026 \times 0.3} = 0.94. \qquad \square$$

1.3 Further reading

Many textbooks on bioinformatics and computational genomics have been published since the mid 1990s. See, for example, Waterman (1995), Baldi and Brunak (2001), Ewens and Grant (2001), Koonin and Galperin (2003), Jones and Pevzner (2004), Deonier, Waterman, and Tavaré (2005). They differ with respect to the selection of covered topics, reflecting the remarkable variety of problems and research directions in today's bioinformatics. We would like also to mention (Brown, 1999b) a text on molecular biology by Brown (1999b) that contains a comprehensive and concise description of genomics by a molecular biologist.

2

Pairwise alignment

The notion of sequence similarity is perhaps the most fundamental concept in biological sequence analysis. In the same way that the similarity of morphological traits served as evidence of genetic and functional relationships between species in classic genetics and biology, biological sequence similarity could frequently indicate structural and functional conservation among evolutionary related DNA and protein sequences. Introduction of the biologically relevant quantitative measure of sequence similarity, the similarity score, is not a trivial task. No simpler is the other task, developing algorithms that would find the alignment of two sequences with the best possible score given the scoring system. Finally, the third necessary component of the computational analysis of sequence similarity is the method of evaluation of statistical significance of an alignment. Such a method, establishing the cut-off values for the observed scores to be statistically significant, works properly as soon as the statistical distribution of similarity scores is determined analytically or computationally.

Chapter 2 of BSA includes twelve problems that require knowledge of the concepts and properties of the pairwise alignment algorithms. This topic is traditionally best known to biologists due to its utmost practical importance. Indeed, an initial characterization of any DNA or protein sequence starts with the BLAST analysis, utilization of a highly efficient heuristic pairwise alignment algorithm for searching for homologous sequences in a database.

Additional nine problems provide more information for understanding the protein evolution theory behind the log-odds scores of amino acid substitutions, as well as the models involved in the assessment of the statistical significance of the observed sequence similarity scores.

2.1 Original problems

Problem 2.1 Amino acids D, E, and K are all charged; V, I, and L are all hydrophobic. What is the average BLOSUM50 score within the charged group of three? What is it within the hydrophobic group? What about between two groups? Suggest reasons for the observed pattern.

Solution The BLOSUM50 substitution scores within the group of charged amino acids and within the group of hydrophobic amino acids were drawn from Table 2.1 and are shown in Table 2.2.

The average substitution score within the charged group is $S_{ch} = 2.66$ and that within the hydrophobic group is $S_h = 3.22$.

The scores of substitution between amino acids from the different groups are given in Table 2.2(c) (with average $S = -3.44$).

As expected, the scores of substitution between amino acids with similar physico-chemical properties (within the charged or the hydrophobic group) are higher than the scores of substitution between the amino acids with different properties (between the groups). The sequence alignment algorithms maximizing the total score will maximize the number of aligned pairs of amino acids with similar physico-chemical properties. □

Problem 2.2 Show that the probability distributions $f(g)$ of the length of gap g that correspond to the linear and affine gap penalty schemes are both geometric distributions of the form $f(g) = ke^{-\lambda g}$.

Solution A gap penalty $\gamma(g)$ is defined as the log-probability of a gap of a given length g: $\gamma(g) = \log(f(g))$. Thus, if a gap penalty $\gamma(g)$ is scored linearly as $\gamma(g) = -gd$ with the gap-open penalty d, then

$$\exp(\gamma(g)) = e^{-gd} = f(g).$$

This equality corresponds to the formula $f(g) = ke^{-\lambda g}$, if $k = 1, \lambda = d$.

If a gap penalty is defined by the affine score $\gamma(g) = -d - (g-1)e$ with gap-open penalty d and gap-extension penalty e, then

$$\exp(\gamma(g)) = f(g) = \exp(e - d)\exp(-ge).$$

Now again $f(g) = ke^{-\lambda g}$ for $k = \exp(e - d)$ and $\lambda = e$.

Remark Note that in both cases described above the gap penalty function $f(g)$ does not define a properly normalized probability distribution of gap lengths. Indeed, for the linear scheme with any positive gap-open penalty d we have

$$\sum_{g=0}^{\infty} f(g) = \sum_{g=0}^{\infty} \exp(-gd) = \frac{1}{1 - \exp(-d)} \neq 1.$$

Table 2.1. *The BLOSUM50 substitution matrix*

The log-odds values are scaled and rounded to the nearest integer.

	A	R	N	D	C	Q	E	G	H	I	L	K	M	F	P	S	T	W	Y	V
A	5	-2	-1	-2	-1	-1	-1	0	-2	-1	-2	-1	-1	-3	-1	1	0	-3	-2	0
R	-2	7	-1	-2	-4	1	0	-3	0	-4	-3	3	-2	-3	-3	-1	-1	-3	-1	-3
N	-1	-1	7	2	-2	0	0	0	1	-3	-4	0	-2	-4	-2	1	0	-4	-2	-3
D	-2	-2	2	8	-4	0	2	-1	-1	-4	-4	-1	-4	-5	-1	0	-1	-5	-3	-4
C	-1	-4	-2	-4	13	-3	-3	-3	-3	-2	-2	-3	-2	-2	-4	-1	-1	-5	-3	-1
Q	-1	1	0	0	-3	7	2	-2	1	-3	-2	2	0	-4	-1	0	-1	-1	-1	-3
E	-1	0	0	2	-3	2	6	-3	0	-4	-3	1	-2	-3	-1	-1	-1	-3	-2	-3
G	0	-3	0	-1	-3	-2	-3	8	-2	-4	-4	-2	-3	-4	-2	0	-2	-3	-3	-4
H	-2	0	1	-1	-3	1	0	-2	10	-4	-3	0	-1	-1	-2	-1	-2	-3	2	-4
I	-1	-4	-3	-4	-2	-3	-4	-4	-4	5	2	-3	2	0	-3	-3	-1	-3	-1	4
L	-2	-3	-4	-4	-2	-2	-3	-4	-3	2	5	-3	3	1	-4	-3	-1	-2	-1	1
K	-1	3	0	-1	-3	2	1	-2	0	-3	-3	6	-2	-4	-1	0	-1	-3	-2	-3
M	-1	-2	-2	-4	-2	0	-2	-3	-1	2	3	-2	7	0	-3	-2	-1	-1	0	1
F	-3	-3	-4	-5	-2	-4	-3	-4	-1	0	1	-4	0	8	-4	-3	-2	1	4	-1
P	-1	-3	-2	-1	-4	-1	-1	-2	-2	-3	-4	-1	-3	-4	10	-1	-1	-4	-3	-3
S	1	-1	1	0	-1	0	-1	0	-1	-3	-3	0	-2	-3	-1	5	2	-4	-2	-2
T	0	-1	0	-1	-1	-1	-1	-2	-2	-1	-1	-1	-1	-2	-1	2	5	-3	-2	0
W	-3	-3	-4	-5	-5	-1	-3	-3	-3	-3	-2	-3	-1	1	-4	-4	-3	15	2	-3
Y	-2	-1	-2	-3	-3	-1	-2	-3	2	-1	-1	-2	0	4	-3	-2	-2	2	8	-1
V	0	-3	-3	-4	-1	-3	-3	-4	-4	4	1	-3	1	-1	-3	-2	0	-3	-1	5

Table 2.2. *Substitution scores for the three groups of aminoacids*

(a)			(b)			(c)		
	D	E	K		V	I	L		V	I	L

(a)	D	E	K
D	8	2	−1
E	2	6	1
K	−1	1	6

(b)	V	I	L
V	5	4	1
I	4	5	2
L	1	2	5

(c)	V	I	L
D	−4	−4	−4
E	−3	−4	−3
K	−3	−3	−3

Similarly, for the affine scheme,

$$\sum_{g=0}^{\infty} f(g) = \sum_{g=0}^{\infty} \exp(e - d)\exp(-ge)$$

$$= \exp(e - d)\sum_{g=0}^{\infty} \exp(-ge) = \frac{\exp(e - d)}{1 - \exp(-e)} \neq 1.$$

However, the situation can be corrected by the introduction of normalizing constants: $C = 1 - \exp(-d)$ for the linear score; $C = 1 - \exp(-e)/\exp(e - d)$ for the affine score. Then the probability distribution of gap lengths can be properly defined:

$$P(\text{the gap length is } g) = P(g) = Cf(g).$$

Note that in the final form the probability of the particular gap length in the affine scheme does not depend on the gap-open penalty d. ☐

Problem 2.3 Typical gap penalty parameters used in practice are $d = 8$ for the linear case and $d = 12$, $e = 2$ for the affine case (the scores are expressed in half bits). A *bit* is the unit obtained when one takes log base 2 of a probability, so in natural log units these gap penalties correspond to $d' = 8\ln 2/2$ and $d' = 12\ln 2/2$, $e' = 2\ln 2/2$, respectively. What are the corresponding probabilities of a gap (of any length), starting at some position, and the distributions of gap length given that there is a gap?

Solution Note that if a quantity x is measured in half bits ($y = 2\log_2 x$) and in natural log units ($z = \ln x$), then y and z are related by the following formula:

$$y = 2\log_2 x = \frac{2\ln x}{\ln 2} = \frac{2z}{\ln 2}.$$

Thus, if in half bits a gap-open penalty $d = 8$, then in the natural log units $d' = 8\ln 2/2$. Similarly, if in half bits a gap-open penalty $d = 12$ and a gap-extension penalty $e = 2$, then in natural log units $d' = 12\ln 2/2$, $e' = 2\ln 2/2$.

In the linear case (see Problem 2.2) the probability distribution of the gap length is defined by the following formula:

$$P(g) = P(\text{the gap length is } g) = (1 - \exp(-d'))\exp(-gd').$$

If $d' = 4\ln 2$, then $P(g) = (15/16)(1/16)^g$. For the probability of a gap (of any length) starting at some position, we have:

$$P(\text{a gap is present}) = 1 - P(\text{no gap}) = 1 - P(0) = 1 - 15/16 = 1/16.$$

Given a gap existence ($g > 0$), the distribution of the gap length is defined by the conditional probabilities

$$P(\text{the gap length is } g|\text{a gap exists}) = P(g|G) = \frac{P(g \cap G)}{P(G)}$$

$$= \frac{P(g)}{P(G)} = 16P(g) = 15\left(\frac{1}{16}\right)^g.$$

In the affine case (Problem 2.2) the probability distribution of the gap lengths is defined by the following formula:

$$P(g) = (1 - \exp(-e'))\exp(-ge').$$

(This distribution does not depend on the gap-open penalty d'.) If $e' = \ln 2$, the probability distribution $P(g)$ becomes $P(g) = (1/2)^{g+1}$. Similarly to the linear case, we first calculate the probability of a gap existence at a given position:

$$P(\text{a gap exists}) = 1 - P(g = 0) = 1 - \frac{1}{2} = \frac{1}{2}.$$

Finally, the probability of a gap of a certain length g given a gap existence is given by

$$P(g|\text{a gap exists}) = \frac{P(g)}{P(G)} = 2P(g) = \left(\frac{1}{2}\right)^g. \qquad \square$$

Problem 2.4 Using the BLOSUM50 matrix (Table 2.1) and an affine gap penalty of $d = 12, e = 2$, calculate the scores of the two given alignments: (a) the human alpha globin and leghaemoglobin from yellow lupin:

```
HBA_HUMAN    GSAQVKGHGKKVADALTNAVAHV---D--DMPNALSALSDLHAHK
             ++ ++++H+ KV    + +A  ++            +L+L+++H+   K
LGB2_LUPLU NNPELQAHAGKVFKLVYEAAIQLQVTGVVVTDATLKNLGSVHVSK
```

(b) the same region of the human alpha globin protein sequence and nematode glutathione S-transferase homologue named F11G11.2:

```
HBA_HUMAN GSAQVKGHGKKVADALTNAVAHVDDMPNALSALSD----LHAHK
          GS+ + G +    +D L  ++ H+ D+  A +AL D     ++AH+
F11G11.2  GSGYLVGDSLTFVDLL--VAQHTADLLAANAALLDEFPQFKAHQ
```

Solution The total score of an alignment is the sum of the substitution scores of aligned amino acids and the penalties for gaps defined by the formula $\gamma(g) = -12 - 2(g-1)$, where g is the gap length. For the first alignment we calculate

$$S_1 = s(G,N) + s(S,N) + \cdots + s(L,G)$$
$$= 0 + 1 - 1 + 2 + 1 + 2 + 0 + 10 + 0 - 2 + 6 + 5 - 3 - 1 - 2 + 1 - 2$$
$$+ 0 + 5 + 0 - 1 + 1 + 1 - 16(gap) - 1 - 14(gap) - 4 - 1 - 1 - 1$$
$$+ 0 + 5 + 0 - 1 + 5 + 0 + 0 + 1 + 10 + 0 - 1 + 6 = 10.$$

For the second alignment

$$S_2 = s(G,G) + s(G,G) + \cdots + s(L,E)$$
$$= 8 + 5 + 0 - 1 + 1 - 3 + 8 - 1 + 0 - 3 - 1 - 1 + 0 + 8 - 2$$
$$+ 5 - 14(gap) + 0 + 0 - 1 + 10 + 0 - 2 + 8 + 3 - 4 - 1 + 5 - 4$$
$$+ 1 + 5 + 5 - 3 + 8 - 18(gap) + 1 + 0 + 5 + 10 + 2 = 39.$$

Interestingly, the structurally and evolutionary plausible alignment of human alpha globin to leghaemoglobin (the first alignment) receives a lower score than the alignment of human alpha globin to an unrelated nematode protein sequence (the second alignment). This example emphasizes the importance of the expert assessment in the homology inference from computer analysis. □

Problem 2.5 Show that the number of ways of intercalating two sequences of lengths n and m to give a single sequence of length $n + m$, while preserving the order of the symbols in each, is $\binom{n+m}{m}$.

Solution A process of intercalating a sequence $x_1, \ldots, x_n$ by another sequence $y_1, \ldots, y_m$ can be described as the filling of a row of $n + m$ empty cells with $n + m$ symbols $x_1, \ldots, x_n, y_1, \ldots, y_m$ (one symbol per cell) while keeping the order of symbols in each sequence. Every selection of n cells for symbols $x_1, \ldots, x_n$ defines only one way to accommodate these symbols which cannot be permutated. As soon as x's are placed, the positions for $y_1, \ldots, y_m$ are also determined unambiguously, as they fill remaining empty cells without permutations. Hence, the total number of distinct intercalated sequences is the same as the number of ways to choose n cells out of $n + m$ cells,

$$\binom{n+m}{n} = \binom{n+m}{m} = \frac{(n+m)!}{m!n!}. \qquad \square$$

Problem 2.6 Is it true that by taking alternating symbols from the upper and lower sequences in an alignment, then discarding the gap characters, one can define a one-to-one correspondence between gapped alignments of the two sequences and intercalated sequences of the type described in problem 2.5?

Solution Note that the discussion of this problem is of a significant length. First we show that the suggested procedure does not establish one-to-one correspondence between gapped alignments of the two sequences and intercalated sequences introduced in Problem 2.5.

As an example, we consider a gapped alignment A of sequences $x = x_1, \ldots, x_4$ and $y = y_1, \ldots, y_5$:

$$A = \begin{pmatrix} - & x_1 & - & x_2 & x_3 & x_4 \\ y_1 & y_2 & y_3 & - & y_4 & y_5 \end{pmatrix}.$$

Then, by unfolding columns of symbols, we rewrite alignment A as one sequence (with gap symbols): $-, y_1, x_1, y_2, -, y_3, x_2, -, x_3, y_4, x_4, y_5$. Omitting gap characters produces the sequence $y_1, x_1, y_2, y_3, x_2, x_3, y_4, x_4, y_5$, the intercalated sequence defined in Problem 2.5. Thus, given alignment A of sequences x and y we construct exactly one intercalated sequence. Obviously, the same is true for any other gapped pairwise alignment. However, it can be shown that any intercalated sequence (except $y_1, y_2, \ldots, y_m, x_1, x_2, \ldots, x_n$) corresponds to more than one gapped alignment. For example, the same intercalated sequence $y_1, x_1, y_2, y_3, x_2, x_3, y_4, x_4, y_5$ could be obtained from eight different gapped alignments A_1–A_8 (and only from them):

$$A_1 = \begin{pmatrix} - & x_1 & - & x_2 & x_3 & x_4 \\ y_1 & y_2 & y_3 & - & y_4 & y_5 \end{pmatrix},$$

$$A_2 = \begin{pmatrix} - & x_1 & - & - & x_2 & x_3 & x_4 \\ y_1 & - & y_2 & y_3 & - & y_4 & y_5 \end{pmatrix},$$

$$A_3 = \begin{pmatrix} - & x_1 & - & x_2 & x_3 & - & x_4 \\ y_1 & y_2 & y_3 & - & - & y_4 & y_5 \end{pmatrix},$$

$$A_4 = \begin{pmatrix} - & x_1 & - & x_2 & x_3 & x_4 & - \\ y_1 & y_2 & y_3 & - & y_4 & - & y_5 \end{pmatrix},$$

$$A_5 = \begin{pmatrix} - & x_1 & - & - & x_2 & x_3 & - & x_4 \\ y_1 & - & y_2 & y_3 & - & - & y_4 & y_5 \end{pmatrix},$$

$$A_6 = \begin{pmatrix} - & x_1 & - & - & x_2 & x_3 & x_4 & - \\ y_1 & - & y_2 & y_3 & - & y_4 & - & y_5 \end{pmatrix},$$

$$A_7 = \begin{pmatrix} - & x_1 & - & x_2 & x_3 & - & x_4 & - \\ y_1 & y_2 & y_3 & - & - & y_4 & - & y_5 \end{pmatrix},$$

$$A_8 = \begin{pmatrix} - & x_1 & - & - & x_2 & x_3 & - & x_4 & - \\ y_1 & - & y_2 & y_3 & - & - & y_4 & - & y_5 \end{pmatrix}.$$

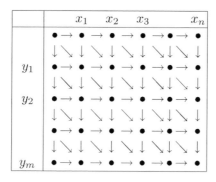

Figure 2.1. The two-dimensional lattice L of the order $(n + 1) \times (m + 1)$.

In general, any intercalated sequence containing k pairs of adjacent symbols x_i, y_j corresponds to 2^k different gapped alignments of sequences x and y. This implies not only that there exists no one-to-one correspondence between pairwise gapped alignments and intercalated sequences, but also that the number $\mathcal{N}_{n,m}$ of possible alignments of sequences $x_1, \ldots, x_n$ and $y_1, \ldots, y_m$ is strictly greater than $\binom{n+m}{m}$, the number of intercalated sequences that could be created from sequences x and y (see Problem 2.5).

The number $\mathcal{N}_{n,m}$ of possible gapped pairwise alignments

To find $\mathcal{N}_{n,m}$, we turn to a two-dimensional lattice L of the order $(n + 1) \times (m + 1)$ (see Figure 2.1).

The columns of L correspond to elements of sequence $x_1, \ldots, x_n$ (a column with index 0 is added to the left side); the rows correspond to elements of sequence $y_1, \ldots, y_m$ (a row with index 0 is added to the top). A diagonal arrow leading to cell (i, j), $i = 1, \ldots, n$, $j = 1, \ldots, m$, corresponds to aligned symbols $\binom{x_i}{y_j}$; a horizontal arrow leading to cell $(i, \bullet)$, $i = 1, \ldots, n$, corresponds to x_i aligned to a gap $\binom{x_i}{-}$; a vertical arrow leading to cell $(\bullet, j)$, $j = 1, \ldots, m$, corresponds to y_j aligned to a gap $\binom{-}{y_j}$. Let cell $(0, 0)$ be an *original corner* and let cell (n, m) be an *end corner* of L. It is easy to see that any continuous path (a broken line of arrows) from the original corner to the end corner uniquely determines a particular gapped alignment of sequences x and y. Conversely, any alignment of x and y corresponds to only one continuous path through the lattice L connecting its original and end corners. Thus, there is one-to-one correspondence between gapped alignments of x and y and continuous paths through L from the original corner to the end corner. Hence, finding the number of such full paths will determine $\mathcal{N}_{n,m}$ at the same time.

We assign horizontal, vertical, and diagonal arrows to vectors $(1, 0)$, $(0, 1)$, and $(1, 1)$, respectively. Every admissible path through L corresponds to a sequence of vectors $(i_1, j_1), \ldots, (i_q, j_q)$, where each (i_l, j_l) is either vector $(1, 0)$, or $(0, 1)$, or

(1, 1), and

$$\sum_{l=1}^{q} (i_l, j_l) = (n, m). \tag{2.1}$$

On the other hand, every such sequence determines one admissible path, i.e. one alignment. Now we have to determine how many such sequences exist. If the number of vectors $(1, 1)$ in the sum in (2.1) is k, $k = 0, \ldots, \min(n, m)$, then the number of vectors $(1, 0)$ and $(0, 1)$ must be equal to $n - k$ and $m - k$, respectively, with the total number q of terms in the sum in (2.1) equal to $n + m - k$. The number of ways to arrange these vectors in sequence $(i_1, j_1), \ldots, (i_{n+m-k}, j_{n+m-k})$ equals the multinomial coefficient $\binom{n+m-k}{k, m-k, n-k}$. Hence, the number of permitted paths through L (and the number of gapped alignments of sequences of lengths n and m) is equal to

$$\mathcal{N}_{n,m} = \sum_{k=0}^{\min(n,m)} \binom{n+m-k}{k, m-k, n-k}. \tag{2.2}$$

Formula (2.2) leads to several corollaries.

Corollary 1: Recurrent formula for $\mathcal{N}_{n,m}$

The reductionist approach recommends the reduction of a complex problem to a combination of simpler ones. We now discuss the implementation of this general principle. The whole set of possible alignments could be divided into three classes: (i) alignments ended with the aligned pair $\binom{x_n}{y_m}$; (ii) alignments ended with a gap in sequence x, $\binom{-}{y_m}$; (iii) alignments ended with a gap in sequence y, $\binom{x_n}{-}$. Therefore, a gapped alignment of sequences $x_1, \ldots, x_n$ and $y_1, \ldots, y_m$ can be obtained from shorter alignments in one of three possible ways depending on a pair of symbols in the alignment last column: from an alignment of sequences $x_1, \ldots, x_{n-1}$ and $y_1, \ldots, y_{m-1}$ by adding the aligned pair $\binom{x_n}{y_m}$; from an alignment of sequences $x_1, \ldots, x_n$ and $y_1, \ldots, y_{m-1}$ by adding the pair $\binom{-}{y_m}$; or from an alignment of sequences $x_1, \ldots, x_{n-1}$ and $y_1, \ldots, y_m$ by adding the pair $\binom{x_n}{-}$. This virtual construction gives the following recurrent formula for the number of gapped alignments of two sequences with lengths n and m:

$$\mathcal{N}_{n,m} = \mathcal{N}_{n-1,m-1} + \mathcal{N}_{n,m-1} + \mathcal{N}_{n-1,m}. \tag{2.3}$$

The initial values of $\mathcal{N}_{n,m}$ ($\mathcal{N}_{n,0}$ and $\mathcal{N}_{0,m}$) are defined by the direct application of formula (2.2):

$$\mathcal{N}_{n,0} = \binom{n}{0, 0, n} = \frac{n!}{0! 0! n!} = 1,$$

$$\mathcal{N}_{0,m} = \binom{m}{0, m, 0} = \frac{m!}{0! m! 0!} = 1. \tag{2.4}$$

Table 2.3. *Results of* $\mathcal{N}_{n,m}$
computations for small n and m
values

	0	1	2	3	4
0	1	1	1	1	1
1	1	3	5	7	9
2	1	5	13	25	41
3	1	7	25	63	129
4	1	9	41	129	321

From Equations (2.4) and (2.3) the values of $\mathcal{N}_{n,m}$ can be computed for any integers n and m. For instance, for $n, m \leq 4$ the values of $\mathcal{N}_{n,m}$ are listed in Table 2.3.

Corollary 2: Number of gapped alignments without aligned pairs of symbols

The term corresponding to $k = 0$ in (2.2) gives the number of alignments containing no aligned pairs of symbols. Although the case $k = 0$ is not of a serious practical interest, it is easy to see that

$$a_0 = \binom{n+m}{0, m, n} = \frac{(n+m)!}{0!m!n!} = \binom{n+m}{m},$$

which means that a_0 is equal to the number of intercalated sequences of lengths n and m (Problem 2.5). One-to-one correspondence between these two sets can be established by the procedure proposed in the statement of Problem 2.5.

Note that for $n, m > 1$ the term for $k = 1$ in formula (2.2) is strictly greater than a_0:

$$a_1 = \binom{n+m-1}{1, m-1, n-1} = \frac{(n+m-1)!}{1!(m-1)!(n-1)!} = \frac{nm}{n+m}\binom{n+m}{m}$$

$$= \frac{nm}{n+m} a_0 > a_0.$$

Corollary 3: Asymptotic behavior of $\mathcal{N}_{n,n}$, lower and upper bounds

For sequences x and y with equal lengths, $n = m$, formula (2.2) becomes

$$\mathcal{N}_{n,n} = \sum_{k=0}^{n} \binom{2n-k}{k, n-k, n-k}. \tag{2.5}$$

As n increases to infinity, an asymptotic behavior of the right-hand side terms in Equation (2.5) could be elucidated. First we note that the term for $k = 0$ (see

Problem 2.7) is given by

$$a_0 = \binom{2n}{0,n,n} = \binom{2n}{n} \simeq \frac{4^n}{\sqrt{\pi n}}.$$

The term for $k = 1$ (see corollary 2) is given by

$$a_1 = \frac{n^2}{2n} a_0 \simeq \frac{4^n \sqrt{n}}{2\sqrt{\pi}}.$$

For $k = n$, the case of an ungapped alignment with n aligned pairs of symbols, we have

$$a_n = \binom{n}{n,0,0} = \frac{n!}{n!} = 1.$$

In general, for $a_k = \binom{2n-k}{k,n-k,n-k}$ the analysis of the ratio

$$\frac{a_{k+1}}{a_k} = \frac{(n-k)^2}{(2n-k)(k+1)}$$

shows that terms of the sum (2.5) increase as k grows from zero to

$$k^* = \left[n - \frac{1}{4} - \frac{\sqrt{8n^2 + 8n + 1}}{4} \right],$$

and decrease as k grows from k^* to n. For large n we assume that $k^* \approx (1-(1/\sqrt{2}))n$ and find by applying Stirling's formula that the maximum term in the sum (2.5) can be approximated by the following expression:

$$a_{k^*} \simeq \frac{(3 + 2\sqrt{2})^n}{2\pi n}.$$

For large values of n we have

$$\max_k a_k = a_{k^*} < \mathcal{N}_{n,n} < (n+1)a_{k^*};$$

and, therefore, the lower and upper bounds for the number $\mathcal{N}_{n,n}$ of gapped alignments of two sequences both of length n are as follows:

$$\frac{(3 + 2\sqrt{2})^n}{2\pi n} < \mathcal{N}_{n,n} < \frac{(3 + 2\sqrt{2})^n}{\pi}.$$

Remark 1 An asymptotic expression for $\mathcal{N}_{n,n}$ for large n,

$$\mathcal{N}_{n,n} \simeq \frac{1}{2^{5/4}\sqrt{\pi}} \frac{(3 + 2\sqrt{2})^{(n+1/2)}}{\sqrt{n}},$$

was derived by Laquer (1981).

Remark 2 Some pairwise alignment algorithms such as the Viterbi algorithm for a pair hidden Markov model (HMM) discussed in Chapter 4 (see Figure 4.3 and recurrence Equations (4.5)–(4.7)) admit to competition for optimality only those alignments of sequences x and y, where x-gap does not follow y-gap and vice versa. Let $\mathcal{A}_{n,m}$ be the set of such pairwise alignments of $x = x_1, \ldots, x_n$ and $y = y_1, \ldots, y_m$. The number $|\mathcal{A}_{n,m}|$ is to be determined.

First, note that the procedure described in Problem 2.6 does not establish a one-to-one correspondence between the set of intercalated sequences of length $n + m$ and set $\mathcal{A}_{n,m}$. While an alignment from $\mathcal{A}_{n,m}$ corresponds to some intercalated sequence, there exist intercalated sequences which do not correspond to any alignment in $\mathcal{A}_{n,m}$. (For example, no alignment from $\mathcal{A}_{3,3}$ corresponds to the intercalated sequence $x_1, y_1, y_2, x_2, x_3, y_3$.)

To find $|\mathcal{A}_{n,m}|$, the scheme used above to calculate the number $\mathcal{N}_{n,m}$ of all possible alignments of x and y has to be modified. Now in the sum (2.1) the order of vectors $(1, 1)$, $(1, 0)$, and $(0, 1)$, corresponding to a pair of aligned symbols, y-gap and x-gap, respectively, must satisfy the following restriction: vectors $(1, 0)$ and $(0, 1)$ cannot be adjacent to each other. The number of sequences of vectors $(1, 1)$, $(1, 0)$, and $(0, 1)$ satisfying this restriction is equal to $|\mathcal{A}_{n,m}|$. If the number of vectors $(1, 1)$ in the sum (2.1) is k, $1 \le k \le \min(n, m)$, they are present in the sum in the form of i runs, $1 \le i \le k$, that correspond to blocks of successive matches in the alignment. The number of ways to divide the set of k elements into i non-empty subsets is equal to $N(k, i) = \binom{k-1}{k-i}$, the number of ways to place k undistinguishable items into i boxes (see Problem 11.14). For each value i, the number of blocks of vectors $(1, 1)$ within the sum (2.1), there are three possibilities: (a) the sum in (2.1) starts and ends with blocks of $(1, 1)$; (b) the sum in (2.1) either starts or ends with block of $(1, 1)$; (c) the sum in (2.1) neither starts nor ends with a block of $(1, 1)$. In case (a), $i - 1$ blocks of $(0, 1)$ or blocks of $(1, 0)$ must be located between i blocks of vectors $(1, 1)$. If $n - k$ vectors $(1, 0)$, corresponding to y-gaps, are divided into l blocks, $l \le i - 1$ (there are $N(n - k, l)$ ways to do so), then $m - k$ vectors $(0, 1)$, corresponding to x-gaps, must be divided into $i - 1 - l$ remaining blocks (there are $N(m - k, i - 1 - l)$ ways to do so). The number of ways to place l blocks of $(1, 0)$ into $i - 1$ possible locations while preserving the order of blocks is $\binom{i-1}{l}$. Now the positions for $i - 1 - l$ blocks of vectors $(0, 1)$ are determined uniquely. When the order of all blocks (and their lengths) in the sum (2.1) is fixed, the chain of vectors is uniquely defined. Therefore, the number N_1 of sequences corresponding to (a) becomes

$$N_1 = \sum_{k=1}^{\min(n,m)} \sum_{i=1}^{k} N(k, i) \sum_{l=0}^{i-1} \binom{i-1}{l} N(n - k, l) N(m - k, i - 1 - l).$$

Similar arguments lead to the formulas for the numbers N_2 and N_3 of possible sequences of vectors in the sum (2.1) for cases (b) and (c), respectively:

$$N_2 = 2 \sum_{k=1}^{\min(n,m)} \sum_{i=1}^{k} N(k,i) \sum_{l=0}^{i} \binom{i}{l} N(n-k,l)N(m-k,i-l);$$

$$N_3 = \sum_{k=1}^{\min(n,m)} \sum_{i=1}^{k} N(k,i) \sum_{l=0}^{i+1} \binom{i+1}{l} N(n-k,l)N(m-k,i+1-l).$$

By combining together the numbers of sequences for cases (a)–(c), we obtain

$$|\mathcal{A}_{n,m}| = N_1 + N_2 + N_3$$

$$= \sum_{k=1}^{\min(n,m)} \sum_{i=1}^{k} \binom{k-1}{k-i} \left[\sum_{l=0}^{i-1} \binom{i-1}{l} N(n-k,l)N(m-k,i-1-l) \right.$$

$$+ 2 \sum_{l=0}^{i} \binom{i}{l} N(n-k,l)N(m-k,i-l)$$

$$\left. + \sum_{l=0}^{i+1} \binom{i+1}{l} N(n-k,l)N(m-k,i+1-l) \right].$$

Here $N(0,0) = 1$; $N(q,0) = N(0,q) = 0$ for $q \geq 1$; $N(q,j) = \binom{q-1}{q-j}$ for $j \geq 1$, $q \geq j$; $N(q,j) = 0$ for $j \geq 1$, $1 \leq q < j$.

For example, for two short sequences $x = x_1, x_2$ and $y = y_1, y_2, y_3$, formula (2.2) yields for the number of all possible gapped alignments of x and y: $N_{2,3} = 25$, while $|\mathcal{A}_{2,3}| = 5$. The five alignments in $\mathcal{A}_{2,3}$ are as follows:

$$\begin{pmatrix} x_1 & x_2 & - \\ y_1 & y_2 & y_3 \end{pmatrix}, \begin{pmatrix} - & x_1 & x_2 \\ y_1 & y_2 & y_3 \end{pmatrix}, \begin{pmatrix} x_1 & x_2 & - & - \\ - & y_1 & y_2 & y_3 \end{pmatrix},$$

$$\begin{pmatrix} - & - & x_1 & x_2 \\ y_1 & y_2 & y_3 & - \end{pmatrix}, \begin{pmatrix} x_1 & - & x_2 \\ y_1 & y_2 & y_3 \end{pmatrix}.$$

By following the same logic as in Corollary 1, one can derive for $|\mathcal{A}_{n,m}|$ the analog of formula (2.3) obtained for $\mathcal{N}_{n,m}$. We leave to the reader to check that the following recurrent equation holds:

$$|\mathcal{A}_{n,m}| = |\mathcal{A}_{n-1,m-1}| + \sum_{i=2}^{m} |\mathcal{A}_{n-1,i-2}| + \sum_{i=2}^{n} |\mathcal{A}_{i-2,m-1}|.$$

On the right-hand side of the equation the first term is the number of alignments in $\mathcal{A}_{n,m}$ with aligned symbols x_n and y_m; the sum $\sum_{i=2}^{m}$ is the number of alignments in $\mathcal{A}_{n,m}$ with y_m aligned to a gap; and the sum $\sum_{i=2}^{n}$ is the number of alignments in $\mathcal{A}_{n,m}$ with x_n aligned to a gap.

Table 2.4. *Results of $\mathcal{A}_{n,m}$ computations for small numbers n and m*

	0	1	2	3	4
0	1	1	1	1	1
1	1	1	2	3	4
2	1	2	3	5	8
3	1	3	5	9	15
4	1	4	8	15	27

Assuming that $|\mathcal{A}_{n,0}| = |\mathcal{A}_{0,m}| = 1$, we calculate the values $\mathcal{A}_{n,m}$ for $n, m \leq 4$ and list them in Table 2.4.

Remark (suggested by Anders Krogh) Let us consider subalignments between aligned pairs of symbols in a gapped pairwise alignment (with alignment ends as aligned pairs). In the example

$$
\begin{array}{ccccccccc}
* & - & x & X & - & - & X & * \\
* & y & - & Y & y & y & Y & *
\end{array}
$$

we put capital letters for aligned pairs and aligned $*$'s at the ends. Let us refer to an alignment section between aligned pairs as an "unalignment." Two subsequences can be "unaligned" in many different ways; for instance, xx and yy can give these "unalignments":

$$
\begin{pmatrix} - & - & x & x \\ y & y & - & - \end{pmatrix}, \begin{pmatrix} - & x & - & x \\ y & - & y & - \end{pmatrix}, \begin{pmatrix} x & - & - & x \\ - & y & y & - \end{pmatrix},
$$
$$
\begin{pmatrix} - & x & x & - \\ y & - & - & y \end{pmatrix}, \begin{pmatrix} x & - & x & - \\ - & y & - & y \end{pmatrix}, \begin{pmatrix} x & x & - & - \\ - & - & y & y \end{pmatrix}.
$$

One may argue that it is senseless to count these as separate alignments, because there is no reasonable way to discriminate between them (unless you have actually witnessed the evolution step by step). With affine gap scoring, only the first and the last would have a chance to survive in the procedure of building the optimal alignment; and with many programs none of them would, because gaps following gaps are not allowed. However, if we count all these unalignments once only, and let them be represented by the alignment with all the y-insertions before x-insertions (the first form in the example above), then it is easy to see that the number of such alignments, $N_{n,m}$, is equal to $\binom{n+m}{n}$ since there is precisely one alignment for any intercalation and vice-versa.

We will show that there is a one-to-one mapping between intercalations and alignments in which we allow insertions in the lower (y) sequence before insertions in the upper (x) sequence, but not vice versa. To show this, we need to show that any alignment maps to one, and only one, intercalation, that any two alignments map to different intercalations, and that there exists an alignment which maps to any intercalation. The first part is easy, because there is exactly one intercalation for a given alignment, which follows from the method of construction of an intercalation from an alignment described in Problem 2.5.

An intercalation is interpreted as follows.

(1) Any pair xy corresponds to an aligned pair in the alignment. Add virtual aligned pairs at the ends of the intercalation.

(2) Between two aligned pairs we can have some y's followed by some x's in the intercalation. If the numbers of y's and x's are k and l, respectively, it means that we have k y-insertions followed by l x-insertions in the alignment. Both k and l may be zero. We cannot have x's followed by y's, because this would infer that we forgot to mark an aligned pair.

This shows that there exists an allowed alignment for any intercalation.

Finally, it is clear that only one allowed alignment maps to a given intercalation. Assume that two alignments give the same intercalation. Then they would have to have the same set of aligned pairs and the same number of y-insertions and x-insertions between all the aligned pairs, so the two alignments would be identical.

This concludes the proof of the one-to-one correspondence; hence, $N_{n,m} = \binom{n+m}{n}$. □

Problem 2.7 Use Stirling's formula ($x! \simeq \sqrt{2\pi}\, x^{x+\frac{1}{2}} e^{-x}$) to prove the following equation:

$$\binom{2n}{n} \simeq \frac{2^{2n}}{\sqrt{\pi n}}.$$

Solution To prove the above equality, we use Stirling's formula for $(2n)!$ and $n!$:

$$\binom{2n}{n} = \frac{(2n)!}{(n!)^2} \simeq \frac{\sqrt{2\pi}\,(2n)^{(2n+\frac{1}{2})} e^{-2n}}{2\pi n^{2n+1} e^{-2n}}$$

$$= \frac{2^{2n}\sqrt{2n}^{(2n+\frac{1}{2})} e^{-2n}}{\sqrt{2\pi}\, n^{2n+1} e^{-2n}} = \frac{2^{2n}}{\sqrt{\pi n}}. \qquad \Box$$

Problem 2.8 Find two equal-scoring optimal alignments in the dynamic programming matrix in Table 2.5.

Table 2.5. *The sequence alignment matrix obtained by dynamic programming for sequences HEAGAWGHEE and PAWHEAE*

	H	E	A	G	A	W	G	H	E	E
0	← −8	← −16	← −24	← −32	← −40	← −48	← −56	← −64	← −72	← −80
P −8	−2	−9	← −17	← −25	−33	← −42	← −49	← −57	−65	−73
A −16	−10	−3	−4	← −12	−20	← −28	← −36	← −44	← −52	← −60
W −24	−18	−11	−6	−7	−15	−5	← −13	← −21	← −29	← −37
H −32	−14	−18	−13	−8	−9	−13	−7	−3	← −11	← −19
E −40	−22	−8	← −16	−16	−9	−12	−15	−7	**3**	−5
A −48	−30	−16	−3	← −11	−11	−12	−12	−15	−5	2
E −56	−38	−24	−11	−6	−12	−14	−15	−12	−9	**1**

Solution The existence of several equally scoring optimal alignments is betrayed by "forks" in the traceback graph spanning the dynamic programming (DP) matrix (Table 2.5). A fork occurs if the traceback procedure recovering the optimal path arrives at a cell (i, j) of the DP matrix whose optimal value $F(i, j)$ was derived from more than one "parent" cells. Alternative use of the parent cells by the traceback procedure generates different paths through the DP matrix and, hence, different optimal alignments.

In the DP matrix for sequences *HEAGAWGHEE* and *PAWHEAE* (Table 2.2), the value of $F(3, 1)$ on the traceback path can be obtained from both $(2, 0)$ and $(2, 1)$ cells:

$$F(3, 1) = F(2, 0) + s(A, P) = -16 - 1 = -17,$$

$$F(3, 1) = F(2, 1) - d = -9 - 8 = -17.$$

These possibilities correspond to two optimal alignments with score 1:

```
H  E  A  G  A  W  G  H  E  -  E
-  -  P  -  A  W  -  H  E  A  E
```

and

```
H  E  A  G  A  W  G  H  E  -  E
-  P  -  -  A  W  -  H  E  A  E
```
□

Problem 2.9 Calculate the DP matrix and an optimal alignment for the DNA sequences *GAATTC* and *GATTA*, scoring +2 for a match, −1 for a mismatch, and with a linear gap penalty of $d = 2$.

Table 2.6. *The sequence alignment matrix obtained by dynamic programming for sequences GATTC and GATTA*

		G	A	A	T	T	C
	0	−2	−4	−6	−8	−10	−12
G	−2	2	0	−2	−4	−6	−8
A	−4	0	4	2	0	−2	−4
T	−6	−2	2	3	4	2	0
T	−8	−4	0	1	5	6	4
A	−10	−6	2	2	3	4	5

Solution Computation of the DP matrix and subsequent determination of back-tracking pointers produce the result shown in Table 2.6.

The traceback algorithm recovers two equally scoring optimal alignments (with score 5):

$$
\begin{array}{ccccccc}
G & A & A & T & T & C \\
G & - & A & T & T & A
\end{array}
$$

and

$$
\begin{array}{ccccccc}
G & A & A & T & T & C \\
G & A & - & T & T & A
\end{array}
$$

Problem 2.10 Calculate the score of the example alignment

$$
\begin{array}{cccccccc}
V & L & S & P & A & D & - & K \\
H & L & - & - & A & E & S & K
\end{array}
$$

with $d = 12$, $e = 2$.

Solution Use of the substitution scores defined by the BLOSUM50 matrix (Table 2.1) allows us to compute

$$S = s(V,H) + s(L,L) - d - e + s(A,A) + s(D,E) - d + s(K,K) = -12.$$

Problem 2.11 Fill the correct values of $c(i,j)$ for the global alignment of the example pair of sequences *HEAGAWGHEE* and *PAWHEAE* for the first pass of the linear space algorithm ($u = 5$).

Remark The linear space algorithm (Myers and Miller, 1988) constructs the optimal alignment of two sequences of lengths n and m using memory space $O(n+m)$ instead of $O(n \times m)$ necessary for the standard DP algorithm. This reduction is important for sequences of large lengths. The major challenge in devising the linear space algorithm is to overcome the requirement to store the DP matrix of scores of the size $n \times m$. It is feasible to find a score of global (or local) pairwise alignment using $O(n+m)$ storage since the standard DP algorithm can be performed column by column (or row by row) while keeping only data needed to compute a new column (row). Similarly, "extended traceback pointers" $c(i,j)$ can be computed column by column and kept upon computation of the scores in each column i, $i \geq u+1$. Thus, the space required for storing extended pointers is also $O(n+m)$. A result of the first pass of the algorithm is finding the cell $(u, c(n,m))$, the intersection of column u, and the optimal path through the DP matrix. This information allows to divide the initial dynamic programming problem into two parts by working on two separate submatrices (with diagonals from $(0,0)$ to $(u, c(n,m))$ and from $(u, c(n,m))$ to (n,m), respectively). Each of the submatrices can be split in the same way, and the procedure continues until the submatrices obtained at the last step contain only one column (or one row). Now, the whole optimal alignment can be determined uniquely through the collection of cells (u', v'), where the optimal path intercepts all m columns of the DP matrix.

Solution The first pass of the algorithm should determine $c(i,j)$ values starting with the fifth column ($u = n/2 = 5$). When we calculate $c(i,j)$, using the values $F(i,j)$ of the DP matrix, we have to remember that, in the real algorithm, after computation of a new column of values $F(i,j)$ and $c(i,j)$ the data in the previous column are erased.

We fill in the cells of matrix C by the values $c(i,j)$: $c(5,j) = j$, for any j; for $i > 5$ $c(i,j) = c(i',j')$ if value $F(i,j)$ in cell (i,j) of the DP matrix (Table 2.5) is derived from value $F(i',j')$ in cell (i',j'). The calculations produce the result shown in Table 2.7.

The value $c(10,7) = 2$ defines the extended traceback pointer to cell $(u,v) = (5,2)$ that belongs to the optimal path. Cell $(5,2)$ is the point of splitting of the DP matrix into two submatrices of sizes 5×2 and 5×5, which should be used in the second pass of the linear space algorithm. $\qquad \square$

Problem 2.12 The linear space algorithm allows us to decrease the memory required to find the optimal alignment of two sequences of lengths n and m from $O(nm)$ (as in the standard DP algorithm) to $O(n+m)$. Show that the time required by the linear space algorithm is only about twice that of the standard $O(nm)$ algorithm.

Table 2.7. *The matrix of*
$c(i,j)$ values obtained by
the linear space algorithm

$c(i,j)$	5	6	7	8	9	10
0	0	0	0	0	0	0
1	1	1	1	1	0	0
2	2	2	2	2	2	2
3	3	2	2	2	2	2
4	4	2	2	2	2	2
5	5	4	2	2	2	2
6	6	5	4	2	2	2
7	7	6	5	4	2	2

Solution First, we estimate the time required to run the standard global alignment algorithm with a linear gap scoring. (An affine gap scoring can be worked out in a similar way.) The algorithm fills in $n \times m$ cells of the score matrix $F(i,j)$, performing four operations for each cell: calculations of values $F(i-1,j-1)+s(i,j)$, $F(i-1,j)-d$, $F(i,j-1)-d$, and their maximum. Hence, the total running time (measured in the number of operations) of a global alignment algorithm is given by $T_{St} \approx 4nm$.

For the linear space algorithm the total running time is the sum of the times required for N passes:

$$T_{LS} = T_1 + T_2 + \cdots + T_N.$$

In the first pass it computes $m \times n$ scores $F(i,j)$, fills in $(m \times n)/2$ values $c(i,j)$, and eventually finds the value $v = c(n,m)$. So, $T_1 \approx 4nm$. In the second pass it computes the elements of two submatrices $F_1(i,j)$ and $F_2(i,j)$ of approximate size $(n/2) \times (m/2)$ (actual orders depend on v), fills in values of $c_1(i,j)$ and $c_2(i,j)$, and finds v_1 and v_2. Then, $T_2 \approx (n/2) \times (m/2) \times 4 \times 2 = 2nm$. As the algorithm continues, in the kth pass it computes elements of 2^{k-1} submatrices of sizes $(n/2^{k-1}) \times (m/2^{k-1})$, thus $T_k \approx nm/2^{k-3}$. Therefore, the time required for the linear space algorithm is estimated as

$$T_{LS} \approx \sum_{i=1}^{N} \frac{nm}{2^{i-3}} = 8nm \sum_{i=1}^{N} \left(\frac{1}{2}\right)^i < 8nm \sum_{i=1}^{\infty} \left(\frac{1}{2}\right)^i = 8nm.$$

Remark The maximum number N of passes of the linear space algorithm can be estimated as follows. The sizes of the submatrices ($n' \times m'$) decrease at least twice during each pass of the algorithm. The "divide and conquer" process continues until the last remaining submatrix is reduced to only one cell (or to the matrix with only one row or one column). For the longest branch of the matrix division process we have

$$\frac{\max(n,m)}{2^N} \approx 1.$$

Thus, $N \approx [\log_2 \max(n,m)]$.   □

2.2 Additional problems and theory

Several problems included in this section deal with the derivation of the substitution score matrix, the cornerstone of a scoring system for pairwise sequence alignments (Problems 2.15–2.17), as well as with the score distributions of high-scoring local alignments of two random sequences, needed for the assessment of statistical significance of the local pairwise alignment (Problems 2.18–2.21).

Since the original work by Dayhoff, Schwartz, and Orcutt (1978) on the derivation of the PAM family of substitution matrices is rather difficult to find, we provide an extended discussion of their method in Section 2.2.1. We also illustrate by a "toy" example (Problem 2.17) the ideas that lead to the derivation of the BLOSUM family of substitution matrices (Henikoff and Henikoff, 1992).

A brief review of the theoretical results on the distribution of the statistics of the scores of high-scoring pairs in the alignment of random sequences (ungapped and gapped cases) is given in the theoretical introductions to Problems 2.18 and 2.19. In the theoretical introduction to Problem 2.20, we review results on the distribution of the length of the longest common word among several unrelated random sequences. The expected number of words of fixed length common to two DNA fragments is calculated in Problem 2.21.

Problem 2.13 Prove that the running time of the DP algorithm for optimal pairwise alignment of two sequences of lengths n with a gap penalty function of a general form is $O(n^3)$.

Solution For a gap scoring function $\gamma(g)$ of a general form the DP equation for the optimal score $F(i,j)$ is given by the following formula:

$$F(i,j) = \max \begin{cases} F(i-1,j-1) + s(x_i,y_j), & \\ F(k,j) + \gamma(i-k), & k = 0,\ldots,i-1; \\ F(i,k) + \gamma(j-k), & k = 0,\ldots,j-1. \end{cases} \qquad (2.6)$$

Here $s(x_i, y_j)$ is the substitution score for the pair of residues (x_i, y_j). To determine $F(i,j)$, Equation (2.6) requires us to perform $2(i+j)+2$ operations. Then, the

Table 2.8. *The sequence alignment matrix obtained by dynamic programming for sequences TACGA and ACTGAC*

		T	A	C	G	A
	0	−2	−4	−6	−8	−10
A	−2	−1	0	−2	−4	−6
C	−4	−3	−2	2	0	−2
T	−6	−2	−4	0	1	−1
G	−8	−4	−3	−2	2	0
A	−10	−6	−2	−4	0	4
C	−12	−8	−4	0	−2	2

number of operations required to compute the whole $n \times n$ matrix is given by

$$T_n = 2 \sum_{i=0}^{n} \sum_{j=0}^{n} (i+j+1) = 2 \left((n+1) \sum_{i=0}^{n} i + (n+1) \sum_{j=0}^{n} j + n^2 \right)$$

$$= 4(n+1) \frac{n(n+1)}{2} + n^2 = 2n^3 + 5n^2 + 2n = O(n^3). \qquad \square$$

Problem 2.14 Find the optimal pairwise global alignment of the sequences *TACGAGTACGA* and *ACTGACGACTGAC* with the condition that **G** nucleotides shown in bold font must be aligned together. The scoring parameters are defined as +2 for match, −1 for mismatch, and $d = -2$ for a linear gap penalty.

Solution It is easy to see that the middle nucleotides **G** divide each sequence into two identical subsequences. Hence, if we find the optimal alignment for subsequences *TACGA* and *ACTGAC*, the optimal global alignment satisfying the defined above condition will be the concatenation of the two "subalignments" with a pair of aligned G's between them. We compute the DP matrix (see Tabe 2.8), determine the score of the optimal "subalignment," and, subsequently, construct the "subalignment" itself by applying a traceback procedure. The score of the optimal subalignment $\tilde{S} = F(5,6) = 2$. The total score of conditional global alignment is the sum of scores, $S = \tilde{S} + s(G,G) + \tilde{S} = 2 + 2 + 2 = 6$, and the alignment itself comes out as

$$
\begin{array}{ccccccccccccc}
T & A & C & - & G & A & - & G & T & A & C & - & G & A & - \\
- & A & C & T & G & A & C & G & - & A & C & T & G & A & C
\end{array}
$$

Remark It turns out that the conditional optimal alignment shown above coincides with one of the two "unconditional" optimal global alignments of the given sequences. The second optimal global alignment (with the same score, 6) is given by

$$
\begin{array}{ccccccccccccc}
T & A & C & - & G & A & G & T & A & C & - & G & A & - \\
- & A & C & T & G & A & C & G & A & C & T & G & A & C
\end{array}
$$

In this alignment two **G**'s are not aligned. Still, the loss of the positive score for the match is compensated for by the decrease in the number of gaps.   □

Problem 2.15 A substitution scoring matrix for alignment of nucleotide sequences is given as follows (with the log-odds scores defined in bits):

	T	C	A	G
T	1	0	−1	−1
C	0	1	−1	−1
A	−1	−1	1	0
G	−1	−1	0	1

(a) Determine the average score per nucleotide pair for DNA sequences described by the independence model with equal probabilities of nucleotides ($\frac{1}{4}$).

(b) Determine the "target frequencies" of nucleotide pairs this matrix is designed to search for in alignments of evolutionary related sequences.

Solution (a) In the ungapped alignment of two random nucleotide sequences the average score H per aligned pair is given by

$$
H = \sum_{i,j} q_i q_j s_{ij}.
$$

Here s_{ij} are the elements of the substitution scoring matrix, the sum is taken over all sixteen possible pairs of nucleotides, and q_i, q_j are the probabilities of nucleotides of types i and j under the uniform independence model. Therefore,

$$
H = \frac{1}{16} \sum_{i,j} s_{ij} = -\frac{4}{16} = -0.25.
$$

(b) By the definition of the substitution score as the log-odds score,

$$s_{ij} = \log_2 \frac{p_{ij}}{q_i q_j},$$

where p_{ij} are the "target frequencies" of the aligned pair (i, j) of nucleotides. Then the value of the target frequency $p_{ij} = (1/q_i q_j)2^{s_{ij}}$. For example, $p_{TT} = 1/16 \times 2 = 1/8$. All the "target frequencies" p_{ij} are shown in the following table:

	T	C	A	G
T	1/8	1/16	1/32	1/32
C	1/16	1/8	1/32	1/32
A	1/32	1/32	1/8	1/16
G	1/32	1/32	1/16	1/8

□

2.2.1 Derivation of the amino acid substitution matrices (PAM series)

Discussion of this method requires knowledge of the key notions of phylogenetic tree construction (Durbin *et al.* (1998), Chap. 7).

The substitution matrix is an important component of the scoring system for a pairwise sequence alignment. Dayhoff *et al.* (1978) offered a solid theoretical approach to defining the elements of substitution matrices, scaled by evolutionary distance, from counts of amino acid substitutions. These counts were calculated as frequencies of aligned residue pairs in carefully crafted alignments of closely related protein sequences from seventy-one families. Multiple alignments of these sequences were reduced to seventy-one ungapped alignment blocks (each sequence in a block had to be at least 85% identical to any other sequence in the block). The most parsimonious (with minimal number of substitutions along edges) phylogenetic tree, or several trees if parsimony was not unique, was constructed for sequences from each block. To illustrate the method, we use the artificial ungapped block considered by Dayhoff *et al.* (1978):

$$\begin{matrix} A & C & G & H \\ D & B & G & H \\ A & D & I & J \\ C & B & I & J \end{matrix} \tag{2.7}$$

The four most parsimonious trees, T_1, T_2, T_3, T_4, for sequences from block (2.7) are shown in Figure 2.2. For each pair of different amino acids (i, j) the total number a_{ij} of substitutions from i to j along the downward directed edges of trees T_k, $k = 1, \ldots, 4$, was calculated, and the matrix A of *accepted point mutations* with elements $A_{ij} = A_{ji} = a_{ij} + a_{ji}, i \neq j$, was produced (see Table 2.9).

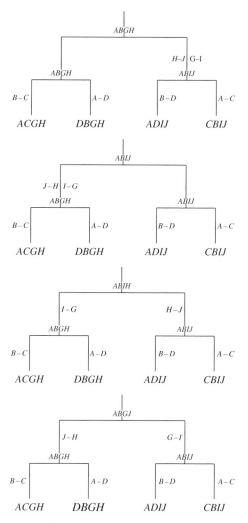

Figure 2.2. The most parsimonious trees T_1, T_2, T_3, and T_4 (in top down order) with the "observed" in block (2.7) amino acid sequences at the leaf nodes; each tree carries a total of six substitutions. The ancestor sequences (unobserved but inferred) are placed at the root and branch nodes; amino acid substitutions are indicated along the edges.

Note that we keep the term "accepted point mutation" and the two which will appear below, "relative mutability" and "mutation probability matrix," that were used by Dayhoff *et al.* (1978), although in contemporary literature the term "mutation" in this context has been replaced by the term "substitution."

Next, the relative mutability m_j for each amino acid j was determined as follows. An edge of a tree T_k, $k = 1, \ldots, 4$, is associated with the ungapped pairwise

Table 2.9. *The matrix A of accepted point mutation counts*

	A	B	C	D	G	H	I	J
A		0	4	4	0	0	0	0
B	0		4	4	0	0	0	0
C	4	4		0	0	0	0	0
D	4	4	0		0	0	0	0
G	0	0	0	0		0	4	0
H	0	0	0	0	0		0	4
I	0	0	0	0	4	0		0
J	0	0	0	0	0	4	0	

Table 2.10. *The relative amino acid mutability values m_j derived from the sequence alignment block (2.7)*

Amino acid	A	B	I	H	G	J	C	D
Changes (substitutions)	8	8	4	4	4	4	8	8
Frequency of occurrence	40	40	24	24	24	24	8	8
Relative mutability m	0.2	0.2	0.167	0.167	0.167	0.167	1	1

alignment of two sequences connected by this edge. Thus, any tree T_k in Figure 2.2 generates six alignments; for example, for T_1 they are as follows:

A	B	G	H		A	B	G	H		A	B	G	H
---	---	---	---		---	---	---	---		---	---	---	---
A	B	G	H		A	B	I	J		A	C	G	H

A	B	G	H		A	B	I	J		A	B	I	J
---	---	---	---		---	---	---	---		---	---	---	---
D	B	G	H		A	D	I	J		C	B	I	J

The relative mutability m_j is defined as the ratio of the total number of times that amino acid j has changed in all twenty-four pairwise alignments to the number of times that j has occurred in these alignments. The values of m_j are listed in Table 2.10.

Now we introduce the effective frequency f_j of an amino acid j. This notion takes into account the difference in variability of the primary structure conservation in proteins with different functional roles (thus, two alignment blocks corresponding to two different families may contribute differently to f_j, even if the number of occurrences of amino acid j in these blocks is the same). The effective frequency f_j is defined as

$$f_j = k \sum_{B_i} q_j^{B_i} N_i,$$

Table 2.11. *The amino acid effective frequencies f_j derived from the*
sequence alignment block (2.7)

Amino acid	A	B	I	H	G	J	C	D
Frequencies f	0.125	0.125	0.125	0.125	0.125	0.125	0.125	0.125

Table 2.12. *The amino acid effective frequencies f_i determined*
for the original alignment data (Dayhoff et al. (1978), Table 22)

Gly (G)	Ala (A)	Leu (L)	Lys (K)	Ser (S)	Val (V)	Thr (T)
0.089	0.087	0.085	0.081	0.070	0.065	0.058
Pro (P)	Glu (E)	Asp (D)	Arg (R)	Asn (N)	Phe (F)	Gln (Q)
0.051	0.050	0.047	0.041	0.040	0.040	0.038
Ile (I)	His (H)	Cys (C)	Tyr (Y)	Met (M)	Trp (W)	
0.037	0.034	0.033	0.030	0.015	0.010	

where the sum is taken over all alignment blocks B_i, $q_j^{B_i}$ is the observed frequency
of amino acid j in block B_i, N_i is the number of substitutions in a tree built for B_i,
and coefficient k is chosen to ensure that the sum of the frequencies f_j is 1. In our
example, with only the one block (2.7), the values of effective frequencies are equal
to the values of compositional frequencies ($f_j = q_j$) and are shown in Table 2.11.
The effective frequencies of twenty amino acids derived from seventy-one original
alignment blocks (Dayhoff *et al.*, 1978), are given in Table 2.12.

The next step is to find elements of the mutation probability matrix $M = (M_{ij})$.
The element M_{ij} defines the probability of an amino acid in column j having been
substituted by an amino acid in row i over a given evolutionary time. The non-
diagonal elements of M are defined by the following formula:

$$M_{ij} = \frac{\lambda m_j A_{ij}}{\sum_k A_{kj}}, \tag{2.8}$$

where λ is a constant to be determined below, m_j is the relative mutability of amino
acid j (Table 2.10), and A_{ij} is an element of the accepted point mutation matrix A
(Table 2.9). For the diagonal elements of M we have

$$M_{jj} = 1 - \lambda m_j. \tag{2.9}$$

Note that M is a non-symmetric matrix if $m_i \neq m_j$ for some $i \neq j$.

The coefficient λ represents a degree of freedom that could be used to connect
the matrix M with an evolutionary time scale. For instance, the coefficient λ could
be adjusted to ensure that a specified (small) number of substitutions would occur

Table 2.13. *The example mutation probability matrix for evolutionary distance 1 PAM calculated from alignment block (2.7)*

The element M_{ij} gives the probability of an amino acid in column j having been substituted by an amino acid in row i over evolutionary time 1 PAM.

	A	B	C	D	G	H	I	J
A	0.9948	0	0.0131	0.0131	0	0	0	0
B	0	0.9948	0.0131	0.0131	0	0	0	0
C	0.0026	0.0026	0.9740	0	0	0	0	0
D	0.0026	0.0026	0	0.9740	0	0	0	0
G	0	0	0	0	0.9957	0	0.0043	0
H	0	0	0	0	0	0.9957	0	0.0043
I	0	0	0	0	0.0043	0	0.9957	0
J	0	0	0	0	0	0.0043	0	0.9957

on average per hundred residues. This adjustment of λ was done by Dayhoff *et al.* (1978) in the following way. The expected number of amino acids that will remain unchanged in a protein sequence one hundred amino acids long is given by the formula $100 \sum_j f_j M_{jj} = 100 \sum_j f_j (1 - \lambda m_j)$. If only one substitution per hundred residues is allowed, then λ is calculated from equation

$$100 \sum_j f_j (1 - \lambda m_j) = 99. \tag{2.10}$$

Subsequently, Equations (2.8) and (2.9) are used for the calculation of all elements of the matrix M. Such a mutation probability matrix is associated with an evolutionary time interval 1 PAM (one accepted point mutation per hundred amino acids), and is called the 1PAM matrix. From the sequence data in alignment block (2.7) we obtain $\lambda = 0.0261$ and the example matrix M (Table 2.13). The actual 1 PAM matrix M derived from the original data in Dayhoff *et al.* (1978) is shown in Table 2.14.

Further, it is assumed that the mutation probability matrix M serves as the matrix of transition probabilities for the stationary (homogeneous) Markov chain x_n, $n \in \mathbf{N}$, the model of the evolutionary change at each site of a protein sequence. Therefore, the stochastic 1 PAM matrix M associated with one unit (1 PAM) of evolutionary distance can be used for derivation of the matrices associated with larger evolutionary distances (multiples of 1 PAM). Then the mutation probability matrix for evolutionary distance n PAM will coincide with the matrix M^n of transition probabilities of the Markov chain x_n for n time units, the nth power of matrix M. (For how to derive a Markov process, which is a continuous-time version of Markov chain x_n, see Problem 8.18.) It is easy to check by direct calculation that M^3, the mutation probability matrix for time units 3 PAM, is strictly positive;

Table 2.14. *Mutation probability matrix M for the evolutionary distance 1 PAM (i.e., one accepted point mutation per hundred amino acids)*

The element of this matrix, M_{ij}, gives the probability of an amino acid in column j having been substituted by an amino acid in row i over the evolutionary time interval of 1 PAM. The actual transition probabilities are multiplied by 10000 to simplify the matrix appearance (Dayhoff *et al.* (1978), Fig. 82).

	Ala	Arg	Asn	Asp	Cys	Gln	Glu	Gly	His	Ile	Leu	Lys	Met	Phe	Pro	Ser	Thr	Trp	Tyr	Val
Ala	9867	2	9	10	3	8	17	21	2	6	4	2	6	2	22	35	32	0	2	18
Arg	1	9913	1	0	1	10	0	0	10	3	1	19	4	1	4	6	1	8	0	1
Asn	4	1	9822	36	0	4	6	6	21	3	1	13	0	1	2	20	9	1	4	1
Asp	6	0	42	9859	0	6	53	6	4	1	0	3	0	0	1	5	3	0	0	1
Cys	1	1	0	0	9973	0	0	0	1	1	0	0	0	0	1	5	1	0	3	2
Gln	3	9	4	5	0	9876	27	1	23	1	3	6	4	0	6	2	2	0	0	1
Glu	10	0	7	56	0	35	9865	4	2	3	1	4	1	0	3	4	2	0	1	2
Gly	21	1	12	11	1	3	7	9935	1	0	1	2	1	1	3	21	3	0	0	5
His	1	8	18	3	1	20	1	0	9912	0	1	1	0	2	3	1	1	1	4	1
Ile	2	2	3	1	2	1	2	0	0	9872	9	2	12	7	0	1	7	0	1	33
Leu	3	1	3	0	0	6	1	1	4	22	9947	2	45	13	3	1	3	4	2	15
Lys	2	37	25	6	0	12	7	2	2	4	1	9926	20	0	3	8	11	0	1	1
Met	1	1	0	0	0	2	0	0	0	5	8	4	9874	1	0	1	2	0	0	4
Phe	1	1	1	0	0	0	0	1	2	8	6	0	4	9946	0	2	1	3	28	0
Pro	13	5	2	1	1	8	3	2	5	1	2	2	1	1	9926	12	4	0	0	2
Ser	28	11	34	7	11	4	6	16	2	2	1	7	4	3	17	9840	38	5	2	2
Thr	22	2	13	4	1	3	2	2	1	11	2	8	6	1	5	32	9871	0	2	9
Trp	0	2	0	0	0	0	0	0	0	0	0	0	0	1	0	1	0	9976	1	0
Tyr	1	0	3	0	3	0	1	0	4	1	1	0	0	21	0	1	1	2	9945	1
Val	13	2	1	1	3	2	2	3	3	57	11	1	17	1	3	2	10	0	2	9901

therefore, the Markov chain with the matrix M of transition probabilities is a regular Markov chain. From the theory of stochastic matrices (Berman and Plemmons, 1979; Meyer, 2000) it is known that for a regular Markov chain there exists a unique strictly positive invariant (stationary) distribution π: $M\pi = \pi$, and for any initial distribution of probabilities of states (amino acids in our case) π_0,

$$\pi_n = M^n \pi_0 \to \pi \tag{2.11}$$

as $n \to +\infty$. It turns out that vector f of the effective frequencies $(f_1, f_2, \ldots, f_{20})$ (Table 2.12), with the order of its components corresponding to the order of rows (columns) of matrix M, satisfies the equation $Mf = f$ for the stationary distribution. Then convergence in (2.11) implies that f is the vector of equilibrium frequencies of M:

$$M^n \to \begin{pmatrix} f_A & f_A & \cdots & f_A \\ f_R & f_R & \cdots & f_R \\ \cdots & \cdots & & \cdots \\ f_V & f_V & \cdots & f_V \end{pmatrix} \tag{2.12}$$

as $n \to +\infty$. The convergence (2.12) was verified in Dayhoff *et al.* (1978) by direct calculation: it was shown that M^{2034} closely approximates the matrix of equilibrium frequencies.

The theory described above was used further to derive the matrices of the log-odds scores for amino acid substitutions. These matrices are critically important for algorithms of protein sequence alignment. Since we have a family of mutation probability matrices M^n, $n = 1, 2, \ldots$, we can derive a family of substitution matrices S_n, $n = 1, 2, \ldots$, with elements $s_n(i,j)$ of log-odds scores as follows. For a pair of amino acids (i,j), $i,j = 1, \ldots, 20$, the log-odds score $s_n(i,j)$ is defined by the following formula:

$$s_n(i,j) = \log \frac{M_{ji}^n}{f_j},$$

where $M_{ji}^n = P(x_n = j | x_0 = i)$ is an element of matrix M^n. The interpretation of the substitution score $s_n(i,j)$ from the standpoint of protein sequence evolution starts with the application of the properties of the Markov chain $\{x_k\}$, $k \in \mathbf{N}$:

$$s_n(i,j) = \log \frac{M_{ji}^n}{f_j} = \log \frac{P(x_n = j | x_0 = i)}{f_j} = \log \frac{f_i P(x_n = j | x_0 = i)}{f_i f_j}$$

$$= \log \frac{\sum_a f_i P(x_n = j | x_{n/2} = a) P(x_{n/2} = a | x_0 = i)}{f_i f_j}. \tag{2.13}$$

Note that for large n we have $M_{ij}^n \approx f_i$, $M_{ji}^n \approx f_j$, since f is the vector of equilibrium frequencies for matrix M due to (2.12). Therefore, even though the Markov chain $\{x_k\}$ does not possess the reversibility property ($f_j M_{ij} \neq f_i M_{ji}$ for some $i \neq j$), for sufficiently large n we can assume that $f_j M_{ij}^n \approx f_i M_{ji}^n$ for all i, j. Then Equation (2.13) becomes

$$s_n(i,j) = \log \frac{\sum_a f_i P(x_n = j | x_{n/2} = a) P(x_{n/2} = a | x_0 = i)}{f_i f_j}$$

$$\approx \log \frac{\sum_a f_a P(x_n = j | x_{n/2} = a) P(x_{n/2} = i | x_0 = a)}{f_i f_j}$$

$$= \log \frac{\sum_a f_a P(x_{n/2} = j | x_0 = a) P(x_{n/2} = i | x_0 = a)}{f_i f_j}$$

$$= \log \frac{P(x_{n/2} = i, x'_{n/2} = j | x_0 = x'_0)}{f_i f_j}, \qquad (2.14)$$

where $\{x'_k\}$ is an independent copy of the Markov chain $\{x_k\}$. The last expression in Equation (2.14) is the log-odds ratio that involves the Markovian evolutionary model with molecular clock property and the independence pair-sequence model R with parameters defined as the product of the effective $f_i f_j$, $i, j = 1, \ldots, 20$.

The numerator in (2.14) is the probability that two aligned protein sequences diverged from a common ancestor $n/2$ PAM time units ago would have at a given site amino acids i and j, assuming that substitutions in proteins are described by the Markov process $\{x_k\}$.

The term $f_i f_j$ in the denominator in (2.14) is the probability of observing amino acids i and j at a given site of two aligned protein sequences under the independence pair-sequence model R. Since the same arguments as in Equations (2.13) and (2.14) are true for the substitution score $s_n(j, i) = \log M_{ij}^n / f_i$, we have $s_n(i,j) \approx s_n(j, i)$. Thus, unlike the mutation probability matrix M^n, the substitution matrix S_n is a symmetric one. For practical convenience, log-odds values $s_n(i,j)$ are rescaled (by multiplying by ten, or by an other scaling factor) and then rounded to the nearest integer.

Among the PAM substitution matrices, the most frequently used is the 250 PAM matrix shown in Table 6.1.

With the rapid growth of the protein data, the update of the matrices of PAM series was undertaken by Jones, Taylor, and Thornton (1992). They used the same technique for counting amino acid substitutions as Dayhoff *et al.* (1978). Another empirical model of protein evolution which combines a parsimony-based counting and the maximum likelihood approach was derived by Whelan and Goldman (2001). Extended discussions of the Dayhoff model can be found in Wilbur (1985) and George, Barker, and Hunt (1990).

The important practical question is how to choose the optimal substitution matrix for the alignment of two given (homologous) protein sequences x and y. The accurate answer would include the estimation of the evolutionary distance (time)

Table 2.15. *Correspondence between the observed percent of*
amino acid differences d between two aligned homologous
sequences and the evolutionary distance n (in PAM) between them

As the evolutionary distance increases, the probability of multiple
substitutions at the same site reversing initial changes becomes greater
and results in a slower growth of observed percent difference.
(Dayhoff *et al.* (1978), Table 23).

d	1	5	10	15	20	25	30	34	40	45
n	1	5	11	17	23	30	38	47	56	67
d	50	55	60	65	70	75	80	85		
n	80	94	112	133	159	195	246	328		

n between x and y, which requires the construction of an alignment of x and y to
determine the percent difference d (the percentage of mismatches in aligned sites).
The correspondence between the percent difference d and evolutionary distance n
(in PAM) could be determined from the following equation:

$$100 \sum_{j} f_j M_{jj}^n = 100 - d,$$

where M_{jj}^n are elements of the mutation probability matrix M^n for the evolutionary
time n PAM. For a given d, this equation allows us to choose the appropriate mutation probability matrix M^n and the substitution matrix S_n (Table 2.15 derived by
Dayhoff *et al.* (1978) lists some pairs of corresponding values n and d). However,
construction of an alignment of x and y to find d requires the substitution matrix
in the first place! A possible, but cumbersome, way to break this circular logic is
to use an iterative approach. Additional iterations, however, become prohibitively
expensive for the database searches.

The problem of choosing an appropriate substitution matrix for local sequence
alignment was studied by Altschul (1991) from the information theory perspective.
It was shown that if a single matrix has to be selected then, for database searches,
the 120 PAM matrix is the most appropriate one, while for comparing two specific
proteins with suspected homology the best choice is the 200 PAM matrix.

Finally, to construct alignments with gaps, the scoring system has to be augmented by the gap scoring scheme. This issue was addressed, for example, by Vingron
and Waterman (1994), Pearson (1996), Mott (1999), and Reese and Pearson (2002).

Problem 2.16 The original 250 PAM substitution matrix (Dayhoff *et al.*, 1978)
scores a substitution of *Gly* by *Arg* by negative score −3 (decimal logarithms

and scaling factor 10 are used, with rounding to the nearest integer). The average frequency of Arg in the protein sequence database is 0.041. Use this information, as well as the method described in Section 2.2.1, to estimate the probability that Gly will be substituted by Arg after a 250 PAM time period.

Solution The element s_{ij} of the 250 PAM substitution matrix and the frequency of amino acid q_j in a protein sequence database (the data set from which the parameters of the background independence model are derived) are connected by the following formula:

$$s_{ij} = \left[10 \lg \frac{P(i \to j \text{ in } 250 \text{ PAM})}{q_j} \right].$$

Therefore, for the probability of substitution of Gly by Arg we have

$$P(Gly \to Arg \text{ in } 250 \text{ PAM}) = 0.041 \times 10^{-0.3} = 0.0205. \qquad \square$$

Problem 2.17 Matrices of the BLOSUM series are frequently used in protein sequence alignment algorithms. The method of derivation of the amino acid substitution scores used in the BLOSUM matrices was introduced by Henikoff and Henikoff (1992). From the following multiple alignment block clustered into three sections at the 75% threshold

$$
\begin{array}{cccc}
A & D & A & D \\
A & D & C & D \\
A & C & C & D \\
\\
D & C & A & A \\
D & C & A & A \\
\\
A & A & C & C \\
D & A & C & C \\
\end{array}
$$

define the BLOSUM-type substitution score matrix (3×3) using half-bit units.

Solution We define the matrix F of counts with elements $f_{ij} = f_{ji}$ equal to the number of (weighted) pairs of amino acids i and j over all columns of the block. We explain how to calculate f_{ij}, taking as an example the weighted count f_{AA}. First, note that every residue from the first cluster (sequences 1–3) has weight $1/3$, and that every residue from the second cluster (sequences 4, 5) and the third cluster (sequences 6, 7) has weight $1/2$, since the total weight of the residues from sequences of the same cluster (per column) must be equal to 1 (each cluster contributes into the alignment as a single sequence). In the first column there are three residues A from the first cluster (each with weight $1/3$) and one residue A from the third cluster (with weight $1/3$). Thus, the total weighted count (per column 1) of A-to-A substitutions between different clusters is $f_{AA}^1 = 1/3 \times 1/2 + 1/3 \times 1/2 + 1/3 \times 1/2 = 1/2$. There are no A-to-A substitutions between clusters in columns 2 and 4. The total count of A-to-A substitutions per column 3 is $f_{AA}^3 = 1/3 \times 1/2 = 1/3$. Finally, $f_{AA} = f_{AA}^1 + f_{AA}^2 + f_{AA}^3 + f_{AA}^4 = 1/3 + 0 + 1/2 + 0 = 5/6$. Repeating this counting scheme for each pair i and j, we fill in the following symmetric matrix F of pair counts:

	A	C	D
A	5/6	13/3	11/3
C	13/3	1	5/3
D	11/3	5/3	1/2

The observed frequency of occurrence of pair (i,j) is defined by the following formula:

$$q_{ij} = \frac{f_{ij}}{\sum_i \sum_{j=1}^{i} f_{ij}},$$

which produces the following values of q_{ij}:

	A	C	D
A	5/72	13/36	11/36
C		1/12	5/36
D			1/24

Next, the expected frequency e_{ij} of occurrence for the amino acid pair (i,j) is defined by $e_{ii} = p_i^2$ and $e_{ij} = 2p_i p_j$, $i \neq j$, where p_i is the probability of occurrence of amino acid i. The observed frequencies p_i are determined by the following formula:

$$p_i = q_{ii} + \frac{1}{2} \sum_{j \neq i} q_{ij}.$$

Thus, we have $p_A = 29/72$, $p_C = 19/72$, $p_D = 1/3$, and the expected frequencies of pairs e_{ij} come out as follows:

	A	C	D
A	0.1622	0.2683	0.2125
C		0.1108	0.1757
D			0.0696

Finally, calculating a log-odds ratios in half-bit units $s_{ij} = 2\log_2 q_{ij}/e_{ij}$ and rounding them to the nearest integer produces the (3×3) substitution score matrix:

	A	C	D
A	−2	1	1
C	1	−1	−1
D	1	−1	−1

□

2.2.2 Distributions of similarity scores

2.2.2.1 Theoretical introduction to Problems 2.19 and 2.20.

The development of comparative and evolutionary genomics would be impossible without efficient similarity search algorithms. However, as soon as a similarity characterized by a high score is found in a database search one needs to make sure that this high score has not occurred by chance. To establish confidence in identifying similarities, it is natural to study the alignments of random sequences and determine the statistically significant thresholds for similarity scores. Several definitions are necessary for an in-depth discussion.

Let a biological (protein) sequence be described by the independence model with probability q_a of occurrence of amino acid a at any position, $\sum_a q_a = 1$. The score of an ungapped pairwise alignment is determined as the sum of scores for amino acid pairs, while the score for amino acid pair (a, b) is defined by an element $s(a, b)$ of the substitution matrix $\mathbf{S}$ (such as BLOSUM or PAM matrix). Local ungapped pairwise alignments whose scores cannot be improved by extension or trimming are called *high-scoring segment pairs* (HSPs).

The elements of the substitution matrix and parameters of the independence model are supposed to satisfy the negative bias condition,

$$\sum_{a,b} q_a q_b s(a, b) < 0,$$

which means that the average score per alignment position is negative. This condition prevents the increase of a total score of a local alignment due to the mere increase of the length of the alignment. On the other hand, some elements of matrix $\mathbf{S}$ must be positive, otherwise all HSPs will have zero length. Note that the negative bias condition implies that the function $f(\lambda) = \sum_{a,b} q_a q_b e^{\lambda s(a,b)} - 1$ has only one positive root λ (see Problems 5.3–5.5).

Distributions of statistics of the HSP scores have been studied in numerous publications, for example Iglehart (1972); Lipman *et al.* (1984); Reich, Drabsch, and Däumler (1984); Smith, Waterman, and Burks (1985); Karlin and Altschul (1990, 1993); Karlin, Dembo, and Kawabata (1990); Dembo, Karlin, and Zeitouni (1994a,b); Altschul and Gish (1996).

It was proved in Dembo, Karlin, and Zeitouni (1994b) that for unrelated (independent) sequences X and Y with sufficiently large lengths n and m the distribution of the number N_S of HSPs with scores greater than S can be closely approximated by the Poisson distribution with parameter $\Lambda = Knme^{-\lambda S}$. Here λ is the positive root of $f(\lambda)$, and the constant K depends only on $\{q_a\}$ and the scoring matrix $\mathbf{S}$. Therefore, the expected number of HSPs with scores greater than S is given by

$$\mathbf{E}N_S \approx \Lambda = Knme^{-\lambda S},$$

and the probability of observing a certain number z of such HSPs is given by

$$P(N_S = z) \approx e^{-\Lambda} \frac{\Lambda^z}{z!}.$$

The distribution of N_S can be used to derive the probability distribution of the maximum score $S_{\max}$ (the score of the optimal ungapped local alignment for sequences X and Y) as follows:

$$P(S_{\max} \geq S) = P(N_S > 0) = 1 - P(N_S = 0) \approx 1 - e^{-\mathbf{E}N_S}. \qquad (2.15)$$

The value $\mathbf{E}N_S$ is called the *E-value* of the score S; the probability value $P(S_{\max} \geq S) = 1 - e^{-\mathbf{E}N_S}$ is called the *P-value* of the score S. The **E**-value is frequently used to characterize the statistical significance of the number of HSPs observed in database searches.

The normalized sum T_r of the scores $S_1, \ldots, S_r$ of r highest-scoring segment pairs,

$$T_r = \lambda \left(\sum_{k=1}^{r} S_k \right) - \ln K^r - \ln(mn)^r,$$

can be of interest as well. It was shown in Karlin and Altschul (1993) that for sufficiently large n and m the probability density function of T_r is approximated by

$$f(t) = \frac{e^{-t}}{r!(r-2)!} \int_0^{\infty} y^{r-2} \exp(-e^{(y-t)/r}) \, dy.$$

The moments of this distribution can be calculated by the Laplace transform. In particular,

$$\mathbf{E}T_r \approx r \left(1 + \gamma - \sum_{k=1}^{r} 1/k \right) \approx r(1 - \ln r) - 1/2,$$

$$\mathbf{Var}\, T_r \approx r^2 \left(\pi^2/6 - \sum_{k=1}^{r} 1/k^2 \right) + r \approx 2r - 1/2,$$

where $\gamma \approx 0.577$ is the Euler constant.

Instead of ungapped alignments and HSPs, Neuhauser (1994) considered local alignments with insertions and deletions but no mismatches (so $x_i = y_j$ in a pair of aligned residues (x_i, y_j)). Neuhauser proposed an algorithm searching for local alignments of sequences X and Y with t matching pairs, k gaps, and a length of each gap at most l ((t, k, l)-alignments). It has been proved that, for independent sequences X and Y both described by the independence model and with sufficiently large lengths n and m, the number W of (t, k, l)-alignments found by this algorithm approximately has the Poisson distribution with parameter

$$\Lambda = C^k mn(1 - p)\binom{t-1}{k} l^k p^{t-k},$$

where the probability of match p and the constant C depend only on the parameters of the independence model. Then for sufficiently large n and m the largest number S^* of matching pairs in any local alignment with k gaps (with length at most l each) has the following distribution:

$$P(S^* < t) \approx e^{-\Lambda}.$$

Finally, the formula for the P-value of an alignment score was derived for the general case of gapped local alignment (with mismatches in aligned pairs allowed) of two independent sequences and a scoring scheme based on the log-odds substitution matrix $\mathbf{S}$. The results were obtained heuristically by Mott (1999, 2000) and Mott and Tribe (1999) and analytically by Siegmund and Yakir (2000, 2003). We outline below the main results on the statistics of the gapped alignments scores (Siegmund and Yakir, 2000).

It is assumed, as it was for HSPs, that the negative bias condition holds for the elements of matrix $\mathbf{S}$ and parameters of the independence model, and that λ is the only positive root of function f. For the linear gap scoring function $\gamma(g) = -\delta g$, for sufficiently large n and m, the tail of the distribution of the maximum score $\mathbf{S}_{\max}^j$ of local alignment with at most j gaps can be approximated by the following formula:

$$P(\mathbf{S}_{\max}^j \geq S) \approx K' nm e^{-\lambda S} \frac{(b\lambda S)^j}{j!} \sum_{l=j}^{\infty} a_l e^{-\lambda \delta l}.$$

Here constants K', b, and a_l depend only on $\{q_a\}$ and the scoring matrix $\mathbf{S}$, and the upper bound for a_l is $1/(\exp(\lambda\delta) - 1)$.

For the scoring function $\gamma(g) = -d - \delta g$ (the affine gap score), and sufficiently large n and m, the score $\mathbf{S}_{\max}$ of optimal local alignment (with any number of gaps) approximately has the following distribution:

$$P(\mathbf{S}_{\max} \geq S) \approx K' nm e^{-\lambda S} \sum_{j=0}^{\infty} \frac{(b\lambda S e^{-d\lambda})^j}{j!} \sum_{l=j}^{\infty} a_l e^{-\lambda \delta l}. \tag{2.16}$$

In Equation (2.16) K', b, and a_l are the same constants as in the case of the linear gap scoring function, and it is assumed that $\lambda d = \ln S + C$ for some constant C. If, additionally, the rates of growth of n, m, and S satisfy the condition $nm \exp(-S) \to v$ for some finite positive v, then the formula for the P-value becomes

$$P(\mathbf{S}_{max} \geq S) \approx 1 - \exp \Lambda, \qquad (2.17)$$

where Λ is defined by the expression on the right-hand side of Equation (2.16). If in the expression for Λ we set $j = 0$ and, therefore, consider ungapped local alignments only, then Equation (2.17) for the P-value is reduced to Equation (2.15) derived by Dembo, Karlin, and Zeitouni (1994b). The P-values used in the original program, BLAST (Altscul *et al.*, 1990) for the similarity scores of ungapped local alignments have been calculated by formula (2.15). Notably, in the gapped version of the BLAST program Altschul *et al.* (1997) use the approximate formula (2.17) for P-values with the values of constants involved in the expression for Λ estimated numerically for each particular search. Shuler, Atschul, and Lipman (1991) used an analog of the distribution defined by Equation (2.15) for the assessment of the statistical significance of the sum-of-pair score of the ungapped block of several amino acid sequences. Further developments and discussions of problems related to the score distributions can be found in Waterman and Vingron (1994), Altschul *et al.* (2001), Webber and Barton (2001), Bailey and Gribskov (2002) and Grossman and Yakir (2004). Statistical methods for sequence analysis were reviewed by Karlin (2005).

The distributions of statistics of pairwise alignment scores N_S, S_{max}, T_r, W, S^*, or $\mathbf{S}_{max}$ provide a rigorous basis for testing hypotheses on the relatedness of sequences X and Y. The following two tests for the null hypothesis,

$$H_0 = \{X \text{ and } Y \text{ are unrelated (independent) sequences}\},$$

versus the alternative hypothesis,

$$H_a = \{X \text{ and } Y \text{ are (evolutionary) related sequences}\},$$

could be considered. In Test 1, based on statistic S_{max}, H_0 is accepted (H_a is rejected) if $S_{max} < S^\circ$; otherwise, H_0 is rejected (H_a is accepted). In Test 2, based on statistic T_r, H_0 is accepted (H_a is rejected), if $|T_r - \mathbf{E}T_r| < \varepsilon^*$; otherwise, H_0 is rejected (H_a is accepted). Note that the cut-off values S° and ε^* should be determined *a priori* (before the actual sequences X and Y are observed).

Problem 2.18 Test 1 and Test 2 are applied to establish relatedness of locally aligned protein sequences X and Y of lengths $n = 100$ and $m = 300$. It is assumed that $K = 0.1$, $\lambda = 0.7$.

(a) Define the cut-off value S° corresponding to the significance level α_1 (the false negative rate) of Test 1 equal to 0.05.

(b) Estimate the significance level α_2 of Test 2 for the cut-off value $\varepsilon^* = 13.8$ and $r = 5$.

(c) Given observed scores $S_1 = 15$, $S_2 = S_3 = 12$, $S_4 = 11$, $S_5 = 10$ of highest-scoring segment pairs, use Tests 1 and 2 with cut-off values defined as in (a) and (b) to test H_0 versus H_a.

Solution (a) From the definition of significance level α_1 of the test and Equation (2.15) we have

$$\alpha_1 = P(\text{type-one error}) = P(H_0 \text{ is rejected}|H_0 \text{ is true})$$

$$= P(S_{\max} \geq S^\circ | X \text{ and } Y \text{ are unrelated (random) sequences})$$

$$\approx 1 - e^{-E N_{S^\circ}}.$$

Thus, the significance level of Test 1 for any cut-off value S° is equal to the P-value of S°. For a given α_1, value S° has to satisfy the equation $-Kmne^{-\lambda S^\circ} = \ln(1-\alpha_1)$. By solving this equation we determine that to test H_0 versus H_a at significance level $\alpha_1 = 0.05$ the cut-off value $S^\circ = 16$ should be used.

(b) To estimate the significance level of Test 2 for a given cut-off value ε^*, we apply Chebyshov's inequality in the form (11.10):

$$\alpha_2 = P(\text{type-one error}) = P(H_0 \text{ is rejected}|H_0 \text{ is true})$$

$$= P(|T_r - \mathbf{E}T_r| \geq \varepsilon^* | X \text{ and } Y \text{ are unrelated (random) sequences})$$

$$= P(|T_r - \mathbf{E}T_r| \geq \varepsilon^* | T_r \text{ has p.d.f. } f(t)\,) \leq \frac{\mathbf{Var}T_r}{(\varepsilon^*)^2}.$$

For $r = 5$ the variance of the normalized sum $\mathbf{Var}T_5 \approx 2 \times 5 - 1/2 = 9.5$; thus for the cut-off value $\varepsilon^* = 13.8$ the significance level of the test $\alpha_2 \leq 0.05$.

(c) To test hypothesis H_0 versus H_a, we calculate statistics $S_{\max}$ (to use in Test 1) and T_5 (to use in Test 2) for the observed alignment scores:

$$S_{\max} = S_1 = 15,$$

$$T_5 = \lambda \left(\sum_{k=1}^{r} S_k \right) - \ln K^r - \ln(mn)^r = 42 - 5\ln 3 - 15\ln 10 = 1.968.$$

Since $S_{\max} = 15 < 16 = S^\circ$, we accept H_0 (reject H_a) in Test 1 with significance level $\alpha_1 = 0.05$. To carry out Test 2, we have to calculate the expectation of the normalized sum: $\mathbf{E}T_5 = 5(1 - \ln 5) - 1/2 = -3.547$. Since $|T_5 - \mathbf{E}T_5| = 5.515 < 13.8 = \varepsilon^*$, the null hypothesis is accepted (and the alternative hypothesis is rejected) in Test 2 with significance level $\alpha_2 \leq 0.05$. Therefore, the observed scores of the five highest-scoring segment pairs do not support the hypothesis of a relation between protein sequences X and Y with significance level of the test at most 0.05.   □

Problem 2.19 A protein sequence X of m amino acids long is used as a query for similarity search against a protein sequence database containing k protein sequences $Y_1, \ldots, Y_k$ with lengths $n_1, \ldots, n_k$. Find the relationships (a) between the E-values of pairwise comparison of sequences X to Y_i, E^i, $i = 1, \ldots, k$, and the E-value of the whole database search; (b) between the P-values of pairwise comparisons, P^i, $i = 1, \ldots, k$, and the P-value of the whole database search.

Solution In a search for similarity between sequence X and a database, the expected number of HSPs with score greater than S is equal to the sum of $\mathbf{E}N_S^i$ for optimal local alignments of pairs of sequences X and Y_i, $i = 1, \ldots, k$. If we assume that all sequences in the database as well as sequence X are generated by the independence model, then for the E-value of the a database search we have

$$E = \mathbf{E}N_S = \sum_{i=1}^{k} \mathbf{E}N_S^i = \sum_{i=1}^{k} E^i = \sum_{i=1}^{k} Kmn_i e^{-\lambda S}$$

$$= Kme^{-\lambda S} \sum_{i=1}^{k} n_i = Kmne^{-\lambda S}, \tag{2.18}$$

where n is the total length of the database sequences. Thus, when calculating the E-value of a database search, one can treat the database as a single sequence of length n.

The database search P-value can be determined as a function of $P_1, P_2, \ldots, P_k$ from Equation (2.18) as follows:

$$P = P(S_{\max} \geq S) \approx 1 - e^{-\mathbf{E}N_S} = 1 - e^{\sum_{i=1}^{k} E^i} = 1 - \prod_{i=1}^{k} e^{E^i} = 1 - \prod_{i=1}^{k} (1 - P^i).$$

□

2.2.3 Distribution of the length of the longest common word among several unrelated sequences

Theoretical introduction to Problem 2.21 Genomic DNA sequences from different species may possess virtually identical contiguous subsequences which are strikingly long. The identification and interpretation of such common subsequences are of substantial biological interest. The related statistical problems can be investigated: what are the properties of the distribution of the length of the longest common word observed in two (or more) random sequences given the type of the sequence model? How will the distribution change when at most k mismatches inside the common word are allowed?

The mathematical study of similar problems was conducted by Erdös and Revesz (1975), who provided almost sure upper and lower bounds for the longest run of heads in a sequence of n tosses of a fair coin, as well as for runs of heads interrupted by at most k tails. For the length of the longest common word $M(n,m)$ of two sequences with lengths n and m, under the assumption that sequences are (independently) generated by independence models (not necessarily the same), Arratia and Waterman (1985) proved that

$$P\left(\lim_{n\to+\infty, m\to+\infty} \frac{M(n,m)}{\log_{1/p}(mn)} = K\right) = 1.$$

Here p is the probability of a match, $p = P(x_i = y_j)$, and the constant K depends on the ratio of the rates of growth of n and m as well as the sequence model parameters. This result was generalized to the case of several sequences described either by independence models or by Markov chains. For the distribution of the longest contiguous run $M_k(n,m)$ of matches between two independent sequences allowing at most k mismatches, Arratia, Gordon, and Waterman (1986) determined the mathematical expectation, the variance, and the limit distribution of $M_k(n,m)$, as $n, m \to +\infty$. If $k = 0$, the uninterrupted run is considered, and both n and m grow at the same rate, then $M_0(n,m) = M(n,m)$ has approximately the same distribution as the maximum of $(1-p)nm$ independent geometrically distributed (with parameter p) random variables. The expectation and the variance of $M_k(n,m)$ are given by the following formulas:

$$\mathbf{E}M_k(n,m) \approx \log_{1/p}((1-p)mn) + k \log_{1/p}\log_{1/p}((1-p)mn)$$
$$+ k \log_{1/p}\frac{1-p}{p} - \log_{1/p}(k!) + \gamma \log_{1/p}(e) - \frac{1}{2}, \qquad (2.19)$$

$$\mathbf{Var}\,M_k(n,m) \approx \frac{(\pi \log_{1/p}(e))^2}{6} + \frac{1}{12}.$$

Here $\gamma \approx 0.577$ is the Euler constant. Further, Karlin and Ost (1987, 1988) derived an asymptotic distribution of the length of the longest common word among r or more out of s independent sequences generated by stationary processes with uniform mixing.

Problem 2.20 Sequences X and Y of lengths 1000 and 10 000 are generated (independently from each other) by the independence model with an equal probability of symbols. What is the expected length of the longest common word between X and Y if mismatches are not allowed ($k = 0$), and if at most two mismatches are allowed ($k = 2$)? (a) Nucleotide and (b) amino acid alphabets are to be considered.

Solution (a) Since the four nucleotide types have equal probabilities $q_a = 1/4$ to appear at a given position of sequences X and Y, the probability $p = P(x_i = y_j)$ of observing matching symbols in positions i of X and j of Y for any pair i and j is given by

$$p = \sum_a q_a q_a = 4 \times \frac{1}{16} = 0.25.$$

Equation (2.19) for the expected value of $M_k(n, m)$ yields

$$EM_0(1000; 10\,000) = \log_4(0.75 \times 1000 \times 10\,000) - \frac{0.577}{\ln 0.25} - \frac{1}{2} = 11.335;$$

$$EM_2(1000; 10\,000) = EM_0(1000; 10\,000) + 2\log_4 \log_4(7\,500\,000)$$

$$+ 2\log_4 3 - \log_4 2 = 15.934.$$

Therefore, the expected length of the longest common word between X and Y is 11 nt, while the average length of the longest common word with at most two mismatches is 15 nt.

(b) Since the twenty types of amino acids have equal probabilities $q_a = 1/20$ to appear at a given position of sequences X and Y, we determine that $p = \sum_a q_a q_a = 20 \times 1/400 = 0.05$. We use Equation (2.19) to find the expected length of longest common word in sequences X and Y:

$$EM_0(1000; 10\,000) = \log_{20}(0.95 \times 1000 \times 10\,000) - \frac{0.577}{\ln 0.05} - \frac{1}{2} = 5.056;$$

$$EM_2(1000; 10\,000) = EM_0(1000; 10\,000) + 2\log_{20} \log_{20}(9\,500\,000)$$

$$+ 2\log_{20} 19 - \log_{20} 2 = 7.912.$$

Naturally, the expected length of the longest common word increases as more mismatches are allowed. Also it would be anticipated that the longest common word in amino acid sequences on average is shorter than the longest common word in nucleotide sequences of the same lengths due to the smaller probability p of a match at a given pair of sequence positions. □

Problem 2.21 A dot-plot sequence comparison algorithm identifies similarities in DNA sequences using the following rule. A pair of DNA fragments of length l starting at positions i and j of two sequences X and Y, respectively, are considered a "matching pair" if at least k nucleotides out of l make identical pairs (no insertions or deletions are allowed) with mismatches permitted only in internal pairs. Determine the expected number of "matching pairs" for 1000 nt long DNA sequences X and Y for $l = 8$, $k = 6$. The sequences are described by the independence model with probabilities of T, C, A, and G equal to 0.3, 0.2, 0.3 and 0.2, respectively.

Solution Nucleotides occupying positions i in X and j in Y are identical with probability

$$
\begin{aligned}
P(x_i = y_j) &= P\{(x_i, y_j) = (T, T), \text{ or } (x_i, y_j) = (C, C) \\
&\quad \text{ or } (x_i, y_j) = (A, A), \text{ or } (x_i, y_j) = (G, G)\} \\
&= 0.3^2 + 0.2^2 + 0.3^2 + 0.2^2 = 0.26.
\end{aligned}
$$

An alignment of two DNA segments generates a sequence of nucleotide pairs associated with a sequence of Bernoulli trials where "success" ("nucleotides in a pair are identical") occurs with probability 0.26. Then a fragment pair starting at positions (i, j) will be a 'matching pair' with probability

$P(\text{matching pair starting at } (i, j)) = P(\text{eight "successes" in eight trials})$

$\quad + P(\text{seven "successes" in eight trials, including start and end})$

$\quad + P(\text{six "successes" in eight trials, including start and end})$

$$
= 0.26^8 + 0.26^7 \times 0.74 \times 6 + 0.26^6 \times 0.74^2 \times \binom{6}{4} = 0.0029.
$$

For two 1000 long DNA sequences there are

$$
N = (1000 - 7)(1000 - 7) = 993^2 = 986049
$$

fragment pairs of length 8. Here the Bernoulli scheme is used again. We consider N Bernoulli trials with "success," defined as "a fragment pair starting at positions (i, j) is a matching pair," taking place with probability $p = 0.0029$. Then the expected number of "matching fragment pairs" is given by

$$
\mathbf{E} = Np = 986049 \times 0.0029 \approx 2860. \qquad \square
$$

2.3 Further reading

The performance of pairwise sequence alignment algorithms has been assessed and the scoring systems discussed by many authors, including Pearson (1995, 1996); Brenner, Chothia, and Hubbard (1998). Shindyalov and Bourne (1998) developed the combinatorial extension (CE) algorithm for pairwise alignment of two protein sequences. An alignment in the CE algorithm was defined by aligned pairs of protein fragments, which confer structure similarity. Pairwise alignment algorithms based on the Bayesian approach were proposed by Zhu, Liu, and Lawrence (1998) and by Webb, Liu, and Lawrence (2002).

Kann, Qian, and Goldstein (2000) developed a new method of derivation for the substitution matrix which maximizes the average confidence of the homology identification verified over a set of homologous and non-homologous pairs of proteins from the COG database (Tatusov, Galperin, and Koonin, 2000). Another method of optimization of the substitution matrix which improves the accuracy of homology searches was proposed by Hourai, Akutsu, and Akiyama (2004). Rost (1999) analyzed the quality of protein sequence alignments in the twilight zone (for pairs of sequences with 20–35% identity).

Algorithms for whole genome comparison were developed by Delcher *et al.* (1999) and Schwartz *et al.* (2000). Brudno *et al.* (2003) developed the LAGAN algorithm for rapid pairwise alignment of homologous genomic sequences: first, the optimal set of non-overlapping local alignments is chosen, then the intermediate sequence fragments are aligned by the Needleman-Wunsch algorithm. Algorithms for aligning cDNA sequences with genomic DNA sequences were developed by Florea *et al.* (1998) (SIM4) and by Kent (2002) (BLAT).

Sequence alignment algorithms were incorporated into the prediction of three-dimensional protein structure (Martí-Renom *et al.*, 2000), and into identification of gene-fusion events in complete genomes (Enright *et al.*, 1999).

Sequence alignment algorithms have been frequently used in comparative genomics projects. One of the most important sources of information about human and other genomes, the Ensembl database (Hubbard *et al.*, 2002; Clamp *et al.*, 2003) uses the automatic pipeline of gene annotation that includes the tools for homology search. The KEGG database by Kanehisa *et al.* (2002) uses a comprehensive search for similarities by the Smith–Waterman algorithm to transfer functional annotation of proteins and establish ortholog/paralog relations of protein-coding genes in complete genomes stored in the SSDB part of KEGG.

3

Markov chains and hidden Markov models

The chapter in *BSA* that introduces Markov chains and hidden Markov models plays a critical role in that book. The sequence comparison algorithms described in Chapter 2 could not be developed without the introduction of the theoretically justified similarity scores and statistical theory of similarity score distributions. These developments, in turn, are not feasible without rational choices of probabilistic models for DNA and protein sequences. Both Markov chains and hidden Markov models are often remarkably good candidates for the sequence models. Moreover, hidden Markov models (HMMs) are potentially a more flexible means for biological sequence analysis because they allow simultaneous modeling of observable and non-observable (hidden) states. The presence of the two types of states perfectly fits the need to model some important additional information existing beyond sequences *per se*, such as the functional meaning of the sequence elements, matches and mismatches of symbols in pairs of aligned sequences, evolutionary conserved regions in multiple sequences, phylogenetic relationships, etc.

Chapter 3 of *BSA* introduces the fundamental algorithms of HMM theory: the Viterbi algorithm, the forward and backward algorithms, as well as the Baum–Welch algorithm. All of these algorithms are amenable for a variety of applications in biological sequence analysis. Of course, some of these HMM constructions exist in parallel with their non-probabilistic counterparts; for example, consider the Viterbi algorithm for a pair HMM and the classic dynamic programming algorithm for pairwise alignment. Both HMM and non-HMM approaches are known for finding conserved domains, building phylogenetic trees, etc.

In this chapter, the *BSA* problems focus on deriving the formulas that support probabilistic modeling and the HMM algorithm construction. The additional problems focus on applications of the basic HMM algorithms, comparing competitive models and estimating model parameters. We also discuss an alternative non-HMM approach to the problem of sequence segmentation into compositionally uniform segments.

3.1 Original problems

Problem 3.1 The sum of the probabilities of all possible sequences of states of length L can be written as follows:

$$\sum_x P(x) = \sum_{x_1} \sum_{x_2} \cdots \sum_{x_L} P(x_1) \prod_{i=2}^{L} a_{x_{i-1}x_i}.$$

Show that this sum is equal to 1.

Solution By changing the order of summation and using the definition of the transition probability, we rewrite the sum as follows:

$$\sum_x P(x) = \sum_{x_1} \sum_{x_2} \cdots \sum_{x_L} P(x_1) \prod_{i=2}^{L} a_{x_{i-1}x_i}$$

$$= \sum_{x_2} \cdots \sum_{x_L} \prod_{i=3}^{L} a_{x_{i-1}x_i} \sum_{x_1} P(x_1) a_{x_1 x_2}$$

$$= \sum_{x_2} \cdots \sum_{x_L} \prod_{i=3}^{L} a_{x_{i-1}x_i} \sum_{x_1} P(x_1) P(x_2|x_1). \tag{3.1}$$

It is obvious that the inner sum (over x_1) gives the probability of x_2, because the summation is made over all possible states at position $i = 1$. Therefore, the last expression in Equation (3.1) becomes

$$\sum_{x_2} \cdots \sum_{x_L} \prod_{i=3}^{L} a_{x_{i-1}x_i} P(x_2) = \sum_{x_3} \cdots \sum_{x_L} \prod_{i=4}^{L} a_{x_{i-1}x_i} \sum_{x_2} P(x_2) P(x_3|x_2).$$

Now we can apply the same argument for the inner sum over x_2, and so on. After $L - 1$ steps we obtain

$$\sum_x P(x) = \sum_{x_L} P(x_L) = 1.$$

The last equality holds since the sum includes the probabilities of all possible states of the Markov chain at position L.   □

Problem 3.2 Assume that the model has an end state and that the transition from any state to the end state has probability τ. Show that the sum of the probabilities over all sequences of length L (and properly terminating by making a transition to the end state) is $\tau(1 - \tau)^{L-1}$.

Solution Assuming that the transition probabilities from any state to the end state ε are equal to τ, we have $a_{x_1\varepsilon} = \tau$ for any state x_1, which implies that

$$\sum_{\{x_2:x_2\neq\varepsilon\}} a_{x_1x_2} = 1 - \tau.$$

Similarly, for any state x_i such that $x_i \neq \varepsilon$ we have

$$\sum_{\{x_{i+1}:x_{i+1}\neq\varepsilon\}} a_{x_ix_{i+1}} = 1 - \tau,$$

with $i = 1, \ldots, L-1$. Then the sum of probabilities of all sequences of length L terminating at the end state ε is given by

$$\Sigma = \sum_{x_1} \sum_{\{x_2:x_2\neq\varepsilon\}} \cdots \sum_{\{x_L:x_L\neq\varepsilon\}} P(x_1)a_{x_1x_2}a_{x_2x_3}\cdots a_{x_{L-1}x_L}a_{x_L\varepsilon}$$

$$= \tau \sum_{x_1} \sum_{\{x_2:x_2\neq\varepsilon\}} \cdots \sum_{\{x_{L-1}:x_{L-1}\neq\varepsilon\}} P(x_1)a_{x_1x_2}\cdots a_{x_{L-2}x_{L-1}} \left(\sum_{\{x_L:x_L\neq\varepsilon\}} a_{x_{L-1}x_L}\right)$$

$$= \tau(1-\tau) \sum_{x_1} \cdots \sum_{\{x_{L-2}:x_{L-2}\neq\varepsilon\}} P(x_1)a_{x_1x_2}\cdots a_{x_{L-3}x_{L-2}} \left(\sum_{\{x_{L-1}:x_{L-1}\neq\varepsilon\}} a_{x_{L-2}x_{L-1}}\right).$$

$$(3.2)$$

After repeating similar regrouping of terms and factorization $(L-1)$ times, the sum Σ in Equation (3.2) becomes

$$\Sigma = \tau(1-\tau)^{L-1} \sum_{x_1} P(x_1) = \tau(1-\tau)^{L-1}. \qquad \square$$

Problem 3.3 Show that the sum of the probabilities over all possible sequences of any length is 1. This proves that the Markov chain really describes a proper probability distribution over the whole space of sequences.

Solution The goal is to verify the following equality:

$$\sum_{L=1}^{\infty} \left(\sum_{x_1} \sum_{\substack{x_2 \\ x_i\neq\varepsilon}} \cdots \sum_{x_L} P(x_1)a_{x_1x_2}a_{x_2x_3}\cdots a_{x_{L-1}x_L}a_{x_L\varepsilon} \right) = 1,$$

where ε is the end state of the Markov chain. It was proved in Problem 3.2 that the sum S_L of probabilities over all sequences of length L (and properly terminating at the end state ε) is $\tau(1-\tau)^{L-1}$, where τ is the probability of transition to the end state. Then

$$\sum_{L=1}^{\infty}\left(\sum_{x_1}\sum_{\substack{x_2\\x_i\neq\varepsilon}}\cdots\sum_{x_L}P(x_1)a_{x_1x_2}a_{x_2x_3}\cdots a_{x_{L-1}x_L}a_{x_L\varepsilon}\right)$$

$$=\sum_{L=1}^{\infty}S_L=\sum_{L=1}^{\infty}\tau(1-\tau)^{L-1}=\frac{\tau}{\tau}=1. \qquad\square$$

Problem 3.4 Let $P(x,\pi)$ be the joint probability of an observed sequence x and a state sequence π. We define the most probable path $\pi^\star$ as $\pi^\star = \mathrm{argmax}_\pi P(x,\pi)$. Show that this definition is equivalent to $\pi^\star = \mathrm{argmax}_\pi P(\pi|x)$.

Solution According to the definition of conditional probability, $P(\pi|x) = P(x,\pi)/P(x)$. Since $P(x)$ does not depend on π, one and the same state sequence $\pi^\star$ will deliver the maximum value for both $P(x,\pi)$ and $P(x,\pi)/P(x)$:

$$\mathrm{argmax}_\pi P(x,\pi) = \mathrm{argmax}_\pi P(\pi|x) = \pi^\star. \qquad\square$$

Problem 3.5 Derive the formula

$$P(\pi_i = k, \pi_{i+1} = l|x, \Theta) = \frac{f_k(i)a_{kl}e_l(x_{i+1})b_l(i+1)}{P(x|\Theta)}, \qquad (3.3)$$

where the forward variable $f_k(i) = P(x_1,\ldots,x_i,\pi_i = k)$ is the joint probability of the observed subsequence $x_1,\ldots,x_i$ and the specific hidden state $\pi_i = k$; a_{kl} and $e_l(x_i)$ are, respectively, transition and emission probabilities; the backward variable $b_k(i) = P(x_{i+1},\ldots,x_L,\pi_i = k)$ is the joint probability of the observed subsequence $x_{i+1},\ldots,x_L$ and the specific hidden state $\pi_i = k$.

Solution The multiplication theorem of probability,

$$P(A\cap B\cap C) = P(A)P(B|A)P(C|A\cap B),$$

allows us to represent the left hand side of Equation (3.3) as follows:

$$P(\pi_i = k, \pi_{i+1} = l|x, \Theta) = \frac{P(\pi_i = k, \pi_{i+1} = l, x|\Theta)}{P(x|\Theta)}$$

$$= \frac{1}{P(x|\Theta)} \times P(x_1,\ldots,x_i,\pi_i = k|\Theta)P(x_{i+1},\pi_{i+1} = l|x_1,\ldots,x_i,\pi_i = k,\Theta)$$

$$\times P(x_{i+2},\ldots,x_L|x_1,\ldots,x_i,x_{i+1},\pi_i = k,\pi_{i+1} = l,\Theta). \qquad (3.4)$$

Since the state π_{i+1} in position $i+1$ of the first order Markov chain depends only on the state π_i in the previous position i, Equation (3.4) becomes

$$P(\pi_i = k, \pi_{i+1} = l|x, \Theta)$$
$$= \frac{P(x_1,\ldots,x_i, \pi_i = k|\Theta)P(x_{i+1}, \pi_{i+1} = l|\pi_i = k, \Theta)P(x_{i+2},\ldots,x_L|\pi_{i+1} = l, \Theta)}{P(x|\Theta)}.$$

Finally, we use the definitions of the variables $f_k(i)$ and $b_l(i)$ as well as a_{kl} and $e_l(x_{i+1})$ to obtain

$$P(\pi_i = k, \pi_{i+1} = l|x, \Theta) = \frac{f_k(i)a_{kl}e_l(x_{i+1})b_l(i+1)}{P(x|\Theta)}. \qquad \square$$

Problem 3.6 Derive the following equation:

$$E_k(b) = \sum_j \frac{1}{P(x^j)} \sum_{\{i:x_i^j = b\}} f_k^j(i)b_k^j(i)$$

for the expected number $E_k(b)$ of times that observed state b is emitted from hidden state k. Here the inner sum is taken over all positions i emitting symbol b and the outer sum is taken over all training sequences x^j.

Solution The formula in question is part of the Baum–Welch algorithm (Baum, 1972), which is used to estimate parameters of a hidden Markov model, including emission probabilities $e_k(b)$, from a set of realizations which serve as a training set. This algorithm works in iterations. For current estimates $e_k'(b)$ and a_{kl}', forward variables $f_k^j(i)$ and backward variables $b_k^j(i)$ are calculated for each training sequence x^j using the forward and the backward algorithm, respectively. A new estimate $e_k''(b)$ is defined as a maximum likelihood estimate:

$$e_k''(b) = \frac{E_k(b)}{\sum_{b'} E_k(b')},$$

where $E_k(b)$ is the expected number of times of symbol b emissions from a hidden state k and the sum $\sum_{b'} E_k(b')$ gives the expected number of occurrences of state k for a given set of training sequences. To derive equations for the expectations $E_k(b)$, we introduce the counting random variable δ_i^j:

$$\delta_i^j = \begin{cases} 1, & \text{if state } k \text{ appears in position } i \text{ of sequence } x^j, \\ 0, & \text{otherwise.} \end{cases}$$

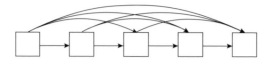

Figure 3.1. Forward connected chain of states.

Then $N(k) = \sum_j \sum_i \delta_i^j$ is the number of occurrences of state k in the paths of hidden states, and the expected value of $N(k)$ is given by

$$\mathbf{E}N(k) = \sum_j \sum_i \mathbf{E}\delta_i^j = \sum_j \sum_i P(\pi_i^j = k, |x^j, \Theta)$$

$$= \sum_j \frac{1}{P(x^j|\Theta)} \sum_i f_k^j(i) b_k^j(i).$$

Here Θ designates the probabilistic model associated with the set of current estimates of parameters. Similarly, we derive the equation for the expected number $E_k(b)$ of times that symbol b is emitted from hidden state k given a set of training sequences. The only difference is that the variables $\delta_i^j(b)$ must now count occurrences of state k only if it emits symbol b. Thus, we complete the derivation as follows:

$$E_k(b) = \sum_{x^j} \sum_{\{i:x_i^j=b\}} \mathbf{E}\delta_i^j(b) = \sum_{x^j} \sum_{\{i:x_i^j=b\}} P(\pi_i^j = k, |x^j, \Theta)$$

$$= \sum_j \frac{1}{P(x^j|\Theta)} \sum_{\{i:x_i^j=b\}} f_k^j(i) b_k^j(i).$$

Note that the updated estimate $e_k''(b)$ of the emission probability in the iterative procedure becomes

$$e_k''(b) = \frac{E_k(b)}{\sum_{b'} E_k(b')} = \frac{E_k(b)}{\mathbf{E}N(k)} = \frac{\sum_j \frac{1}{P(x^j|\Theta)} \sum_{\{i:x_i^j=b\}} f_k^j(i) b_k^j(i)}{\sum_j \frac{1}{P(x^j|\Theta)} \sum_i f_k^j(i) b_k^j(i)}. \qquad \square$$

Problem 3.7 Calculate the total number of transitions needed in a forward connected model with a length L in Figure 3.1. Calculate the same for a model with silent states in Figure 3.2.

Solution In the forward connected chain of L states (Figure 3.1) the total number of transitions is given by

$$N_1 = (L-1) + (L-2) + \cdots + 2 + 1 = \frac{(L-1)+1}{2}(L-1) = \frac{L(L-1)}{2}.$$

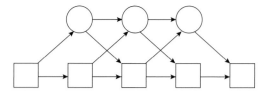

Figure 3.2. Model with silent states, represented by circles.

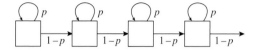

Figure 3.3. Array of states with self-loops.

In the model with silent states (Figure 3.2) there are two transitions from each 'regular' state except for the last two states. There are two transitions from each silent state (there are $L - 2$ of them) except for the last one. So, the total number of transitions is given by

$$N_2 = (L - 2)2 + 1 + (L - 3)2 + 1 = 4L - 8.$$

Models with a smaller number of possible state transitions are preferable. For short chains ($L \leq 6$) N_1 is smaller than N_2. As L increases, N_1 grows faster than N_2, and N_1 becomes bigger than N_2 for $L \geq 7$. □

Problem 3.8 Show that the number of paths through an array of n states (similar to Figure 3.3) is $\binom{l-1}{n-1}$ for length l.

Solution Any path of length l through an array of n states ($l \geq n$) consists of $n - 1$ transitions to the next state (with probability $1 - p$ each) and $l - n$ transitions to the same state (with probability p each). The last transition in a path must be the 'exit' from the nth state. Therefore, the number of different paths is equal to the number of ways to choose $l - n$ transitions returning to the same state out of $l - 1$ internal transitions in the path. This number is equal to the binomial coefficient:

$$\binom{l-1}{l-n} = \binom{l-1}{n-1} = \frac{(l-1)!}{(n-1)!(l-n)!}.$$

Note that the probability distribution of the length L of paths through this model is the negative binomial distribution:

$$P(L = l) = \binom{l-1}{l-n} p^{l-n}(1 - p)^n, \qquad (3.5)$$

where $l \geq n$. □

Problem 3.9   (in the new formulation by Anders Krogh) Consider the model of n states with self-loops (Figure 3.3) giving rise to Equation (3.5). (a) What is the probability of the most likely path through the model that the Viterbi algorithm will find? (b) Is this type of length modeling useful with the Viterbi algorithm?

Solution (by Anders Krogh) (a) Any path of length $l, l \geq n$, through the model has probability $p^{l-n}(1-p)^n$. Thus the most likely path would also have this probability, so the Viterbi algorithm does not 'see' the negative binomial distribution. Then the answer to (b) is 'no.'   □

Problem 3.10   A prokaryotic gene is a continuous sequence of nucleotide triplets, codons. A gene starts with a start codon *ATG* and ends with one of three stop codons: *TAA, TAG, TGA*. Calculate the number of parameters in such a codon model. The data set contains on the order of 300 000 codons. Would it be feasible to estimate a second order Markov chain from this data set?

Solution First, we assume that the sequence of codons is modeled by the first order Markov chain with states defined as sixty-one sense codons. The three stop codons are clumped together into the *end* state of the Markov chain. Since a gene starts with codon *ATG*, the initial probabilities of the states are determined as follows: $P(ATG) = 1$, and zero for all other states. To describe this Markov chain, one needs to define $61 \times 61 = 3721$ probabilities of transitions between sense codons, and $61 \times 3 = 183$ probabilities of transitions from sense codons to the end state. Therefore, the total number of the first order model parameters is given by

$$N_1 = 3721 + 183 = 3904.$$

If the sequence of codons is modeled by the second order Markov chain, the number of parameters increases dramatically. Indeed, there are sixty-one initial probabilities for all possible pairs (ATG, XYZ) of initial codons, $61^3 = 226\,981$ probabilities of transitions between sense codons, and $61^2 \times 3 = 11\,532$ probabilities of transitions from pairs of sense codons $(X_1Y_1Z_1, X_2Y_2Z_2)$ to the end state. Thus, the total number of parameters required for the second order Markov chain of codons is given by

$$N_2 = 61 + 226\,981 + 11\,532 = 238\,205.$$

Could we achieve reliable estimates of these N_2 parameters from a data set comprising 300 000 codons? A maximum likelihood estimate of transition probability $P(X_3Y_3Z_3|X_1Y_1Z_1, X_2Y_2Z_2)$ of this second order Markov chain is the ratio N_1/N_2, where N_1 is a count of the codon triplets $(X_1Y_1Z_1, X_2Y_2Z_2, X_3Y_3Z_3)$ and N_2 is a count of the codon pairs $(X_1 Y_1 Z_1, X_2 Y_2 Z_2)$ in the data set. It is likely that many out of $61^3 = 226\,981$ different codon triplets may never appear in the sample containing about $300\,000$ codon triplets. The observed zero codon triplet counts (unless we use pseudocounts) produce zero transition probabilities, thus leading to overfitting. Clearly, the data set containing on the order of $300\,000$ codons is not sufficient to estimate parameters of the second order Markov model. □

Problem 3.11 Which improvement can be made for the first order Markov chain model of protein-coding ORF with codon states?

Solution Instead of a Markov chain of codons we can consider an inhomogeneous Markov chain (IMC) or a hidden Markov model (HMM), both with states defined as single nucleotides rather than nucleotide triplets. Such IMCs and HMMs have a substantially smaller number of parameters (see Problem 3.12). It can be shown (Borodovsky *et al.*, 1986c; Borodovsky and McIninch, 1993; Krogh, Mian, and Haussler, 1994; Burge and Karlin, 1997) that these models describe nucleotide composition patterns in protein-coding and non-coding regions with accuracy sufficient to design accurate gene finding algorithms. □

Problem 3.12 It is well known that in protein-coding genes the three codon positions have different nucleotide frequency statistics and, therefore, it is natural to use the inhomogeneous three-periodic Markov chain to model protein-coding regions (Borodovsky *et al.*, 1986b). The three phases of the model are numbered 1 to 3 according to the position in the codon. Assuming that x_1 is in codon position 1, the probability of $x_1, x_2, x_3, \ldots$ will be

$$a^1_{x_1 x_2} a^2_{x_2 x_3} a^3_{x_3 x_4} a^1_{x_4 x_5} a^2_{x_5 x_6} \ldots,$$

where a^k are the elements of the transition probability matrix of phase k. Describe the HMM that corresponds to this first order inhomogeneous Markov chain.

Solution The set of hidden states of such an HMM could be defined as

$$\{A_1, A_2, A_3, C_1, C_2, C_3, G_1, G_2, G_3, T_1, T_2, T_3\},$$

where for X_i index i designates the codon position of nucleotide X. The observed states, the nucleotide symbols A, C, G, T, are emitted from hidden states with probabilities $e_{X_\bullet}(X) = 1$ and $e_{X_\bullet}(Y) = 0$ for $Y \neq X$, where $X, Y \in \{A, C, G, T\}$.

To define this HMM, seventy-two parameters are required: twelve initial probabilities a_{0X_1}, a_{0X_2}, a_{0X_3}; forty-eight probabilities of transitions between hidden states, $a_{X_1 Y_2}$, $a_{X_2 Y_3}$, $a_{X_3 Y_1}$; and twelve probabilities of transitions to the *end* state, $a_{X_1 0}$, $a_{X_2 0}$, $a_{X_3 0}$.

Note that some transitions between hidden states are impossible:

$$a_{X_1 Y_1} = a_{X_2 Y_2} = a_{X_3 Y_3} = a_{X_1 Y_3} = a_{X_2 Y_1} = a_{X_3 Y_2} = 0.$$

These zero value parameters are not included in the above counts. $\square$

Problem 3.13 Prove that $P(x) = \prod_{j=1}^{L} s_j$ if scaling variables s_j are defined by the following equations:

$$\tilde{f}_l(i) = \frac{f_l(i)}{\prod_{j=1}^{i} s_j}, \quad \sum_l \tilde{f}_l(i) = 1.$$

Solution With the scaling variables s_i, $i = 1, \ldots, L$, chosen to satisfy equations $\sum_l \tilde{f}_l(i) = 1$ for any i, the forward algorithm equation for the probability $P(x)$ of sequence $x = (x_1, \ldots, x_L)$ becomes

$$P(x) = \sum_k f_k(L) a_{k0} = \left(\sum_k \tilde{f}_k(L) a_{k0} \right) \prod_{j=1}^{L} s_j = \prod_{j=1}^{L} s_j.$$

Here the probabilities of transition from any state k to the end state, a_{k0}, are equal to 1 at position L, thus all sequences have a fixed length L.

For such a choice of s_i, the following recurrent equations hold:

$$\tilde{f}_l(i + 1) = \frac{1}{s_{i+1}} e_l(x_{i+1}) \sum_k \tilde{f}_k(i) a_{kl},$$

$$s_{i+1} = \sum_l e_l(x_{i+1}) \sum_k \tilde{f}_k(i) a_{kl}.$$

For the backward variables rescaled with the same set of s_j, the recurrence equation becomes

$$\tilde{b}_k(i) = \frac{1}{s_i} \sum_l a_{kl} \tilde{b}_l(i + 1) e_l(x_{i+1}).$$

Use of the rescaled variables $\tilde{f}_l(i)$ ($\tilde{b}_k(i)$) allows us to avoid an underflow error when running the forward (the backward) algorithm for long sequences. $\square$

Problem 3.14 Use the result of Problem 3.13 to show that the following equations:

$$A_{kl} = \sum_j \frac{1}{P(x^j)} \sum_i f_k^j(i) a_{kl} e_l(x_{i+1}^j) b_l^j(i+1),$$

$$E_k(b) = \sum_j \frac{1}{P(x^j)} \sum_{\{i:x_i^j=b\}} f_k^j(i) b_k^j(i)$$

used in the Baum–Welch algorithm for the estimation of parameters of an HMM simplify when using the scaled variables $\tilde{f}$ and $\tilde{b}$.

Solution With scaling variables s_i, $i = 1, \ldots, L$, chosen as defined in Problem 3.13, the expected number of transitions from a hidden state k to a hidden state l is given by the following formula:

$$A_{kl} = \sum_j \frac{1}{P(x^j)} \sum_i \tilde{f}_k^j(i) \prod_{q=1}^i s_q^j a_{kl} e_l(x_{i+1}^j) \tilde{b}_l^j(i+1) \prod_{q=i+1}^L s_q^j$$

$$= \sum_j \frac{\prod_{q=1}^L s_q^j}{P(x^j)} \sum_i \tilde{f}_k^j(i) a_{kl} e_l(x_{i+1}^j) \tilde{b}_l^j(i+1) = \sum_{j,i} \tilde{f}_k^j(i) a_{kl} e_l(x_{i+1}^j) \tilde{b}_l^j(i+1).$$

Here the sum is taken over all training sequences x^j and all positions i. The expected number of emissions of symbol b in state k becomes

$$E_k(b) = \sum_j \frac{1}{P(x^j)} \sum_{\{i:x_i^j=b\}} \tilde{f}_k^j(i) \tilde{b}_k^j(i) \prod_{q=1}^L s_q^j s_i^j = \sum_{\{j,i:x_i^j=b\}} \tilde{f}_k^j(i) \tilde{b}_k^j(i) s_i^j,$$

where the sum is taken over all training sequences x^j and positions i occupied by state k emitting symbol b. $\qquad\square$

3.2 Additional problems and theory

Practice with classic algorithms for a simple HMM has been our target in Problems 3.15–3.18: the Viterbi algorithm (Problems 3.15 and 3.16), the forward algorithm (Problem 3.17), the backward algorithm (Problem 3.17), the forward algorithm with scaled forward variables (Problem 3.17), and the posterior decoding algorithm (Problem 3.18).

Other problems from this section are related to the general topic of model selection for biological sequences (of course, an HMM is one of such models). In Problem 3.19, two stochastic models (an independence model and an HMM) are compared; the comparison is based on the model likelihoods calculated first as full probabilities (over all paths) and then as probabilities of the Viterbi paths.

The theoretical introduction to Problems 3.20 and 3.21 describes the basic ideas of model selection for a biological sequence, model testing, the maximum likelihood approach to parameter estimation, and the basic properties of the maximum likelihood estimates of parameters of an independence model and a first order stationary Markov chain. A test on the selection of an appropriate model for a DNA fragment is carried out in Problem 3.21, the estimation of transition probabilities of the Markov chain in Problems 3.20 and 3.21. The procedure of sampling the paths of hidden states from the posterior distribution is described in Problem 3.22. Yet another model for an empirical sequence, a basic segmentation model, is described in the theoretical introduction to Problem 3.23, along with a Bayesian approach to the estimation of parameters of the model. Using these ideas, we find the posterior distributions of unknown parameters and missing data for a given sequence generated by the segmentation model (Problem 3.23).

Problem 3.15 Hidden Markov models can be used in algorithms of protein secondary structure prediction. One rather straightforward approach uses the secondary structure conformations *α-helix*, *β-strand*, and *turn* as the hidden states emitting observable amino acids. It is assumed that the frequencies of appearance of each of twenty amino acids in either conformation have been determined from analysis of the proteins with the three-dimensional structures known from experiment. Draw the state of the HMM and describe the Viterbi and the posterior decoding algorithms that could be used for predicting the protein secondary structure.

Solution The state diagram of the HMM is shown in Figure 3.4. There are three hidden 'structural' states H, E, and T corresponding to *α-helix*, *β-strand*, and *turn*, respectively. These states emit symbols of amino acid residues with probabilities $e_H(j)$, $e_E(j)$, and $e_T(j)$, estimated from the frequencies of occurrence of amino acid $j, j = 1, \ldots, 20$, in corresponding conformations. A *begin* state B and an *end* state $\mathcal{E}$ do not emit any symbols. Other parameters of the HMM are: initial probabilities $a_{BE}, a_{BT}, a_{BH}, a_{B\mathcal{E}}$; transition probabilities $a_{EE}, a_{EH}, a_{ET}, a_{TE}, a_{TH}, a_{TT}, a_{HE}, a_{HH}, a_{HT}$; termination probabilities $a_{E\mathcal{E}}, a_{T\mathcal{E}}, a_{H\mathcal{E}}$. The total number of parameters of this HMM is seventy-six.

For a given protein sequence $x = (x_1, x_2, \ldots, x_n)$ the Viterbi algorithm determines the most probable sequence $\pi = (\pi_1, \pi_2, \ldots, \pi_n)$ of hidden structural states. For each sequence position $i, i = 1, 2, \ldots, n$, it maximizes the probability

$$v_l(i) = e_l(x_i) \max_k (v_k(i-1)a_{kl})$$

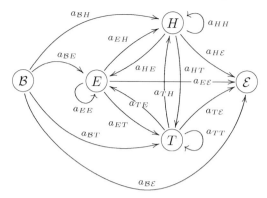

Figure 3.4. The state diagram of an HMM that can be used in the algorithms of protein secondary structure prediction. Hidden states H (*α-helix*), E (*β-strand*), and T (*turn*) emit amino acid symbols (observable states of the HMM).

and essentially determines the subsequence of hidden states up to the sequence position i which, together with the observed amino acid subsequence $(x_1, \ldots, x_i)$, has the maximum joint probability. At the last step, the Viterbi algorithm calculates the joint probability of the full sequence x and the optimal path π:

$$P(x, \pi) = \max_k (v_k(n)a_{k\mathcal{E}}).$$

Then the traceback procedure recovers the path π itself.

The posterior decoding algorithm determines the positional posterior distributions of the structural hidden states given protein sequence x. For sequence position i it finds the hidden state k for which, given sequence x, the posterior probability $P(\pi_i = k|x)$ is maximal.

Both the Viterbi and the posterior decoding algorithms are suitable for finding the sequence of structural states $\{H, E, T\}$: as the optimal or the Viterbi path (by the Viterbi algorithm), and as the sequence of hidden states singled out by their maximal posterior probabilities at consecutive positions of the protein sequence (by the decoding algorithm). These sequences of structural states may not coincide since at a given position the hidden state with the maximum posterior probability may not belong to the optimal path. Still, both sequences could be interpreted as the predicted protein secondary structure. □

Remark We are not aware of the direct use of the described HMM for protein secondary structure prediction. However, a more general HMM with the same hidden state diagram has been used for protein secondary structure prediction by Schmidler, Liu, and Brutlag (2000).

Problem 3.16 Real DNA sequences are inhomogeneous and can be described by a hidden Markov model with hidden states representing different types of nucleotide composition. Consider an HMM that includes two hidden states H and L for higher and lower $C + G$ content, respectively. Initial probabilities for both H and L are equal to 0.5, while transition probabilities are as follows: $a_{HH} = 0.5$, $a_{HL} = 0.5$, $a_{LL} = 0.6$, $a_{LH} = 0.4$. Nucleotides T, C, A, G are emitted from states H and L with probabilities 0.2, 0.3, 0.2, 0.3, and 0.3, 0.2, 0.3, 0.2, respectively. Use the Viterbi algorithm to define the most likely sequence of hidden states for the 'toy' sequence $x = GGCACTGAA$.

Solution The Viterbi algorithm calculations are performed here in logarithmic mode (base 2), which is recommended in general to avoid an underflow error. We use a tilde to designate new values of parameters, so $\tilde{a}_{kl} = \log_2 a_{kl}$, etc. As the recursion equations we have

$$V_l(i + 1) = \tilde{e}_l(x_{i+1}) + \max_k (V_k(i) + \tilde{a}_{kl}),$$

where V is the logarithm of v. The Viterbi algorithm is initialized at the *begin* state designated as 0, and scores $V_H(i)$, $V_L(i)$ are calculated in turn:

$i = 0$, $V_0(0) = 0$, $V_H(0) = -\infty$, $V_L(0) = -\infty$;

$i = 1$, $V_H(1) = \tilde{e}_H(G) + \tilde{a}_{0H} = -2.736966$,

$V_L(1) = \tilde{e}_L(G) + \tilde{a}_{0L} = -3.321981$;

$i = 2$, $V_H(2) = \tilde{e}_H(G) + V_H(1) + \tilde{a}_{HH} = -5.473931$,

$V_L(2) = \tilde{e}_L(G) + V_H(1) + \tilde{a}_{HL} = -6.058893$;

$i = 3$, $V_H(3) = \tilde{e}_H(C) + V_H(2) + \tilde{a}_{HH} = -8.210897$,

$V_L(3) = \tilde{e}_L(C) + V_H(2) + \tilde{a}_{HL} = -8.795859$;

$i = 4$, $V_H(4) = \tilde{e}_H(A) + V_H(3) + \tilde{a}_{HH} = -11.532825$,

$V_L(4) = \tilde{e}_L(A) + V_H(3) + \tilde{a}_{HL} = -10.947862$;

$i = 5$, $V_H(5) = \tilde{e}_H(C) + V_L(4) + \tilde{a}_{LH} = -14.006756$,

$V_L(5) = \tilde{e}_L(C) + V_L(4) + \tilde{a}_{LL} = -14.006756$;

$i = 6$, $V_H(6) = \tilde{e}_H(T) + V_H(5) + \tilde{a}_{HH} = -17.328684$,

$V_L(6) = \tilde{e}_L(T) + V_L(5) + \tilde{a}_{LL} = -16.480673$;

$i = 7$, $V_H(7) = \tilde{e}_H(G) + V_L(6) + \tilde{a}_{HH} = -19.539581$,

$V_L(7) = \tilde{e}_L(G) + V_L(6) + \tilde{a}_{LL} = -19.539581$;

$i = 8,$ $V_H(8) = \tilde{e}_H(A) + V_H(7) + \tilde{a}_{HH} = -22.861509,$

$V_L(8) = \tilde{e}_L(A) + V_L(7) + \tilde{a}_{LL} = -22.013512;$

$i = 9,$ $V_H(9) = \tilde{e}_H(A) + V_L(8) + \tilde{a}_{LH} = -25.657368,$

$V_L(9) = \tilde{e}_L(A) + V_L(8) + \tilde{a}_{LL} = -24.487443.$

The most probable path π^* is determined by the traceback procedure: $\pi^* = $ *HHHLLLLLL*, while $P(x, \pi^*) = 2^{V_L(9)} = 4.251528 \times 10^{-8}$. For sequence x the Viterbi path π^* is unique, because at each step i of the algorithm the maximum score $V_\bullet(i)$ was derived unambiguously from only one of the two previous values $V_\bullet(i-1)$.   □

Problem 3.17 For the hidden Markov model defined in Problem 3.16 and the DNA sequence fragment $x = GGCA$ find $P(x)$ by both the forward algorithm and the backward algorithm. Repeat the computation by the forward algorithm with use of scaling variables (Problem 3.13).

Solution The forward algorithm proceeds as follows. Initialization:

$i = 0,$ $f_0(0) = 1, f_H(0) = 0, f_L(0) = 0;$

$i = 1,$ $f_H(1) = e_H(x_1)a_{0H} = 0.3 \times 0.5 = 0.15,$

$f_L(1) = e_L(x_1)a_{0L} = 0.2 \times 0.5 = 0.1;$

$i = 2,$ $f_H(2) = e_H(G)(f_H(1)a_{HH} + f_L(1)a_{LH})$

$= 0.3(0.15 \times 0.5 + 0.1 \times 0.4) = 0.0345,$

$f_L(2) = e_L(G)(f_H(1)a_{HL} + f_L(1)a_{LL})$

$= 0.2(0.15 \times 0.5 + 0.1 \times 0.6) = 0.027;$

$i = 3,$ $f_H(3) = e_H(C)(f_H(2)a_{HH} + f_L(2)a_{LH})$

$= 0.3(0.0345 \times 0.5 + 0.027 \times 0.4) = 0.008415,$

$f_L(3) = e_L(C)(f_H(2)a_{HL} + f_L(2)a_{LL})$

$= 0.2(0.0345 \times 0.5 + 0.027 \times 0.6) = 0.00669;$

$i = 4,$ $f_H(4) = e_H(A)(f_H(3)a_{HH} + f_L(3)a_{LH})$

$= 0.2(0.008415 \times 0.5 + 0.00669 \times 0.4) = 0.0013767,$

$f_L(4) = e_L(A)(f_H(3)a_{HL} + f_L(3)a_{LL})$

$= 0.3(0.008415 \times 0.5 + 0.00669 \times 0.6) = 0.00246645.$

At the termination step we have

$$P(x) = P(GGCA) = f_H(4) + f_L(4) = 0.00384315.$$

Here we assume that the probabilities of transitions to the *end* state a_{H0} and a_{L0} are equal to 1 at position $L = 4$ since we are dealing with sequences of length 4.

The backward algorithm, starting with position 4, proceeds as follows:

$i = 4,$ $b_H(4) = 1,\ b_L(4) = 1;$

$i = 3,$ $b_H(3) = a_{HH}e_H(A)b_H(4) + a_{HL}e_L(A)b_L(4)$

$\qquad\qquad = 0.5 \times 0.2 + 0.5 \times 0.3 = 0.25,$

$\qquad b_L(3) = a_{LH}e_H(A)b_H(4) + a_{LL}e_L(A)b_L(4)$

$\qquad\qquad = 0.4 \times 0.2 + 0.6 \times 0.3 = 0.26;$

$i = 2,$ $b_H(2) = a_{HH}e_H(C)b_H(3) + a_{HL}e_L(C)b_L(3)$

$\qquad\qquad = 0.5 \times 0.3 \times 0.25 + 0.5 \times 0.2 \times 0.26 = 0.0635,$

$\qquad b_L(2) = a_{LH}e_H(C)b_H(3) + a_{LL}e_L(C)b_L(3)$

$\qquad\qquad = 0.4 \times 0.3 \times 0.25 + 0.6 \times 0.2 \times 0.26 = 0.0612;$

$i = 1,$ $b_H(1) = a_{HH}e_H(G)b_H(2) + a_{HL}e_L(G)b_L(2)$

$\qquad\qquad = 0.5 \times 0.3 \times 0.0635 + 0.5 \times 0.2 \times 0.0612 = 0.015645,$

$\qquad b_L(1) = a_{LH}e_H(G)b_H(2) + a_{LL}e_L(G)b_L(2)$

$\qquad\qquad = 0.4 \times 0.3 \times 0.0635 + 0.6 \times 0.2 \times 0.0612 = 0.014964.$

At the termination step we have

$$P(x) = P(GGCA) = e_H(G)b_H(1)a_{0H} + e_L(G)b_L(1)a_{0L}$$

$$= 0.3 \times 0.015645 \times 0.5 + 0.2 \times 0.014964 \times 0.5 = 0.00384315.$$

As expected, the probability $P(x)$ determined by both the forward and the backward algorithms is the same.

Now we calculate the probability $P(x)$ again by using the forward algorithm with rescaled variables $\tilde{f}$ and scaling variables s_i defined in Problem 3.13. Since variables $s_i,\ i = 1, \ldots, 4$, were defined to make $\sum_l \tilde{f}_l(i) = 1$ for any i, then $P(x) = \prod_{j=1}^{4} s_j$.

The recurrence equations are as follows:

$$s_{i+1} = \sum_l e_l(x_{i+1}) \sum_k \tilde{f}_k(i)a_{kl},$$

$$\tilde{f}_l(i+1) = \frac{1}{s_{i+1}} e_l(x_{i+1}) \sum_k \tilde{f}_k(i)a_{kl},$$

with termination $P(x) = \prod_{j=1}^{4} s_j$. The iterations proceed as follows (the PERL and C++ programs are included in the Web Supplemental Materials available at opal.biology.gatech.edu/PSBSA). At the initialization step we have

$i = 0$, $\tilde{f}_0(0) = 1, \tilde{f}_H(0) = 0, \tilde{f}_L(0) = 0$;

$i = 1$, $s_1 = 0.25, \tilde{f}_H(1) = 0.6, \tilde{f}_L(1) = 0.4$;

$i = 2$, $s_2 = 0.246, \tilde{f}_H(2) = 0.560976, \tilde{f}_L(2) = 0.439024$;

$i = 3$, $s_3 = 0.245610, \tilde{f}_H(3) = 0.557100, \tilde{f}_L(3) = 0.442810$;

$i = 4$, $s_4 = 0.254429, \tilde{f}_{H(4)} = 0.358222, \tilde{f}_L(3) = 0.641778$.

The termination step gives

$$P(x) = P(GGCA) = s_1 \times s_2 \times s_3 \times s_4 = 0.00384315.$$

The advantage of using rescaled variables $\tilde{f}$ is in avoiding underflow errors that could occur in computations of vanishingly small f-values for sequences with large lengths. Even for the short sequence $GGCA$ one can see the rapid decrease of values of forward variables f and relatively stable behavior of rescaled variables $\tilde{f}$. ☐

Remark Since sequence $x = GGCA$ is a prefix of the sequence $GGCACTGAA$ considered in Problem 3.16, the Viterbi path π^* for x can be immediately determined as a prefix of the Viterbi path found in Problem 3.16. Thus, $\pi^* = HHHL$. The joint probability value is $P(x, \pi^*) = \upsilon(4) = 0.15^4 = 0.00050625$. Note that the contribution of the Viterbi path to the full probability of x is about 13%.

Problem 3.18 Find the posterior probability of states H and L at position 4 of the DNA sequence $x = GGCA$. Consider the hidden Markov model described in Problem 3.16.

Solution The posterior probabilities can be found by the following formula:

$$P(\pi_i = k|x) = \frac{f_k(i)b_k(i)}{P(x)}.$$

The forward and backward variables $f(4)$ and $b(4)$ were determined upon solving Problem 3.17. Therefore, we have

$$P(\pi_4 = H|GGCA) = \frac{f_H(4)b_H(4)}{P(GGCA)} = \frac{0.0013767 \times 1}{0.00384315} = 0.35822.$$

Similarly,

$$P(\pi_4 = L|GGCA) = \frac{f_L(4)b_L(4)}{P(GGCA)} = \frac{0.002466457 \times 1}{0.00384315}$$

$$= 0.64178 = 1 - P(\pi_4 = H|GGCA).$$

We see that state L is more likely to appear at position 4 than state H. Incidentally, state L appears at position 4 in the most probable (Viterbi) path $\pi^* = HHHL$ (see the remark to Problem 3.17). Still, we should remember that states of the Viterbi path and the most probable states obtained by the posterior decoding algorithm do not necessarily coincide. □

Problem 3.19 On a particular day a casino either uses a fair die all the time (mode F) or uses a fair die most of the time, but occasionally switches to a loaded die (mode L). While in mode L, the probability of switching from a fair to a loaded die before each roll is ξ and the probability of switching back is ψ. Given the following sequence x of 300 rolls observed on a particular day:

```
31511624644664424531132163116415213362514454363165662656666
65116645313265124563666463163666316232645523626666625151631
22255544166656656356432436413151346514635341112641462625356
36616366646623253441366166116325256246225526525226643535336
23312162536441443233516324363366556246666263266661235245242
```

describe the procedure of statistical inference discriminating between two modes F and L of the casino operation. Assume that the parameters of mode L are $\psi = 0.1$ and $\xi = 0.05$ and that the loaded die has probability 0.5 of a six and probability 0.1 for each of the other numbers.

Solution To discriminate between the modes F and L, the following statistical rule is suggested: if the ratio $S = \log P(L|x)/P(F|x)$ is positive, then the casino is more likely to operate in mode L; if S is negative, the casino is more likely to operate in mode F; finally, if $S = 0$, no decision is made.

Under the assumption that the prior probabilities of modes F and L are equal, S becomes the log-odds ratio:

$$S = \log \frac{P(L|x)}{P(F|x)}$$

$$= \log \left(\frac{P(x|L)P(L)}{P(x|L)P(L) + P(x|F)P(F)} \frac{P(x|L)P(L) + P(x|F)P(F)}{P(x|F)P(F)} \right)$$

$$= \log \frac{P(x|L)}{P(x|F)}.$$

For mode F (using log base 10)

$$\log P(x|F) = \log \left(\frac{1}{6^{300}} \right) = -234 + \log 3.586.$$

Computation of the likelihood $P(x|L)$ is more complicated. Use of a fair or a loaded die defines a particular hidden state (1 or 2). Transitions between states and emissions of observed numbers are controlled by the transition probabilities $a_{11} = 1 - \xi = 0.95$, $a_{12} = \xi = 0.05$, $a_{21} = \psi = 0.1$, $a_{22} = 1 - \psi = 0.9$ and the emission probabilities $e_1(i) = 1/6$, for $i = 1, \ldots, 6$, $e_2(1) = e_2(2) = e_2(3) = e_(4) = e_1(5) = 1/10$, $e_2(6) = 1/2$. Now, $P(x|L)$ is defined as the sum of probabilities of x generated by all possible sequences π of hidden states:

$$P(x|L) = \sum_{\pi} P(x, \pi|L).$$

The full probability $P(x|L)$ is determined by the forward (backward) algorithm.

The forward algorithm starts with initialization: $f_0(0) = 1$, $f_k(0) = 0$ for $k > 0$; continues with recursion for $i = 1, \ldots, L$: $f_l(i) = e_l(x_i) \sum_k f_k(i-1)a_{kl}$; and terminates with calculation of the full probability of x:

$$P(x) = \sum_k f_k(L)a_{k0} = P(x|L).$$

The general forward algorithm has been implemented in both PERL and C++ languages (see the Web Supplemental Materials at opal.biology.gatech.edu/PSBSA). It was found computationally that

$$\log P(x|L) = -225 + \log 3.294.$$

Therefore, the log-odds ratio becomes

$$S = 8 + \log 9.186 = 8.963145 > 0.$$

Thus, the statistical inference suggests that the given sequence of die rolls x was observed when the casino operated in mode L.

Remark Yet another approach could be suggested for discrimination between modes F and L using statistical inference. Instead of the full probabilities of x given modes F and L, we could compare the Viterbi probabilities, i.e. the probabilities $P(x, \pi_L)$ and $P(x, \pi_F)$ of the most likely paths for x given modes L and F. The log-odds ratio S^* is then defined by the following formula:

$$S^* = \log \frac{P(x, \pi_L)}{P(x, \pi_F)}.$$

In mode F (using the fair die only) there exists only one possible path of hidden states (consisting of fair die states) for any observed sequence of rolls, thus $P(x, \pi_F)$ is the same as the full probability $P(x|F)$:

$$\log P(x, \pi_F) = \log P(x|F) = -234 + \log 3.586.$$

In mode L the probability $P(x, \pi_L)$ of the most likely path is calculated by the Viterbi algorithm (see Problem 3.16 and the computer programs in PERL and C++ included

in the Web Supplementary Materials available at opal.biology.gatech.edu/PSBSA).
For an observed sequence x we obtain

$$P(x, \pi_L) = \text{argmax}_\pi P(x, \pi) = 1.585 \times 10^{-236},$$

$$S^* = \log \frac{P(x, \pi_L)}{P(x, \pi_F)} = \log 0.004 < 0.$$

Interestingly, the statistical inference using the Viterbi paths has led to the conclusion that the casino operated in mode F. This conclusion is in contradiction to what was inferred with the full probabilistic model!

The reason for switching the decision-making outcome is that the Viterbi path delivers only a fraction of the full probability of the observed sequence. Therefore, there is a risk of making incorrect decisions when the Viterbi path is used for the inference. This issue will be further illustrated by Problems 4.4–4.6.   □

3.2.1 Probabilistic models for sequences of symbols: selection of the model and parameter estimation

Theoretical introduction to Problems 3.20 and 3.21

To describe approaches to model selection, we will follow Reinert, Schbath, and Waterman (2000a). A biological sequence $x = x_1, \ldots, x_N$ can be considered as a realization of a random sequence of symbols from a finite alphabet $\mathcal{A}$, $|\mathcal{A}| = 4$ for DNA and $|\mathcal{A}| = 20$ for proteins. Sequence x can be described by a variety of probabilistic models.

The simplest model, a homogeneous independence model M, assumes that the occurrences of symbols at different sites of the sequence are independent events and a symbol α, $\alpha \in \mathcal{A}$, has probability p_α of appearing at any site of the sequence, $\sum_{\alpha \in \mathcal{A}} p_\alpha = 1$. If we assume that the observed sequence x is generated by model M with unknown parameters $\{p_\alpha\}$, the maximum likelihood estimates of p_α are given by ratios

$$\hat{p}_\alpha = \frac{N(\alpha)}{N}, \tag{3.6}$$

where $N(\alpha)$ is the number of symbols α observed in sequence x.

A more general model, the model $M(m)$, is a stationary mth order Markov chain with transition probabilities

$$p_{\alpha_1, \ldots, \alpha_m, \alpha_{m+1}} = P(x_i = \alpha_{m+1} | x_{i-1} = \alpha_m, \ldots, x_{i-m} = \alpha_1),$$

$\alpha_k \in \mathcal{A}$, $i = m + 1, \ldots, N$. Obviously, the independence model M is a Markov chain of order $m = 0$, $M = M(0)$. It is known that for a given sequence x generated by the first order Markov model $M(1)$ with unknown parameters, the maximum likelihood estimates of the transition probabilities $p_{\alpha,\beta}$, $\alpha, \beta \in \mathcal{A}$, are delivered by the ratios of counts:

$$\hat{p}_{\alpha,\beta} = \frac{N(\alpha\beta)}{N(\alpha\bullet)}. \qquad (3.7)$$

Here $N(\alpha\beta)$ is the number of occurrences of the pair of adjacent symbols (α, β) in sequence x and $N(\alpha\bullet) = \sum_{\gamma \in \mathcal{A}} N(\alpha\gamma)$. Note that $N(\alpha\bullet) = N(\alpha)$ if $x_N \neq \alpha$ and $N(\alpha\bullet) = N(\alpha) - 1$ otherwise. Similarly, for sequence x generated by model $M(m)$ with unknown parameters the maximum likelihood estimates of the transition probabilities $p_{\alpha_1,\dots,\alpha_m,\alpha_{m+1}}$ are given by

$$\hat{p}_{\alpha_1,\dots,\alpha_m,\alpha_{m+1}} = \frac{N(\alpha_1 \cdots \alpha_m \alpha_{m+1})}{N(\alpha_1 \cdots \alpha_m \bullet)}.$$

Here $N(\alpha_1 \cdots \alpha_k)$ designates the number of occurrences of a k-letter word $(\alpha_1, \dots, \alpha_k)$ in sequence x and $N(\alpha_1 \cdots \alpha_m \bullet) = \sum_{\gamma \in \mathcal{A}} N(\alpha_1 \cdots \alpha_m \gamma)$.

Selection of the most appropriate model for a sequence x among models $M(m)$, $m \geq 0$, can be achieved on the basis of the χ^2-test. First we test whether the independence model $M(0)$ is an appropriate model for the sequence x. The null hypothesis of independence,

$$H_0 : P(x_i = \alpha, x_{i+1} = \beta) = p_\alpha p_\beta, \ i = 1, \dots, N-1, \qquad \alpha, \beta \in \mathcal{A},$$

must be tested versus the alternative hypothesis H_a stating that the only restriction on the joint probabilities $P(x_i = \alpha, x_{i+1} = \beta)$ is that they do not depend on i. If H_0 is true, the maximum likelihood estimate of $P(x_i = \alpha, x_{i+1} = \beta), i = 1, \dots, N-1$, is given by

$$\hat{P}_{H_0}(\alpha, \beta) = \frac{N(\alpha\bullet)}{N-1} \frac{N(\bullet\beta)}{N-1}.$$

Here $N(\bullet\beta) = \sum_{\gamma \in \mathcal{A}} N(\gamma\beta)$. Note that

$$\frac{N(\alpha\bullet)}{N-1} \frac{N(\bullet\beta)}{N-1} = \hat{p}_\alpha \hat{p}_\beta,$$

where $\hat{p}_\alpha$ is the maximum likelihood estimate of p_α based on sequence $x_1, \dots, x_{N-1}$, and $\hat{p}_\beta$ is the maximum likelihood estimate of p_β based on sequence $x_2, \dots, x_N$. If the alternative hypothesis is true, the maximum likelihood estimate of $P(x_i = \alpha, x_{i+1} = \beta), i = 1, \dots, N-1$, is given by

$$\hat{P}_{H_a}(\alpha, \beta) = \frac{N(\alpha\beta)}{N-1}.$$

Then,

$$\chi^2 = \sum_{\alpha \in A} \sum_{\beta \in A} \frac{((N-1)\hat{P}_{H_a}(\alpha, \beta) - (N-1)\hat{P}_{H_0}(\alpha, \beta))^2}{(N-1)\hat{P}_{H_0}(\alpha, \beta)}$$

$$= \sum_{\alpha \in A} \sum_{\beta \in A} \frac{((N-1)N(\alpha\beta) - N(\alpha\bullet)N(\bullet\beta))^2}{(N-1)N(\alpha\bullet)N(\bullet\beta)} \qquad (3.8)$$

defines the Pearson χ^2-statistic that under the null hypothesis asymptotically has a χ^2-distribution with $(|A| - 1)^2$ degrees of freedom. Thus we reject H_0 when the sample value of the χ^2-statistic (3.8) is larger than a critical value of the χ^2-distribution corresponding to a specified significance level; otherwise, we accept H_0. We have to use the χ^2-distribution with nine degrees of freedom for DNA sequences, and the χ^2-distribution with 361 degrees of freedom for protein sequences.

If H_0 is rejected, a test for a higher order dependence is due. At the next step, the null hypothesis stating that the first order Markov chain $M(1)$ is the relevant model for sequence x can be formulated as follows:

$$H_0 : P(x_i = \alpha, x_{i+1} = \beta, x_{i+2} = \gamma)$$

$$= P(x_i = \alpha)P(x_{i+1} = \beta | x_i = \alpha)P(x_{i+2} = \gamma | x_i = \alpha, x_{i+1} = \beta)$$

$$= P(x_i = \alpha)p_{\alpha,\beta}p_{\beta,\gamma} = \sum_{x_1, x_2, \ldots, x_{i-1}} \pi(x_1)p_{x_1, x_2} \times \cdots \times p_{x_{i-1}, \alpha}p_{\alpha,\beta}p_{\beta,\gamma}$$

$$= \pi(\alpha)p_{\alpha,\beta}p_{\beta,\gamma},$$

for $i = 1, \ldots, N-2$, $\alpha, \beta, \gamma \in A$. Here π is the stationary distribution of the Markov chain $M(1)$. The null hypothesis should be tested against the alternative hypothesis H_a, which assumes that the only restriction on the probabilities $P(x_i = \alpha, x_{i+1} = \beta, x_{i+2} = \gamma)$ is that they do not depend on i.

Under the null hypothesis, the maximum likelihood estimate of $P(x_i = \alpha, x_{i+1} = \beta, x_{i+2} = \gamma)$, $i = 1, \ldots, N-2$, is given by

$$\hat{P}_{H_0}(\alpha, \beta, \gamma) = \hat{\pi}(\alpha)\hat{p}_{\alpha,\beta}\hat{p}_{\beta,\gamma} = \frac{N(\alpha \bullet \bullet)}{N-2} \frac{N(\alpha\beta\bullet)}{N(\alpha \bullet \bullet)} \frac{N(\bullet\beta\gamma)}{N(\bullet\beta\bullet)} = \frac{N(\alpha\beta\bullet)}{N-2} \frac{N(\bullet\beta\gamma)}{N(\bullet\beta\bullet)}.$$

Here $N(\alpha\beta\bullet)$ is the number of occurrences of triplets starting with symbols (α, β); $N(\alpha \bullet \bullet) = \sum_{\beta \in A} N(\alpha\beta\bullet)$; $N(\bullet\beta\gamma)$ is the number of occurrences of triplets ending with symbols (β, γ); $N(\bullet\beta\bullet) = \sum_{\gamma \in A} N(\bullet\beta\gamma)$.

Under H_a, the maximum likelihood estimate of $P(x_i = \alpha, x_{i+1} = \beta, x_{i+2} = \gamma)$, $i = 1, \ldots, N-2$, is given by

$$\hat{P}_{H_a}(\alpha, \beta, \gamma) = \frac{N(\alpha\beta\gamma)}{N-2}.$$

Then under H_0 the Pearson χ^2-statistic,

$$\chi^2 = \sum_{\alpha \in \mathcal{A}} \sum_{\beta \in \mathcal{A}} \sum_{\gamma \in \mathcal{A}} \frac{((N-2)\hat{P}_{H_a}(\alpha,\beta,\gamma) - (N-2)\hat{P}_{H_0}(\alpha,\beta,\gamma))^2}{(N-2)\hat{P}_{H_0}(\alpha,\beta,\gamma)}$$

$$= \sum_{\alpha \in \mathcal{A}} \sum_{\beta \in \mathcal{A}} \sum_{\gamma \in \mathcal{A}} \frac{(N(\alpha\beta\gamma)N(\bullet\beta\bullet) - N(\alpha\beta\bullet)N(\bullet\beta\gamma))^2}{N(\alpha\beta\bullet)N(\bullet\beta\gamma)N(\bullet\beta\bullet)}, \qquad (3.9)$$

asymptotically has a χ^2-distribution with $(|\mathcal{A}|^2-1)(|\mathcal{A}|-1)-l$ degrees of freedom, where l is the number of triplets (α,β,γ), $\alpha,\beta,\gamma \in \mathcal{A}$, such that the number of expected counts $(N-2)\hat{P}_{H_0}(\alpha,\beta,\gamma)$ is equal to zero (thus the corresponding l terms

$$\frac{((N-2)\hat{P}_{H_a}(\alpha,\beta,\gamma) - (N-2)\hat{P}_{H_0}(\alpha,\beta,\gamma))^2}{(N-2)\hat{P}_{H_0}(\alpha,\beta,\gamma)}$$

cannot be properly defined and they will be absent from the sum for the χ^2-statistic). Application of this test to DNA sequences at a specified significance level requires us to compare the observed value of the χ^2-statistic with the critical value of the χ^2-distribution with $45 - l$ degrees of freedom, while for protein sequences we need the χ^2-distribution with $7581 - l$ degrees of freedom. If the null hypothesis is rejected, the test for the second order Markov chain should be carried out in the analogous way. Detailed descriptions of the statistical tests for the independence, homogeneity, Markovity, and the order of the Markov chain can be found in the classical publications by Goodman (1959), Billingsley (1961a,b) and Kullback, Kupperman, and Ku (1962).

Now we come back to the independence model $M(0)$ and assume that it has been established by the χ^2-test that sequence $x = x_1, \ldots, x_N$ is generated by $M(0)$ with the unknown parameters p_α, $\alpha \in \mathcal{A}$. Then, we have the maximum likelihood estimates in Equation (3.6) for the model parameters and we face a logical question: how close are these estimates to the true values of parameters? The maximum likelihood estimates $\hat{p}_\alpha$ are unbiased:

$$\mathbf{E}\hat{p}_\alpha = \mathbf{E}\frac{N(\alpha)}{N} = \frac{Np_\alpha}{N} = p_\alpha.$$

The consistency ($\hat{p}_\alpha \to^P p_\alpha$) and the strong consistency ($\hat{p}_\alpha \to p_\alpha$ with probability 1) of the estimates $\hat{p}_\alpha$ follow from the law of large numbers and the strong law of large numbers, respectively. These properties imply that the estimate $\hat{p}_\alpha$, $\alpha \in \mathcal{A}$, becomes closer to the true value of p_α as the sample size (the sequence x length) N increases.

We designate $P_i = P_{\alpha_i}$, $i = 1, \ldots, k$, for an alphabet of size k, $\mathcal{A} = \{\alpha_1, \ldots, \alpha_k\}$. The asymptotic normality property of a multidimensional maximum likelihood estimator (Cox and Hinkley (1974), Sect. 9.2) implies that a random vector $Y = (\hat{\mathbf{P}} - \mathbf{P})\sqrt{N}$ with components $Y_i = (\hat{p}_i - p_i)\sqrt{N}$, $i = 1, \ldots, k$, is asymptotically a centered Gaussian vector with the covariance matrix $R = (r_{ij}) = (cov(Y_i, Y_j))$, $i, j = 1, \ldots, k$, with elements

$$r_{ii} = p_i - p_i^2, \quad r_{ij} = -p_i p_j, \quad i \neq j.$$

In particular, this means that each component $Y_i = (\hat{p}_i - p_i)\sqrt{N}$ of the vector Y is asymptotically a centered normal random variable.

Next, for the first order ergodic stationary Markov chain with unknown transition probabilities estimated by (3.7) and stationary distribution π estimated by $\hat{\pi}(\alpha) = N(\alpha)/N$, it is also easy to show that the maximum likelihood estimates $\hat{\pi}(\alpha)$ are unbiased:

$$\mathbf{E}\hat{\pi}(\alpha) = \frac{1}{N}\mathbf{E}N(\alpha) = \frac{1}{N}\sum_{i=1}^{N} P(x_i = \alpha)$$

$$= \frac{1}{N}\sum_{i=1}^{N}\sum_{x_1, x_2, \ldots, x_{i-1}} \pi(x_1)p_{x_1, x_2} \times \ldots \times p_{x_{i-1}, \alpha} = \frac{N}{N}\pi(\alpha) = \pi(\alpha).$$

The estimates $\hat{\pi}(\alpha)$ are consistent, and vector ξ with components $\xi_i = (\hat{\pi}(\alpha_i) - \pi(\alpha_i))\sqrt{N}$, $i = 1, \ldots, k$, is asymptotically a centered Gaussian vector (Billingsley (1961b), Lemma 3.2, Theorem 3.3).

For an alphabet of size k, $\mathcal{A} = \{\alpha_1, \ldots, \alpha_k\}$, we designate $p_{ij} = p_{\alpha_i \alpha_j}$, $i, j = 1, \ldots, k$. It is also known that the maximum likelihood estimates of transition probabilities are consistent: $\hat{p}_{ij} \to^P p_{ij}$, $i, j = 1, \ldots, k$ (Billingsley (1961b), Theorem 4.1). The k^2-dimensional vector $\eta = (\eta_{11}, \ldots, \eta_{1k}, \ldots, \eta_{k1}, \ldots, \eta_{kk})$ with components

$$\eta_{ij} = \frac{N(\alpha_i \alpha_j) - N(\alpha_i)p_{ij}}{\sqrt{N(\alpha_i)}},$$

$i, j = 1, \ldots, k$, is an asymptotically centered normal vector (Billingsley, 1961b) with covariance matrix $R = (r_{ij,kl}) = (\mathbf{E}\eta_{ij}\eta_{sl})$, $i, j, s, l = 1, \ldots, k$: $r_{ij,sl} = 0$ for $i \neq s$, $j, l = 1, \ldots, k$; $r_{ij,ij} = p_{ij} - p_{ij}^2$ for $i, j = 1, \ldots, k$; $r_{ij,il} = -p_{ij}p_{il}$ for $i, j, l = 1, \ldots, k, j \neq l$.

These theoretical results on the asymptotic properties of the maximum likelihood estimates can be useful for the evaluation of the possible errors of the estimates of parameters of the statistical models used in bioinformatic algorithms.

Further discussion of the stochastic models for DNA sequences can be found in: Gatlin (1972); Almagor (1983); Fitch (1983); Karlin and Ghandour (1985); Borodovsky *et al.* (1986a,b); Churchill (1989); Tavaré and Song (1989); Cowan (1991); Karlin and Macken (1991); Karlin and Brendel (1992); Karlin and Dembo (1992);

Karlin, Burge, and Campbell (1992); Kleffe and Borodovsky (1992); Schbath (2000); Robin and Schbath (2001); Ekisheva and Borodovsky (2006).

Problem 3.20 A 4200 nt long DNA sequence is used as a training set for parameter estimation of the DNA statistical model. The observed counts of sixteen dinucleotides, N_{XY}, are as follows:

	T	C	A	G
T	510	380	210	190
C	240	170	360	230
A	370	200	220	210
G	190	170	220	220

Find the maximum likelihood estimates of (a) the transition probabilities P_{TT}, P_{AG} of the first order Markov model of the DNA sequence; (b) the transition probabilities P_{TT}, P_{AG} of the first order Markov model for the DNA sequence complementary to the given training sequence.

Solution (a) For the first order Markov model the maximum likelihood estimates of transition probabilities are determined by formula (3.7):

$$\hat{P}_{TT} = \frac{N_{TT}}{\sum_{X=A,G,C,T} N_{TX}} = \frac{51}{129} = 0.395;$$

$$\hat{P}_{AG} = \frac{N_{AG}}{\sum_{X=A,G,C,T} N_{AX}} = \frac{21}{100} = 0.21.$$

(b) The dinucleotide *TT* in the complementary strand corresponds to the dinucleotide *AA* in the direct strand. Therefore, we use the dinucleotide counts known for the direct strand:

$$\hat{P}_{TT} = \frac{N_{AA}}{\sum_{X=A,G,C,T} N_{XA}} = \frac{22}{101} = 0.218.$$

Similarly, for the dinucleotide AG ($5' \rightarrow 3'$ direction) situated in the complementary strand we use the count of the dinucleotide CT ($3' \rightarrow 5'$ direction) situated in the direct strand. Therefore, we have

$$\hat{P}_{AG} = \frac{N_{CT}}{\sum_{X=A,G,C,T} N_{XT}} = \frac{24}{136} = 0.176. \qquad \square$$

Problem 3.21   The 100 nt long sequence x shown below was observed in a non-coding region of the human genome:

TGACTTCAGTCTGTTCTGCAGAAGTGATTCTGATGTCATGAAACTGCCTG
CACTTGGCTGAGAGGATGATAGGGGCAGAGAAGGGTTGTTTAAGCCATAT

Determine the simplest probabilistic model for sequence x among models $M(m)$, $m \geq 0$, that would pass the fitness test described above at significance level 0.05. Provide the maximum likelihood estimates of the parameters of this identified model.

Solution   First we check whether the independence model fits sequence x by using the independence test described above. We find counts of nucleotides and dinucleotides in sequence x as follows:

$$N(A\bullet) = 25, \quad N(C\bullet) = 15, \quad N(G\bullet) = 31, \quad N(T\bullet) = 28,$$

$$N(\bullet A) = 25, \quad N(\bullet C) = 15, \quad N(\bullet G) = 31, \quad N(\bullet T) = 28;$$

$$N(AA) = 6, \quad N(CA) = 5, \quad N(GA) = 12, \quad N(TA) = 2,$$

$$N(AC) = 3, \quad N(CC) = 2, \quad N(GC) = 5, \quad N(TC) = 5,$$

$$N(AG) = 9, \quad N(CG) = 0, \quad N(GG) = 8, \quad N(TG) = 14,$$

$$N(AT) = 7, \quad N(CT) = 8, \quad N(GT) = 6, \quad N(TT) = 7.$$

Next, the value of the χ^2-statistic is calculated by Equation (3.8):

$$\chi^2 = \sum_{\alpha,\beta \in \{A,C,G,T\}} \frac{((N-1)N(\alpha\beta) - N(\alpha\bullet)N(\bullet\beta))^2}{(N-1)N(\alpha\bullet)N(\bullet\beta)} = 19.22.$$

Since all expected counts $(N-1)\hat{P}_{H_0}(\alpha,\beta) = N(\alpha\bullet)N(\bullet\beta)/(N-1)$, $\alpha, \beta \in \{A, C, G, T\}$, are positive, the null hypothesis is tested by comparing the value of the χ^2-statistic with a critical value of the χ^2-distribution with nine degrees of freedom corresponding to significance level 0.05. We have $\chi^2 = 19.22 > 16.92 = \chi^2_{0.05,9}$; therefore, H_0 must be rejected at the 5% significance level. In other words, the independence model M does not fit sequence x according to the testing procedure, with a probability of 0.05 of rejecting M when it is true.

Next, we test the null hypothesis that x is generated by the first order Markov model $M(1)$. To apply the test based on the χ^2-statistic (3.9), we need the counts of the triplets from alphabet $\{A, C, G, T\}$, which are as follows:

$$N(AAT) = N(ACA) = N(ACC) = N(ACG) = N(ATC) = N(CAA)$$
$$= N(CAC) = N(CCC) = N(CCG) = N(CGA) = N(CGC) = N(CGG)$$
$$= N(CGT) = N(CTA) = N(CTC) = N(GCG) = N(GTA) = N(TAC)$$
$$= N(TAG) = N(TCC) = N(TCG) = 0;$$

$$N(AAA) = N(AAC) = N(AGC) = N(ATT) = N(CCA) = N(CCT)$$
$$= N(GCT) = N(GGT) = N(GTG) = N(TAA) = N(TAT) = N(TTA)$$
$$= N(TTT) = 1;$$

$$N(AGG) = N(AGT) = N(ATA) = N(CAT) = N(CTT) = N(GAC)$$
$$= N(GAG) = N(GCA) = N(GCC) = N(GGA) = N(GGC) = N(GTC)$$
$$= N(TCA) = N(TGC) = N(TTG) = 2;$$

$$N(ACT) = N(ATG) = N(CAG) = N(GGG) = N(GTT) = N(TCT)$$
$$= N(TGG) = N(TGT) = N(TTC) = 3;$$

$$N(AAG) = N(AGA) = N(GAA) = N(GAT) = 4;$$

$$N(CTG) = N(TGA) = 6.$$

We calculate the value of the χ^2-statistic (3.9) as follows:

$$\chi^2 = \sum_{\alpha,\beta,\gamma \in \{A,C,G,T\}} \frac{(N(\alpha\beta\gamma)N(\bullet\beta\bullet) - N(\alpha\beta\bullet)N(\bullet\beta\gamma))^2}{N(\alpha\beta\bullet)N(\bullet\beta\gamma)N(\bullet\beta\bullet)} = 45.32$$

and compare it with the critical value of the χ^2-distribution with $45 - 7 = 38$ degrees of freedom (because seven of the expected counts are equal to zero). Since $\chi^2 = 45.32 < 58.38 = \chi^2_{0.05,38}$, H_0 should be accepted at the 5% significance level. Therefore, the first order Markov model has passed the fitness test for the sequence x. The the maximum likelihood estimates (3.7) of the transition probabilities of the Markov chain are as follows:

$$\hat{p}_{A,A} = 0.24, \quad \hat{p}_{C,A} = 0.33, \quad \hat{p}_{G,A} = 0.39, \quad \hat{p}_{T,A} = 0.07,$$
$$\hat{p}_{A,C} = 0.12, \quad \hat{p}_{C,C} = 0.13, \quad \hat{p}_{G,C} = 0.16, \quad \hat{p}_{T,C} = 0.18,$$
$$\hat{p}_{A,G} = 0.36, \quad \hat{p}_{C,G} = 0.00, \quad \hat{p}_{G,G} = 0.26, \quad \hat{p}_{T,G} = 0.50,$$
$$\hat{p}_{A,T} = 0.28, \quad \hat{p}_{C,T} = 0.54, \quad \hat{p}_{G,T} = 0.19, \quad \hat{p}_{T,T} = 0.25.$$

Note that for a sequence of rather small length (100 nt) the estimates $\hat{p}_{ij}$ are not very reliable in the sense that the probability that the true values of parameters p_{ij} lie in a close neighborhood of estimates $\hat{p}_{ij}$ for all i, j is small. The increase of the length of the training sequence ensures a better approximation to the true values of the transition probabilities.

Nevertheless, we should remember that real biological sequences cannot be consistently treated as realizations of some theoretical random model. This modeling is always an approximation, with higher or lower degrees of accuracy depending on the size and the nature of the real sequence. □

Problem 3.22 To sample from the posterior distribution of hidden states paths $\{\pi\}$ of an HMM, given the realization $x = x_1, \ldots, x_L$, the following algorithm is proposed. First, a matrix of forward variables $F = (f_j(i))$ is determined by the forward algorithm for sequence x. Secondly, the path π is constructed from hidden states π_i recursively chosen by the stochastic traceback algorithm. The recursion goes as follows:

- given the *end* state, the state π_L is chosen with probability

$$\frac{a_{\pi_L 0} f_{\pi_L}(L)}{P(x)};$$

- given the π_L state, the state π_{L-1} is chosen with probability

$$\frac{e_{\pi_L}(x_L) a_{\pi_{L-1}\pi_L} f_{\pi_{L-1}}(L-1)}{f_{\pi_L}(L)};$$

- given the π_{L-1} state, state π_{L-2} is chosen with probability

$$\frac{e_{\pi_{L-1}}(x_{L-1}) a_{\pi_{L-2}\pi_{L-1}} f_{\pi_{L-2}}(L-2)}{f_{\pi_{L-1}}(L-1)};$$

...

- given the π_2 state, state π_1 is chosen with probability

$$\frac{e_{\pi_2}(x_2) a_{\pi_1 \pi_2} f_{\pi_1}(1)}{f_{\pi_2}(2)};$$

- given the π_1 state, the *begin* state is chosen with probability

$$\frac{e_{\pi_1}(x_1) a_{0\pi_1}}{f_{\pi_1}(1)} = 1.$$

Show that the algorithm indeed samples from the posterior distribution of the paths of hidden states.

Solution The posterior probability of a path $\pi = \pi_1, \ldots, \pi_L$ given the observed sequence x is defined by the following formula:

$$P(\pi|x) = \frac{P(x, \pi)}{P(x)} = \frac{a_{0\pi_1} e_{\pi_1}(x_1) a_{\pi_1 \pi_2} e_{\pi_2}(x_2) \cdots a_{\pi_{L-1}\pi_L} e_{\pi_L}(x_L) a_{\pi_L 0}}{P(x)}.$$

The algorithm of stochastic sampling described above will select path π with a probability P equal to the product of probabilities P_i of choosing its consecutive states π_i:

$$P = \prod_{i=L}^{1} P_i$$

$$= \frac{a_{\pi_L 0} f_{\pi_L}(L)}{P(x)} \frac{e_{\pi_L}(x_L) a_{\pi_{L-1} \pi_L} f_{\pi_{L-1}}(L-1)}{f_{\pi_L}(L)}$$

$$\times \frac{e_{\pi_{L-1}}(x_{L-1}) a_{\pi_{L-2} \pi_{L-1}} f_{\pi_{L-2}}(L-2)}{f_{\pi_{L-1}}(L-1)} \cdots$$

$$\times \frac{e_{\pi_2}(x_2) a_{\pi_1 \pi_2} f_{\pi_1}(1)}{f_{\pi_2}(2)} \frac{e_{\pi_1}(x_1) a_{0 \pi_1}}{f_{\pi_1}(1)},$$

We see that in the last ratio the same forward variables $f_{\pi_i}(i)$ appear in both the numerator and the denominator. After canceling out all $f_{\pi_i}(i)$, we obtain

$$P = \frac{a_{0 \pi_1} e_{\pi_1}(x_1) a_{\pi_1 \pi_2} e_{\pi_2}(x_2) \cdots a_{\pi_{L-1} \pi_L} e_{\pi_L}(x_L) a_{\pi_L 0}}{P(x)} = P(\pi | x).$$

Thus, the proposed stochastic traceback algorithm is a recurrent implementation of the sampling from the posterior distribution of the hidden state trajectories given the sequence x. $\square$

3.2.2 Bayesian approach to sequence composition analysis: the segmentation model by Liu and Lawrence

Theoretical introduction to Problem 3.23

In biological sequence analysis we frequently have to select a particular probabilistic model (such as an independence model, a Markov chain, an HMM) with statistical properties as close as possible to those observed in a real sequence. Having selected the model type, traditional statistical inference relies on the point estimates of unknown model parameters, which are routinely obtained by the maximum likelihood method (see the theoretical introduction to Problems 3.20 and 3.21).

The Bayesian approach to sequence composition analysis also starts with the selection of the model type. After selecting a probabilistic model, Bayesian statistics treats all unknown parameters of the model, along with observed data and missing data, as random variables and assigns appropriate *prior* distributions to all of them. The goal of this approach is to use the data to obtain the *posterior* distributions of all unknown parameters rather than their point estimates, as in the traditional frequentist approach. Therefore, Bayesian inference relies on the posterior distributions of parameters.

Liu and Lawrence (1999) developed a Bayesian method for sequence composition analysis based on the *basic segmentation model* (see also Braun and Müller, 1998) which can be described as follows. We consider a set of M independence models for an alphabet of size D; the ith model has unknown parameters $(\theta_{1,i}, \theta_{2,i}, \ldots, \theta_{D,i}) = \Theta_i$ $(\Theta_i \neq \Theta_{i+1}, i = 1, \ldots, M - 1)$. The basic segmentation model can serve as a model for heterogeneity in the composition of DNA or protein sequences with $D = 4$ or $D = 20$, respectively. A sequence $x = x_1, \ldots, x_N$ of symbols from the alphabet is generated by the basic segmentation model as follows: the first C_1 symbols are generated by the first model, the next C_2 symbols are generated by the second model, and so on. Finally, the last $C_{\mu+1}$ symbols of sequence x are generated by the model $\mu + 1$, $\mu \leq M - 1$. A vector of change points A_m, $m = 1, \ldots, \mu$, consists of integers $A_m = \sum_{i=1}^{m} C_i + 1$. The number of change points μ and the vector of change points $A = (A_1, A_2, \ldots, A_\mu)$ are unknown. We assume that the observed data (sequence x), the missing data (the number of change points μ and the random vector of change points $A = (A_1, A_2, \ldots, A_\mu)$), and the parameters of the independence models are random variables, and that *a priori* Θ $(\Theta_1, \ldots, \Theta_M) = \Theta$ and μ are independent. Given the prior probabilities $P(\mu)$, $P(A|\mu)$, $P(\Theta_i)$, Bayes' theorem is applied to determine the posterior probability of random variable ψ $(\psi = \mu, A_m$ $(m = 1, \ldots, \mu)$, parameters of the independence model $\Theta(j)$ at position $j, j = 1, \ldots, N)$:

$$P(\psi|x) = \frac{P(x, \psi)}{P(x)}.$$

The probabilities $P(x)$ and $P(x, \psi)$ can be calculated from the joint distribution of all variables $P(x, A, \Theta, \mu)$ by integration (or summation). Since the joint distribution can be written as

$$P(x, A, \Theta, \mu) = P(x|A, \Theta)P(A|\mu)P(\mu)P(\Theta),$$

for $P(x)$ we have

$$P(x) = \sum_{m=1}^{M} P(\mu = m) \sum_{A:||A||=m} P(A|\mu = m) \int_{\Theta} P(x|A, \Theta) \, d\Theta. \tag{3.10}$$

Here the inner sum is taken over all possible sets of m change points. A natural choice of a prior distribution for parameter $\Theta_i = (\theta_{1,i}, \theta_{2,i}, \ldots, \theta_{D,i})$ of the independence model i is the conjugate prior $\mathcal{D}(\alpha_i)$, the Dirichlet distribution with parameter $\alpha_i = (\alpha_{1,i}, \alpha_{2,i}, \ldots, \alpha_{D,i})$, such that the mean of $\theta_{k,i}$ is $\alpha_{k,i}/\sum_l \alpha_{l,i}$. Then, from

Equation (3.10) we derive

$$P(x|\mu = m) = \sum_{A:||A||=m} P(A|\mu = m) \int_{\Theta} P(x|A, \Theta)d\Theta$$

$$= \sum_{A:||A||=m} P(A|\mu = m) \prod_{i=1}^{m} \frac{\Gamma(\sum_d \alpha_{d,i}) \prod_d \Gamma(n_{d,i} + \alpha_{d,i})}{\prod_d \Gamma(\alpha_{d,i})\Gamma(C_i + \sum_d \alpha_{d,i})}. \quad (3.11)$$

Here $\Gamma(y)$ is the gamma function; $n_{d,i}$ is the count of symbols of type d in the ith segment $x[A_{i-1} : A_i] = (x_{A_{i-1}}, \ldots, x_{A_i} - 1)$ for the set $(A_1, A_2, \ldots, A_m, A_{m+1})$, $A_{m+1} = N + 1$; and $\sum_d n_{d,i} = C_i$. Dynamic programming is used to complete the summation in Equation (3.11). If $P(x[i : j]|m)$ denotes the probability of observing the subsequence $x[i : j] = (x_i, \ldots, x_j)$ given that it consists of m segments, then for the recursion equation we have

$$P(x[1 : j]|m) = \sum_{l<j} P(x[1 : l]|m - 1)P(x[l : j]|1). \quad (3.12)$$

The posterior distribution of A_m can be determined from the following equation:

$$P(\text{for some } m \; A_m = l|x) = \frac{1}{P(x)} \sum_{\mu \leq M} \sum_{m=1}^{\mu} P(x[1 : l]|m)P(x(l : j]|\mu - m). $$

$$(3.13)$$

Since the change points $A_1, A_2, \ldots, A_\mu$ are mutually dependent random variables, the posterior distribution of the random vector A cannot be determined as a product of distributions (3.13) of components A_m. However, an approximate distribution of A can be obtained by applying Monte Carlo sampling: we take a draw from the posterior distribution of $\mu = m$, fix A_m, and find the change points $A_1, A_2, \ldots, A_{m-1}$ by recursively sampling backward from the distribution:

$$P(A_{i-1} = j|x, A_i = k) = \frac{P(x[1 : j]|i - 1)P(x(j : k]|1)}{P(x[1 : k]|i)}.$$

Averaging over the draws yields a distribution which converges to the posterior distribution of the random vector of change points as the number of draws increases. Finally, the posterior distribution of the symbol probabilities at each position can also be obtained from the symbol frequencies given a segmentation by averaging over all possible segmentations.

Problem 3.23 To illustrate the Bayesian approach to DNA sequence analysis, we consider a 'toy' nucleotide sequence $x = GGCA$ (the same as in Problem 3.17, where x was assumed to be generated by an HMM with two hidden states). Now we consider x in the context of the basic segmentation model with $M = 4$.

The prior probabilities are as follows: $P(\mu = m) = 1/4$, $m = 0, 1, 2, 3$; $P(A :$ $||A|| = m|\mu = m) = \binom{N}{m}^{-1} = \binom{4}{m}^{-1}$ (thus, all possible segmentations with m change points are equally likely given $\mu = m$); for all independence models Θ_i, $i = 1, \ldots, 4$, the prior distribution of parameters (frequencies of nucleotides) is $\mathcal{D}(1, 1, 1, 1)$, the Dirichlet distribution with parameters $(\alpha_{A,i}, \alpha_{C,i}, \alpha_{G,i}, \alpha_{T,i}) = (1, 1, 1, 1)$.

Compute the posterior distributions of the number μ of change points, of the change points A_m, $m = 1, \ldots, \mu$, and of the probabilities of nucleotides $\Theta(j) = (\theta_A(j), \theta_C(j), \theta_G(j), \theta_T(j))$ at all positions j of the sequence.

Remark We realize that x is too short to demonstrate anything but the algorithm itself. Also, since x is a short sequence, we do not use dynamic programming recursions defined by Equations (3.12) and (3.13) to compute the posterior distributions; instead, we carry out direct computations by considering all possible segmentations.

Solution To calculate $P(x|\mu = m)$, $m = 0, 1, 2, 3$, we apply Equation (3.11). If $\mu = 0$, sequence x is described by the only independence model with parameters Θ_1, and we have

$$P(x|\mu = 0) = P((GGCA) \sim^d \Theta_1) = \frac{\Gamma(4)}{(\Gamma(1))^4} \frac{\Gamma(3)(\Gamma(2))^2\Gamma(1)}{\Gamma(8)} = \frac{1}{420},$$

$$P(x|\mu = 0)P(\mu = 0) = \frac{1}{420} \times \frac{1}{4} = 0.0006. \tag{3.14}$$

For $\mu = 1$, three segmentations are possible: with $A = 2$, $A = 3$, or $A = 4$. Using the independence of the distributions, we calculate

$$P(x|A = 2) = P((G) \sim^d \Theta_1)P((GCA) \sim^d \Theta_2) = \frac{(3!)^2}{4!6!} = 0.0021,$$

$$P(x|A = 3) = P((GG) \sim^d \Theta_1)P((CA) \sim^d \Theta_2) = \frac{(3!)^2 2!}{(5!)^2} = 0.005,$$

$$P(x|A = 4) = P((GGC) \sim^d \Theta_1)P((A) \sim^d \Theta_2) = \frac{(3!)^2 2!}{4!6!} = 0.0042;$$

$$P(x|\mu = 1)P(\mu = 1) = P(x|A = 2, \mu = 1)P(A = 2|\mu = 1)P(\mu = 1)$$
$$+ P(x|A = 3, \mu = 1)P(A = 3|\mu = 1)P(\mu = 1)$$
$$+ P(x|A = 4, \mu = 1)P(A = 4|\mu = 1)P(\mu = 1)$$
$$= \frac{1}{12}(P(x|A = 2) + P(x|A = 3) + P(x|A = 4)) = 0.0009. \tag{3.15}$$

Similarly, for $\mu = 2$ and $\mu = 3$ we obtain

$$P(x|\mu = 2)P(\mu = 2) = \frac{1}{12}(P(x|A = (2,3), \mu = 2) + P(x|A = (2,4), \mu = 2)$$

$$+ P(x|A = (3,4), \mu = 2)) = 0.001,$$

$$P(x|\mu = 3)P(\mu = 3) = \frac{1}{4}P(x|A = (2,3,4), \mu = 3)$$

$$= \frac{1}{4}\frac{(3!)^4}{(4!)^4} = 0.001. \tag{3.16}$$

Equations (3.10) and (3.14)–(3.16) yield the total probability of the sequence x under the segmentation model:

$$P(x) = \sum_{m=0}^{3} P(x|\mu = m)P(\mu = m) = 0.0036.$$

The posterior distribution of the number of change points μ can be found immediately from Bayes' theorem as follows:

$$P(\mu = 0|x) = \frac{P(x|\mu = 0)P(\mu = 0)}{P(x)} = 0.1676,$$

$$P(\mu = 1|x) = \frac{P(x|\mu = 1)P(\mu = 1)}{P(x)} = 0.264,$$

$$P(\mu = 2|x) = \frac{P(x|\mu = 2)P(\mu = 2)}{P(x)} = 0.2933,$$

$$P(\mu = 3|x) = \frac{P(x|\mu = 3)P(\mu = 3)}{P(x)} = 0.275.$$

For the posterior distribution of change point A_m, Bayes' theorem gives

$$P(A_m = j \text{ for some } m|x) = \sum_{i=0}^{3} P(A_m = j \text{ for some } m, \mu = i|x)$$

$$= \frac{1}{P(x)} \sum_{i=0}^{3} P(x|A_m = j \text{ for some } m, \mu = i)$$

$$\times P(A_m = j \text{ for some } m|\mu = i)P(\mu = i);$$

$P(A_m = 2$ for some $m|x)$

$$= \frac{1}{P(x)} \left(\frac{1}{12} P(x|A = (2), \kappa = 1) + \frac{1}{12} P(x|A = (2,3), \mu = 2) \right.$$

$$\left. + \frac{1}{12} P(x|A = (2,4), \mu = 2) + \frac{1}{4} P(x|A = (2,3,4), \mu = 3) \right) = 0.470578;$$

$P(A_m = 3$ for some $m|x)$

$$= \frac{1}{P(x)} \left(\frac{1}{12} P(x|A = (3), \mu = 1) + \frac{1}{12} P(x|A = (2,3), \mu = 2) \right.$$

$$\left. + \frac{1}{12} P(x|A = (3,4), \mu = 2) + \frac{1}{4} P(x|A = (2,3,4), \mu = 3) \right) = 0.612363;$$

$P(A_m = 4$ for some $m|x)$

$$= \frac{1}{P(x)} \left(\frac{1}{12} P(x|A = (4), \mu = 1) + \frac{1}{12} P(x|A = (2,4), \mu = 2) \right.$$

$$\left. + \frac{1}{12} P(x|A = (3,4), \mu = 2) + \frac{1}{4} P(x|A = (2,3,4), \mu = 3) \right) = 0.592806;$$

$$P \text{ (no change points}|x) = P(\mu = 0|x) = 0.167627.$$

Finally, the posterior probability density function $f_j(\theta_A, \theta_C, \theta_G, \theta_T|x)$ of the probabilities of nucleotides at site $j, j = 1, 2, 3, 4$, is given by

$$f_j(\theta_A, \theta_C, \theta_G, \theta_T|x) = \sum_{\text{segment covering } j} P(\text{segment}|\Theta_j)P(\Theta_j)P(\text{segment}). \tag{3.17}$$

Formula (3.17) allows us to calculate the posterior density functions:

$$f_1(\theta_A, \theta_C, \theta_G, \theta_T|x) = \frac{1}{4} \mathcal{D}(2,2,3,1) + \frac{1}{2} \mathcal{D}(1,1,2,1) + \frac{1}{6} \mathcal{D}(1,1,3,1)$$

$$+ \frac{1}{12} \mathcal{D}(1,2,3,1);$$

$$f_2(\theta_A, \theta_C, \theta_G, \theta_T|x) = \frac{1}{4} \mathcal{D}(2,2,3,1) + \frac{1}{12} \mathcal{D}(2,2,2,1) + \frac{1}{6} \mathcal{D}(1,1,3,1)$$

$$+ \frac{1}{12} \mathcal{D}(1,2,3,1) + \frac{1}{3} \mathcal{D}(1,1,2,1) + \frac{1}{12} \mathcal{D}(1,2,2,1);$$

$$f_3(\theta_A, \theta_C, \theta_G, \theta_T | x) = \frac{1}{4}\mathcal{D}(2,2,3,1) + \frac{1}{12}\mathcal{D}(2,2,2,1) + \frac{1}{6}\mathcal{D}(2,2,1,1)$$

$$+ \frac{1}{12}\mathcal{D}(1,2,3,1) + \frac{1}{12}\mathcal{D}(1,2,2,1) + \frac{1}{3}\mathcal{D}(1,2,1,1);$$

$$f_4(\theta_A, \theta_C, \theta_G, \theta_T | x) = \frac{1}{4}\mathcal{D}(2,2,3,1) + \frac{1}{12}\mathcal{D}(2,2,2,1) + \frac{1}{6}\mathcal{D}(2,2,1,1)$$

$$+ \frac{1}{2}\mathcal{D}(2,1,1,1). \tag{3.18}$$

Note that each posterior density function of probabilities of nucleotides is a mixture of the Dirichlet distributions. From (3.18) we determine the posterior mean estimate (PME) of probability of each nucleotide at site j, $j = 1, 2, 3, 4$, as follows:

$$\mathbf{E}_1\theta_A = 0.2022, \quad \mathbf{E}_2\theta_A = 0.2065,$$

$$\mathbf{E}_1\theta_C = 0.2141, \quad \mathbf{E}_2\theta_C = 0.2323,$$

$$\mathbf{E}_1\theta_G = 0.4128, \quad \mathbf{E}_2\theta_G = 0.429,$$

$$\mathbf{E}_1\theta_T = 0.1709; \quad \mathbf{E}_2\theta_T = 0.1634;$$

$$\mathbf{E}_3\theta_A = 0.2343, \quad \mathbf{E}_4\theta_A = 0.3487,$$

$$\mathbf{E}_3\theta_C = 0.3268, \quad \mathbf{E}_4\theta_C = 0.2419,$$

$$\mathbf{E}_3\theta_G = 0.2755, \quad \mathbf{E}_4\theta_G = 0.2453,$$

$$\mathbf{E}_3\theta_T = 0.1634; \quad \mathbf{E}_4\theta_T = 0.1709.$$

Remark Use of the HMM defined in Problem 3.16 to the model sequence $x = GGCA$ allows us to compute the posterior distribution of the 'change points' via the forward and backward variables determined in Problem 3.17. Actually, we should calculate the posterior probabilities of events $A_j =$(hidden states at sites $j - 1$ and j are not the same), $j = 2, 3, 4$. Note that $P(x)$ was determined in Problem 3.17 and is equal to 0.0038. We obtain

$$P(A_2|x) = P(\pi_1 = H, \pi_2 = L|x) + P(\pi_1 = L, \pi_2 = H|x)$$

$$= \frac{P(\pi_1 = H, \pi_2 = L, x)}{P(x)} + \frac{P(\pi_1 = L, \pi_2 = H, x)}{P(x)}$$

$$= \frac{1}{P(x)}(P(x_1, x_2, x_3, x_4, \pi_1 = H, \pi_2 = L)$$

$$+ P(x_1, x_2, x_3, x_4, \pi_1 = L, \pi_2 = H))$$

$$= \frac{1}{P(x)} [P(x_1, x_2, \pi_1 = H, \pi_2 = L) P(x_3, x_4 | x_1, x_2, \pi_1 = H, \pi_2 = L)$$

$$+ P(x_1, x_2, \pi_1 = L, \pi_2 = H) P(x_3, x_4 | x_1, x_2, \pi_1 = L, \pi_2 = H)]$$

$$= \frac{1}{P(x)} (a_{0H} e_H(G) a_{HL} e_L(G) b_H(2) + a_{0L} e_L(G) a_{LH} e_H(G) b_L(2))$$

$$= 0.2389 + 0.1983 = 0.4372.$$

Similarly, for other posterior probabilities we have

$$P(A_3 | x) = \frac{1}{P(x)} (P(x_1, x_2, x_3, x_4, \pi_2 = H, \pi_3 = L)$$

$$+ P(x_1, x_2, x_3, x_4, \pi_2 = L, \pi_3 = H))$$

$$= 0.2334 + 0.2108 = 0.4442;$$

$$P(A_4 | x) = \frac{1}{P(x)} (P(x_1, x_2, x_3, x_4, \pi_3 = H, \pi_4 = L)$$

$$+ P(x_1, x_2, x_3, x_4, \pi_3 = L, \pi_4 = H))$$

$$= 0.3279 + 0.1392 = 0.4671. \qquad \square$$

3.3 Further reading

Since the publication of *BSA* a remarkably large number of publications have appeared on the subject of the applications of Markov chains and HMMs for biological sequences. We are able to provide just a few references to frequently cited papers published since 1997.

Markov chains and hidden Markov models provide a useful formalism for genomic sequence modeling and are frequently employed in gene prediction methods and algorithms. Burge and Karlin (1997) in their gene finding computer program GENSCAN used a three-periodic (inhomogeneous) fifth order Markov chain to model coding regions of DNA sequences. The GeneMark.hmm algorithm (Lukashin and Borodovsky, 1998) uses the HMM framework with gene boundaries modeled as transitions between hidden states; GLIMMER, an algorithm for gene finding in microbial genomes (Salzberg *et al.*, 1998), uses the interpolated Markov models.

Karplus, Barrett, and Hughey (1998) described the SAM-T98 method for finding remote homologs of protein sequences; Tusnády and Simon (1998) proposed a new method to search maximum likelihood transmembrane topology among all possible topologies of a given protein; Bystroff, Thorsson, and Baker (2000) introduced the HMMSTR, a hidden Markov model for the three-dimensional protein structure prediction based on the I-sites, the sequence-structure motifs; Gough *et al.* (2001)

suggested a method of assignment of homologs to protein products encoded in genomes by using a library of HMMs representing all proteins of known structure; Frith, Hansen, and Weng (2001) described a new method for detecting regulatory regions in DNA sequences by searching for clusters of *cis*-elements; Krogh *et al.* (2001) designed the TMHMM method for membrane protein topology prediction; Löytynoja and Milinkovitch (2003) proposed a new algorithm for multiple sequence alignment that combines an HMM approach, a progressive alignment algorithm, and a probabilistic model of evolution describing substitutions of nucleotides or amino acids.

4

Pairwise alignment using HMMs

In the *BSA* Chapter 3 we learned that a DP algorithm for pairwise sequence alignment allows a probabilistic interpretation. Indeed, the equivalent equations appear in the logarithmic form of the Viterbi algorithm for the hidden Markov model of a gapped sequence alignment. The hidden states of such a model, called a pair HMM, correspond to the alignment match, the x-gap, and the y-gap positions. The pair HMM state diagram is topologically similar to the diagram of the finite state machine (Durbin *et al.* (1998), Fig. 4.1), although the pair HMM parameters have clear probabilistic meanings. The optimal finite state machine alignment found by standard DP is equivalent to the most probable path through the pair HMM determined by the Viterbi algorithm. Both global and local optimal DP alignment algorithms have Viterbi counterparts for suitably defined HMMs. Interestingly, the HMM has an advantage over the finite state machine because the HMM can compute the full probability that sequences X and Y could be generated by a given pair HMM; thus, a probabilistic measure can be introduced to help establish evolutionary relationships. This full probabilistic model also defines (i) the posterior distribution over all possible alignments given sequences X and Y and (ii) the posterior probability that a particular symbol x of sequence X is aligned to a given symbol y of sequence Y. However, real biological sequences cannot be considered to be exact realizations of probabilistic models. This explains the difficulties met by the HMM-based alignment methods for the similarity search (Durbin *et al.* (1998), Sect. 4.5), while more simplistic finite state machine methods perform sufficiently well.

The *BSA* problems in Chapter 4 illustrate the properties of pair HMMs and elaborate on the comparison of an HMM and an independence model, both emitting the symbols from a given alphabet.

The additional problems are concerned with the practical use of pair HMMs for sequence alignment, computation of the full probability of two aligned sequences, and computation of either the posterior probability of optimal path or the posterior probability of a given pair of symbols to be aligned. The relationship between the

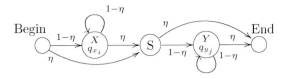

Figure 4.1. Independence random model R emitting a pair of sequences.

parameters of the log-odds score version of the Viterbi algorithm for pair HMM and the parameters of the classic DP alignment algorithm with affine gap scores is considered last.

4.1 Original problems

Problem 4.1 An independence pair-sequence model, R, is shown in Figure 4.1. The main states X and Y emit two full sequences x and y in turn independently of each other. A silent state S between X and Y, as well as *begin* and *end* states, does not emit any symbols. Each main state has a loop onto itself with probability $1 - \eta$ and transition probabilities from Y to *end* state and from S to *end* state are equal to η. What is the probability that sequence x has length t under the independence pair-sequence model R?

Solution Let L_x, L_y be the lengths of sequences x and y. First, we will find the probability that $L_x = t$ under conditions $L_x \leq n$, $L_y \leq m$ for some integers n, m. Since we are interested in the length distributions only, we can leave out the emission probabilities. For integer t such that $t > n$,

$$P(L_x = t | L_x \leq n, L_y \leq m) = 0.$$

If $t \leq n$, we obtain

$$
\begin{aligned}
P(L_x = t | L_x \leq n, L_y \leq m) &= \frac{P(L_x = t, L_x \leq n, L_y \leq m)}{P(L_x \leq n, L_y \leq m)} \\
&= \frac{P(L_x = t, L_y \leq m)}{P(L_x \leq n, L_y \leq m)} \\
&= \frac{\sum_{l=0}^{m} P(L_x = t, L_y = l)}{\sum_{k=0}^{n} \sum_{l=0}^{m} P(L_x = k, L_y = l)} \\
&= \frac{\sum_{l=0}^{m} \eta^2 (1 - \eta)^{t+l}}{\sum_{k=0}^{n} \sum_{l=0}^{m} \eta^2 (1 - \eta)^{k+l}} = \frac{(1 - \eta)^t}{\sum_{k=0}^{n} (1 - \eta)^k} \\
&= \frac{\eta (1 - \eta)^t}{(1 - (1 - \eta)^n)}.
\end{aligned}
\tag{4.1}
$$

Note that the conditional probability $P(L_x = t | L_x \leq n, L_y \leq m)$ does not depend on m due to the independence of sequences x and y. If the restriction on the sequence x length is removed (so $n = +\infty$) then Equation (4.1) becomes

$$P(L_x = t) = \eta(1 - \eta)^t. \qquad \Box$$

Problem 4.2 What is the expected length of sequences from the independence pair-sequence model R introduced in Problem 4.1? How should the parameter η be set?

Solution The formula for the expected length of sequences x and y under conditions $L_x \leq n$ and $L_y \leq m$ for some integers n, m, is derived as a direct continuation of Problem 4.1. For the conditional expectation of the length L_x we obtain from Equation (4.1) the following:

$$\mathbf{E}(L_x | L_x \leq n, L_y \leq m) = \sum_{k=0}^{n} k P(L_x = k | L_x \leq n, L_y \leq m)$$

$$= \sum_{k=0}^{n} k \frac{\eta(1-\eta)^k}{(1-(1-\eta)^n)} = \frac{\eta}{(1-(1-\eta)^n)} \sum_{k=0}^{n} k(1-\eta)^k$$

$$= \frac{\eta}{(1-(1-\eta)^n)} \frac{(1-\eta)(1-(n+1)(1-\eta)^n + n(1-\eta)^{n+1})}{\eta^2}$$

$$= \frac{(1-\eta)(1-(n+1)(1-\eta)^n + n(1-\eta)^{n+1})}{\eta(1-(1-\eta)^n)}.$$

Similarly, for the conditional expectation of the length L_y we have

$$\mathbf{E}(L_y | L_x \leq n, L_y \leq m) = \frac{(1-\eta)(1-(m+1)(1-\eta)^m + m(1-\eta)^{m+1})}{\eta(1-(1-\eta)^m)}.$$

Finally, if the lengths of the sequences generated by model R are not restricted, the expected lengths become

$$EL_x = EL_y = \frac{1-\eta}{\eta}.$$

Therefore, if the characteristic length L of sequences x and y is known, it is reasonable to set the parameter η in the independence pair-sequence model R equal to $1/(L+1)$. $\qquad \Box$

Problem 4.3 An example of uncertainty in positioning a gap is shown below (three significantly different gap placements in the globin alignments π_1, π_2, π_3

with very similar scores):

```
HBA_HUMAN  KVADALTNAVAHVD-----DMPNALSALSDLH
           KV    + +A   ++            +L+ L+++H
LGB2_LUPLU KVFKLVYEAAIQLQVTGVVVTDATLKNLGSVH

HBA_HUMAN  KVADALTNAVAHVDDM-----PNALSALSDLH
           KV    + +A   ++            +L+ L+++H
LGB2_LUPLU KVFKLVYEAAIQLQVTGVVVTDATLKNLGSVH

HBA_HUMAN  KVADALTNA-----VAHVDDMPNALSALSDLH
           KV    + +A     V  V       +L+ L+++H
LGB2_LUPLU KVFKLVYEAAIQLQVTGVVVTDATLKNLGSVH
```

The scores of alignments π_1, π_2, and π_3 calculated using BLOSUM50 (see Table 2.1) with gap-open penalty -12 and gap-extension penalty -2 are 3, 3 and 6, respectively. The relative scores for gap position variants depend only on the substitution scores, not the gap scores. Why is this, and what are the consequences for alignment accuracy using DP algorithms?

Solution Each of the three given alignments has one gap of length 5. Since the gap penalty depends on the gap length only, the three gaps make equal contributions to the alignment scores. Hence, the difference between the total alignment scores is due only to the differences in the substitution scores associated with each alignment.

The first two alignments have equal total scores and, hence, equal sums of substitution scores, while the sum of substitution scores for the last alignment is larger by 3 than for the first two alignments.

Such an outcome is natural for the gap penalty function independent of the gap position in the alignment. The gap score with position dependence could be introduced for the protein sequence alignment by assignment of higher gap-open and gap-extension penalties for the polypeptide regions with a regular secondary structure (α-*helix* or β-*strand*). This option, though, comes at a price since at the pre-processing step it requires the determination of protein secondary structures for both sequences to be aligned.

To summarize, the sequence alignment DP algorithm with position-independent gap penalties generates as many optimal alignments as there are possibilities of shifting gaps around the best alignment without changing the sum of substitution scores. Such a set of alignments with equal scores has to be reviewed by experts in order to select the variant adequate from structural, functional, and evolutionary points of view. □

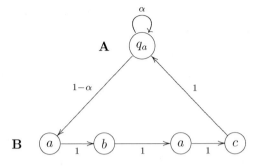

Figure 4.2. Example HMM.

Problem 4.4 Figure 4.2 shows an example of a simple HMM: state A emits symbols with probabilities q_a and the sequential block B of four states emits a fixed string *abac* with probability 1. Show that using the full probabilistic independence model with this example HMM allows discrimination between model and random data.

Solution This problem is not a simple one. We consider two stochastic models generating sequences of symbols from the same alphabet: the HMM shown in Figure 4.2. (model M) and an independence model (model R), which generates symbols with the same emission probabilities q_a as defined for state A of model M. For an observed sequence $x = x_1, \ldots, x_n$ we have to identify the most likely model, M or R, that generates sequence x. To make a decision, we compare the posterior probabilities of the models, $P(M|x)$ and $P(R|x)$. Assuming that the prior probabilities of the models are equal, the ratio of posterior probabilities becomes the odds ratio as follows:

$$\frac{P(M|x)}{P(R|x)} = \frac{P(x|M)}{P(x|M) + P(x|R)} \frac{P(x|M) + P(x|R)}{P(x|R)} = \frac{P(x|M)}{P(x|R)}.$$

Thus, all we need is to compare the likelihoods of the models, $P(x|M)$ and $P(x|R)$. To solve this problem step by step, we start with two special types of sequence x.

(1) If the sequence $x = x_1, \ldots, x_n$ contains no strings *abac*, then $P(x|R) = \prod_{i=1}^{n} q_{x_i}$, $P(x|M) = \alpha^n \prod_{i=1}^{n} q_{x_i}$, and $P(x|R) > P(x|M)$ for any value of transition probability α. Therefore, a sequence x with no strings *abac* is more likely to be produced by the independence model R.

(2) If $x = abac$, then $P(x|R) = q_a^2 q_b q_c$, while $P(x|M)$ is determined by the formula of total probability

$$P(x|M) = P(x, A|M) + P(x, B|M).$$

Here $P(x, A|M)$ is the probability that *abac* is generated by state A (in four steps) and $P(x, B|M)$ is the probability that *abac* is generated by block B. Therefore,

$$P(x|M) = \alpha^4 q_a^2 q_b q_c + (1 - \alpha).$$

To compare the conditional probabilities, it is convenient to analyze the value Δ:

$$\Delta = P(x|M) - P(x|R) = \alpha^4 \eta + (1 - \alpha) - \eta = (\alpha - 1)(\eta(\alpha + 1)(\alpha^2 + 1) - 1),$$

where η stands for $q_a^2 q_b q_c$. Since $\alpha - 1$ is always negative, the sign of Δ is determined by the sign of the second factor. For the function η of three variables, $\eta = \eta(x, y, z) = x^2 yz$, with x, y, z all from interval $[0, 1]$ and $x + y + z = a \leq 1$, we have

$$\eta(x, y, z) = \eta(x, y) = x^2 y(a - x - y) = ax^2 y - x^3 y - x^2 y^2.$$

To determine the maximum value of η for $x, y \in [0, 1]$, we solve the following system of equations:

$$\frac{\partial \eta}{\partial x} = 2axy - 3x^2 y - 2xy^2 = xy(2a - 3x - 2y) = 0,$$

$$\frac{\partial \eta}{\partial y} = ax^2 - x^3 - 2x^2 y = x^2(a - x - 2y) = 0.$$

The values of the variables which maximize η are $x^* = a/2$, $y^* = a/4$, $z^* = a/4$, and the maximum value itself is $\eta(x^*, y^*, z^*) = a^4/64$. Thus, the parameter η satisfies the inequality $0 \leq \eta \leq 1/64$ for any choice of emission probabilities q, and the term $\eta(\alpha + 1)(\alpha^2 + 1) - 1$ is negative. Therefore, Δ is positive, $P(x|M) > P(x|R)$, and the odds ratio r becomes

$$r = \frac{P(x|M)}{P(x|R)} = \frac{\alpha^4 q_a^2 q_b q_c + (1 - \alpha)}{q_a^2 q_b q_c} = \frac{\alpha^4 \eta + (1 - \alpha)}{\eta} > 1, \qquad (4.2)$$

with the log-odds ratio $S(x) = \log P(x|M)/P(x|R) > 0$. Thus, the observation of sequence *abac* allows us to identify model M as the source of the sequence regardless of the values of emission and transition probabilities.

(3) Finally, we consider the general case of sequence $x = x_1, \ldots, x_n$ containing k, $0 \leq k \leq n/4$, strings *abac*. In this case, the likelihood of the independence model R is given by

$$P(x|R) = (q_a^2 q_b q_c)^k \prod_{x_i : x_i \notin X} q_{x_i} = \eta^k \prod_{x_i : x_i \notin X} q_{x_i},$$

where X is a subsequence of sequence x that makes up k strings *abac* in x (note that strings *abac* cannot overlap). The likelihood of model M is calculated as $P(x|M) = \sum_\pi P(x, \pi|M)$, where the sum is taken over all paths π generating the observed sequence x. If path π emits l strings *abac* from state A (each string in four consecutive steps) and $k - l$ strings *abac* from block B, the joint probability of x and π is given by

$$P(x, \pi|M) = (\alpha^4 q_a^2 q_b q_c)^l (1 - \alpha)^{k-l} \alpha^{n-4k} \prod_{x_i : x_i \notin X} q_{x_i}.$$

The number of such paths π is equal to $\binom{k}{l}$ and the likelihood of model M becomes:

$$P(x|M) = \left(\sum_{l=0}^{k}\binom{k}{l}(\alpha^4 q_a^2 q_b q_c)^l(1-\alpha)^{k-l}\alpha^{n-4k}\right)\prod_{x_i:x_i\notin X}q_{x_i}$$

$$= \left(\sum_{l=0}^{k}\binom{k}{l}\alpha^{4l}\eta^l(1-\alpha)^{k-l}\alpha^{n-4k}\right)\prod_{x_i:x_i\notin X}q_{x_i}$$

$$= \alpha^n\eta^k\left(\sum_{l=0}^{k}\binom{k}{l}\left(\frac{1-\alpha}{\alpha^4\eta}\right)^{k-l}\right)\prod_{x_i:x_i\notin X}q_{x_i}$$

$$= \alpha^n\eta^k\left(1+\frac{1-\alpha}{\alpha^4\eta}\right)^k\prod_{x_i:x_i\notin X}q_{x_i}.$$

Then the odds ratio is given by

$$\frac{P(x|M)}{P(x|R)} = \alpha^n\left(\frac{\alpha^4\eta+(1-\alpha)}{\alpha^4\eta}\right)^k = \alpha^{n-4k}\left(\frac{\alpha^4\eta+(1-\alpha)}{\eta}\right)^k = \alpha^{n-4k}r^k,$$

where r is defined in Equation (4.2) and $r>1$. Therefore, for the log-odds ratio we have

$$S(x) = \log\frac{P(x|M)}{P(x|R)} = (n-4k)\log\alpha + k\log r.$$

Obviously, the value of $P(x|M)/P(x|R)$ depends on parameters n and k. If $k=0$, then sequence x with no strings *abac* is more likely to be generated by the independence model R (see case (1) above).

If $k=n/4$, then

$$\frac{P(x|M)}{P(x|R)} = r^{n/4} > 1$$

and x is more likely to be generated by model M.

As k increases from zero to $n/4$, the odds ratio $P(x|M)/P(x|R)$ increases from $\alpha^n < 1$ to $r^{n/4} > 1$ with $P(x|M) = P(x|R)$ when $k = k^*$, where

$$k^* = \frac{n\log\alpha}{4\log\alpha - \log r} = \frac{n\log\alpha}{\log(\alpha^4 q_a^2 q_b q_c) - \log(\alpha^4 q_a^2 q_b q_c + 1 - \alpha)}.$$

Therefore, we conclude that for $k < k^*$ sequence x is identified as one generated by model R; for $k > k^*$ as one generated by model M; for $k = k^*$ no model is selected. As expected, a sequence x with a high 'density' of strings *abac* is identified as one generated by model M.   □

Problem 4.5 Compare using the full probabilistic model in Problem 4.4 with using the Viterbi path in the model M, in which the transition probability to B has been raised to τ such that $\tau > P_A(abac)$.

Solution We need to return to the solution of Problem 4.4 and use the probability of the Viterbi path instead of the full probability of the observed sequence.

(1) If sequence $x = x_1, \ldots, x_n$ contains no strings *abac*, then both models M and R have only one path of hidden states to generate this sequence, with $P(x|R) = \prod_{i=1}^n q_{x_i}$, $P(x|M) = \alpha^n \prod_{i=1}^n q_{x_i}$. Then $P(x|R) > P(x|M)$ regardless of the value of transition probability α, and sequence x is identified as one generated by the independence model R.

(2) If $x = abac$, then the likelihood of model R is $P(x|R) = q_a^2 q_b q_c = \eta$. Given the model M and the condition $\tau > P_A(abac)$, the most probable path π^* goes through block B. Thus,

$$P(x, \pi^*|M) = P(x, B|M) = 1 - \alpha.$$

Then the odds ratio is given by

$$r = \frac{P(x, \pi^*|M)}{P(x|R)} = \frac{1 - \alpha}{q_a^2 q_b q_c} = \frac{1 - \alpha}{\eta}. \tag{4.3}$$

Hence, if $\alpha < 1 - \eta$, then $r > 1$ and model M is identified as the source of sequence x; if $\alpha > 1 - \eta$, then $r < 1$ and the more likely source of x is the independence model R.

(3) Now, if sequence $x = x_1, \ldots, x_n$ contains k, $0 \le k \le n/4$, strings *abac*, the likelihood of model R (with only one path corresponding to x) is the same as it was in Problem 4.4:

$$P(x|R) = (q_a^2 q_b q_c)^k \prod_{x_i : x_i \notin X} q_{x_i} = \eta^k \prod_{x_i : x_i \notin X} q_{x_i}.$$

To generate x on the Viterbi path π^* through model M, all k strings *abac* have to be emitted from block B. Therefore, the joint probability of x and π^* is given by

$$P(x, \pi^*|M) = (1 - \alpha)^k \alpha^{n-4k} \prod_{x_i : x_i \notin X} q_{x_i}.$$

Then we find

$$\frac{P(x, \pi^*|M)}{P(x|R)} = \alpha^{n-4k} \left(\frac{1 - \alpha}{\eta} \right)^k = \alpha^{n-4k} r^k,$$

with r defined as in Equation (4.3). Then, the log-odds ratio becomes

$$S(x) = \log \frac{P(x, \pi^*|M)}{P(x|R)} = (n - 4k) \log \alpha + k \log r.$$

If $\alpha < 1 - \eta = 1 - q_a^2 q_b q_c$, then $r > 1$ and the value of the odds ratio $P(x, \pi^*|M)/P(x|R)$ depends on parameters n and k. From Equation (4.3), we derive that $P(x, \pi^*|M) = P(x, R)$ if

$$k = \frac{n \log \alpha}{4 \log \alpha - \log r} = \frac{n \log \alpha}{\log(\alpha^4 q_a^2 q_b q_c) - \log(1 - \alpha)} = k^*.$$

Therefore, if string *abac* occurs in sequence x k times and $k < k^*$, then model R is identified as the source of sequence x; for $k > k^*$, model M is identified as the source of x.

If $\alpha > 1 - \eta = 1 - q_a^2 q_b q_c$, then $r < 1$ and

$$\frac{P(x, \pi^*|M)}{P(x|R)} = \alpha^{n-4k} r^k < 1.$$

In this case, regardless of the number k of occurrences of string *abac* in sequence x, the independence model R is identified as the source of sequence x.

Remark Here we see again that the use of the Viterbi path probabilities for the model comparison may lead to a different conclusion when compared with the use of the full likelihoods of the models (considered in Problem 4.4). Since the full probabilistic approach is more accurate and produces a more reliable identification of the model behind the data, we see here again the warning that a shortcut to the Viterbi path may end up in erroneous decisions in the model selection. □

Problem 4.6 We can modify the model further by setting all the emission probabilities at A to the same value, $1/A$, where A is the alphabet size. The difference between this model and the independence model with the same emission probabilities is then precisely the number of strings *abac* in the data. Is the quality of discrimination for the modified model the same as for the full probabilistic model?

Solution We have to go through the solution of Problem 4.5 and make modifications if necessary.

(1) The case of sequence $x = x_1, \ldots, x_n$ with no strings *abac* is not affected. The log-odds ratio is, as before,

$$S(x) = \log \frac{P(x|M)}{P(x|R)} = \log \frac{\alpha^n \prod_{i=1}^n q_{x_i}}{\prod_{i=1}^n q_{x_i}} = n \log \alpha < 0.$$

Hence, sequence x is always identified as one generated by independence model R.

(2) If $x = abac$, then, as proved in Problem 4.5, for the Viterbi path π^* the odds ratio r is given by

$$r = \frac{P(x, \pi^*|M)}{P(x|R)} = \frac{1 - \alpha}{q_a^2 q_b q_c}.$$

If the transition probability α satisfies the condition

$$\alpha < 1 - q_a^2 q_b q_c = 1 - \frac{1}{A^4},$$

then $r > 1$ and sequence x is identified as one generated by model M; if $\alpha > 1 - (1/A^4)$, then $r < 1$ and x is identified as a sequence generated by independence model R.

(3) In the general case of sequence $x = x_1, \ldots, x_n$ containing k, $0 \le k \le n/4$, strings *abac*, the odds ratio involving the Viterbi path π^* through model M, found in Problem 4.5, becomes

$$\frac{P(x, \pi^*|M)}{P(x|R)} = \alpha^{n-4k} \left(\frac{1-\alpha}{q_a^2 q_b q_c} \right)^k = \alpha^{n-4k} \left(\frac{1-\alpha}{A^4} \right)^k.$$

If $\alpha < 1 - (1/A^4)$, then the odds ratio depends on parameters n and k. For

$$k = k^* = \frac{n \log \alpha}{4 \log A\alpha - \log(1-\alpha)}$$

no decision can be made since $P(x, \pi^*|M) = P(x|R)$. If $k < k^*$, model R is identified as the source of sequence x; if $k > k^*$, model M is identified as the source of sequence x.

If $\alpha > 1 - (1/A^4)$, then the odds ratio $P(x, \pi^*|M)/P(x|R) < 1$ for any k, and x is identified as a sequence generated by independence model R.

Here is a summary of the decision-making rules derived from the full probabilistic approach (Problem 4.4) and the approach using the Viterbi path. We assume that all emission probabilities q are equal to $1/A$. The critical values k_1^* and k_2^* are given by

$$k_1^* = \frac{n \log \alpha}{4 \log(\alpha A) - \log((\alpha A)^4 + 1 - \alpha)}$$

for the full probabilistic test T_1 and

$$k_2^* = \frac{n \log \alpha}{4 \log(\alpha A) - \log(1-\alpha)}$$

for the test T_2 based on the Viterbi path. Note that $k_1^* < k_2^*$.

The source of sequence x of length n with k strings *abac* is identified as follows. If $\alpha < 1 - (1/A^4)$ there are three possible cases.

(1) For $k < k_1^*$ both tests identify the independence model R.
(2) For $k_1^* < k < k_2^*$, the tests give different results: test T_1 identifies model M, while test T_2 identifies model R.
(3) For $k > k_2^*$, both tests identify model M.

If $\alpha > 1 - (1/A^4)$, test T_2 always identifies model R; however, test T_1 identifies model R for $k < k_1^*$, and model M otherwise. ☐

4.2 Additional problems

The four problems included in this section are practical exercises on using pair HMMs: constructing the optimal alignment of two sequences by using the Viterbi

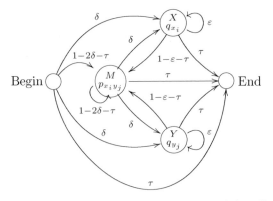

Figure 4.3. Full probabilistic version of the pair HMM emitting aligned sequences
x and y.

algorithm (Problem 4.7); calculating the probability that these sequences are related
according to the pair HMM by using the forward algorithm (Problem 4.8); finding
the posterior probabilities of an alignment, an aligned pair of symbols, as well as
the expected accuracy of a given alignment (Problems 4.9 and 4.10).

Derivation of the Viterbi algorithm equations in log-odds terms is the goal of
Problem 4.11. The log-odds scores that replace probabilistic functions naturally
originate from the likelihood ratios of a pair HMM and an independence pair-
sequence model.

Problem 4.7 The pair HMM shown in Figure 4.3. generates two aligned DNA
sequences x and y. State M emits aligned pairs of nucleotides with emission
probabilities $P_{x_i y_j}$ defined as follows:

$$P_{TT} = P_{CC} = P_{AA} = P_{GG} = 0.5, \quad P_{CT} = P_{AG} = 0.05,$$

$$P_{AT} = P_{GC} = 0.3, \quad P_{GT} = P_{AC} = 0.15.$$

The insert states X and Y emit (aligned with gaps) symbols from sequences x and
y, respectively. The emission probabilities are the same for both insert states:

$$q_A = q_C = q_G = q_T = 0.25.$$

No symbols are emitted by *begin* and *end* states. The values of transition
probabilities are as follows:

$$\pi_{BX} = \pi_{BY} = \pi_{MX} = \pi_{MY} = \delta = 0.2,$$

$$\pi_{BH} = \pi_{MH} = \pi_{XH} = \pi_{YO} = \tau = 0.1,$$

$$\pi_{XX} = \pi_{YY} = \varepsilon = 0.1, \quad \pi_{XM} = \pi_{YM} = 1 - \varepsilon - \tau = 0.8,$$

$$\pi_{BM} = \pi_{MM} = 1 - 2\delta - \tau = 0.5.$$

Use the Viterbi algorithm for pair HMM to find the optimal alignment of DNA sequences $x = TTACG$ and $y = TAG$.

Solution An optimal alignment is associated with the most probable path through the pair HMM generating the aligned sequences x and y. The Viterbi algorithm starts with the following initialization:

$$v^M(0,0) = 1, \quad \text{all other } v^\bullet(i,0), v^\bullet(0,j) = 0. \tag{4.4}$$

Let $v^\bullet(i,j)$ be the probability of emission of the aligned subsequences $x_1, \ldots, x_i$ and $y_1, \ldots, y_j$ by the pair HMM with the subalignment ending with (a) aligned pair x_i and y_j ($v^\bullet(i,j) = v^M(i,j)$), (b) residue x_i aligned to a gap ($v^\bullet(i,j) = v^X(i,j)$), and (c) residue y_j aligned to a gap ($v^\bullet(i,j) = v^Y(i,j)$). Then the recurrence equations required to compute the probability values $v^\bullet(i,j)$ are as follows:

$$v^M(i,j) = P_{x_i,y_j} \max \begin{cases} \pi_{MM} v^M(i-1,j-1), \\ \pi_{XM} v^X(i-1,j-1), \\ \pi_{YM} v^Y(i-1,j-1); \end{cases} \tag{4.5}$$

$$v^X(i,j) = q_{x_i} \max \begin{cases} \pi_{MX} v^M(i-1,j), \\ \pi_{XX} v^X(i-1,j); \end{cases} \tag{4.6}$$

$$v^Y(i,j) = q_{y_j} \max \begin{cases} \pi_{MY} v^M(i,j-1), \\ \pi_{YY} v^X(i,j-1). \end{cases} \tag{4.7}$$

We start the calculations as follows:

$$v^M(1,1) = P_{TT} \pi_{MM} v^M(0,0) = 0.25, \quad v^X(1,1) = 0, \quad v^Y(1,1) = 0;$$

and continue by using Equations (4.5)–(4.7), filling the computed probability values $v^\bullet(i,j)$ in the cells of the DP matrix (Table 4.1). At the termination step we have:

$$v = \tau \max(v^M(5,3), v^X(5,3), v^Y(5,3)) = 10^{-5}.$$

The traceback through the DP matrix determines the optimal path:

$$v^M(5,3) \rightarrow v^X(4,2) \rightarrow v^M(3,2) \rightarrow v^X(2,1) \rightarrow v^M(1,1).$$

The highest-scoring alignment associated with the most probable path π^* through the states of the pair HMM is given by

$$\begin{array}{ccccc} T & T & A & C & G \\ T & - & A & - & G \end{array}$$

The probability of the optimal path π^* (and the optimal alignment) is found at the termination step of the Viterbi algorithm: $P(x, y, \pi^*) = v = 10^{-5}$.

Table 4.1. *The matrix of probability values $v^\bullet(i,j)$ determined by the Viterbi algorithm for sequences TTACG and TAG*

Each cell (i,j) contains three values, $v^M(i,j)$, $v^X(i,j)$, and $v^Y(i,j)$, written in top down order. Entries on the optimal path are shown in bold.

	$i = 0$	$i = 1$	$i = 2$	$i = 3$	$i = 4$	$i = 5$
$j = 0$	1 0 0	0 0 0	0 0 0	0 0 0	0 0 0	0 0 0
$j = 1$	0 0 0	**0.25** 0 0	0 **0.0125** 0	0 3.125×10^{-4} 0	0 7.813×10^{-6} 0	0 1.953×10^{-7} 0
$j = 2$	0 0 0	0 0 0.0125	0.0375 0 0	**0.005** 0.001875 0	3.75×10^{-5} **2.5×10^{-4}** 0	3.125×10^{-7} 6.25×10^{-6} 0
$j = 3$	0 0 0	0 0 3.125×10^{-4}	0.0015 0 0.001875	9.375×10^{-4} 7.5×10^{-5} 2.5×10^{-4}	7.5×10^{-4} 4.688×10^{-5} 1.875×10^{-6}	**10^{-4}** 3.75×10^{-5} 1.563×10^{-8}

The PERL and C++ implementations of the Viterbi algorithm for the pair HMM alignment are available in the Web Supplemental Materials at opal.biology.gatech.edu/PSBSA.

Remark With the initialization conditions of the Viterbi algorithm for the pair HMM as suggested in Equations (4.4) (Durbin *et al.*, 1998, p 84), the resulting alignment of two sequences will always start with a matched pair x_1, y_1 for any two sequences x and y. Hence, the alignment generated by a pair HMM with such a restriction on the initialization step may not be the optimal one.

To allow the discovery of an optimal path which starts with a gap, it suffices to modify the initialization Equations (4.4) of the algorithm as follows:

$$v^M(0,0) = 1, \quad v^X(0,0) = 0, \quad v^Y(0,1) = 0;$$
$$v^M(i,0) = 0, \quad v^Y(i,0) = 0, \quad i \geq 1;$$
$$v^M(0,j) = 0, \quad v^X(0,j) = 0, \quad j \geq 1.$$ □

Problem 4.8 Use the forward algorithm to determine the probability that a pair of DNA sequences $x = TTACG$ and $y = TAG$ could be generated by the pair HMM described in Problem 4.7.

Solution The full probability is determined by the formula

$$P(x, y) = \sum_{\pi} P(x, y, \pi),$$

where the sum is taken over all possible paths π through the pair HMM generating pair of sequences x and y. The value of the full probability can be determined by the forward algorithm for a pair HMM. For each pair (i, j) the algorithm calculates the total probability $f^k(i, j)$ of all aligned subsequences $(x_1, \ldots, x_i)$ and $(y_1, \ldots, y_j)$ with the final pair of symbols emitted by the hidden state k (M, X, or Y). The forward algorithm starts with initialization:

$$f^M(0, 0) = 1, \quad f^X(0, 0) = f^Y(0, 0) = 0, \quad \text{all } f(i, -1), f(-1, j) = 0.$$

The values $f^k(i, j)$ for $i = 0, \ldots, 5$, and $j = 0, \ldots, 3$ are determined by the following formulas:

$$f^M(i, j) = P_{x_i y_j}(\pi_{MM} f^M(i - 1, j - 1) + \pi_{XM} f^X(i - 1, j - 1)$$
$$+ \pi_{YM} f^Y(i - 1, j - 1));$$
$$f^X(i, j) = q_{x_i}(\pi_{MX} f^M(i - 1, j) + \pi_{XX} f^X(i - 1, j));$$
$$f^Y(i, j) = q_{y_j}(\pi_{MY} f^M(i, j - 1) + \pi_{YY} f^Y(i, j - 1)).$$

The values of forward variables appear in the cells of matrix F shown in Table 4.2. At the termination step we have:

$$f^E(5, 3) = \tau(f^M(5, 3) + f^X(5, 3) + f^Y(5, 3)) = 3.632 \times 10^{-5}.$$

Hence, the probability that sequences x and y are generated by the pair HMM with all variants of alignment is given by

$$P(x, y) = f^E(5, 3) = 3.632 \times 10^{-5}.$$

Remark The forward algorithm implementation as computer programs in PERL and C++ are included in the Web Supplemental Materials available at opal.biology.gatech.edu/PSBSA.   □

Problem 4.9 For sequences $x = TTACG$ and $y = TAG$ find the posterior probability $P(\pi^*|x, y)$ of the optimal alignment obtained by the Viterbi algorithm for the pair HMM as described in Problem 4.7.

Solution The posterior probability of path π^* is given by the formula

$$P(\pi^*|x, y) = \frac{P(x, y, \pi^*)}{P(x, y)}.$$

Table 4.2. Forward variables $f^\bullet(i,j)$ found by the forward algorithm for a pair of sequences TTACG and TAG, and the pair HMM defined in Problem 4.7

Each cell (i,j) contains three values, $f^M(i,j)$, $f^X(i,j)$, and $f^Y(i,j)$, written in top down order.

	$i=0$	$i=1$	$i=2$	$i=3$	$i=4$	$i=5$
$j=0$	1 0 0	0 0.5 0	0 0.00125 0	0 3.125×10^{-5} 0	0 7.813×10^{-7} 0	0 1.953×10^{-9} 0
$j=1$	0 0 0.05	0.25 0 0	0.02 0.0125 0	0.0003 0.0013125 0	1.25×10^{-6} 4.781×10^{-5} 0	9.375×10^{-8} 1.258×10^{-6} 0
$j=2$	0 0 0.00125	0.012 0 0.0125	0.0375 0.0006 0.001	0.01 0.00189 1.5×10^{-5}	1.8×10^{-4} 5.473×10^{-4} 6.25×10^{-8}	1.944×10^{-6} 2.268×10^{-5} 4.688×10^{-9}
$j=3$	0 0 3.125×10^{-5}	1.5×10^{-4} 0 9.125×10^{-4}	0.0024 7.5×10^{-6} 2.5×10^{-4}	0.0010015 1.202×10^{-4} 5.004×10^{-4}	0.00196 5.308×10^{-5} 9.002×10^{-6}	2.639×10^{-4} 9.919×10^{-5} 9.730×10^{-8}

The optimal alignment

$$
\begin{array}{ccccc}
T & T & A & C & G \\
T & - & A & - & G
\end{array}
$$

was constructed in Problem 4.7. This alignment is associated with the most probable path π^* through the pair HMM. It was found that its probability is $P(x, y, \pi^*) = 10^{-5}$. The probability that aligned sequences x and y are generated by the HMM is given by $P(x, y) = 3.632 \times 10^{-5}$ (Problem 4.8). Thus,

$$
P(\pi^* | x, y) = \frac{10^{-5}}{3.632 \times 10^{-5}} = 0.275.
$$

Note that there is another optimal alignment π^{**},

$$
\begin{array}{ccccc}
T & T & A & C & G \\
- & T & A & - & G
\end{array}
$$

with the same probability $P(x, y, \pi^{**}) = P(x, y, \pi^*)$. (The remark to Problem 4.7 explains why this second optimal alignment was not recovered by the traceback procedure of the Viterbi algorithm.) A combined contribution of these two alignments to the full probability value exceeds 55%. This relatively large share can be explained as follows. First, the number of possible alignments (contributing to the sum of probabilities) of short sequences x and y is rather small. Indeed, the pair HMM (Figure 4.3) generates gapped alignments of sequences x and y with no x-gap following a y-gap and vice versa. In the second remark for Problem 2.6 we calculated the number $|\mathcal{A}_{n,m}|$ of such alignments for sequences of lengths n and m. For sequences of lengths 5 and 3, $|\mathcal{A}_{5,3}| = 24$. Some of these twenty-four gapped alignments have very small probabilities. For example, the alignment π'

$$
\begin{array}{cccccc}
T & T & A & C & G & - \\
- & - & T & - & A & G
\end{array}
$$

has the probability 3×10^{-9}, which is 333 times smaller than the probability of the optimal alignment and contributes less than 1% to the total probability $P(x, y)$. And π' is not the "worst" alignment in this case! □

Problem 4.10 For the sequences x and y defined in Problem 4.7 determine (a) the posterior probabilities of all aligned pairs in the optimal alignment π^* and (b) the expected accuracy of π^*.

Solution (a) We use the notation $x_i \diamond y_j$ for an aligned pair of symbols in an alignment of sequences x and y. The posterior probability of an aligned pair is given by

$$
P(x_i \diamond y_j | x, y) = \frac{P(x_i \diamond y_j, x, y)}{P(x, y)}.
\tag{4.8}
$$

Table 4.3. *Backward variables $b^\bullet(i,j)$ determined by the backward algorithm*

Each cell (i,j) contains three values, $b^M(i,j)$, $b^X(i,j)$, and $b^Y(i,j)$, written in top down order.

	$i = 5$	$i = 4$	$i = 3$	$i = 2$	$i = 1$
$j = 3$	0.1 0.1 0.1	0.005 0.0025 0	1.25×10^{-4} 6.25×10^{-5} 0	3.125×10^{-6} 1.563×10^{-6} 0	7.813×10^{-8} 3.91×10^{-8} 0
$j = 2$	0.005 0 0.0025	0.25 0.04 0.04	0.0275 0.0022 0.0012	1.131×10^{-4} 6×10^{-5} 5×10^{-6}	3.234×10^{-6} 1.875×10^{-6} 3.75×10^{-7}
$j = 1$	1.25×10^{-4} 0 6.25×10^{-5}	0.00213 0.0002 0.0012	0.00195 0.00301 0.00303	8.38×10^{-4} 0.001175 0.0011	7.574×10^{-5} 5.653×10^{-5} 2.716×10^{-5}

The joint probability on the right hand side of Equation (4.8) can be calculated by using forward and backward variables $f(i,j)$ and $b(i,j)$ defined for forward and backward algorithms for a pair HMM as follows:

$$P(x, y, x_i \diamond y_j) = f^M(i,j)b^M(i,j).$$

The variables $f(i,j)$ were already calculated (Problem 4.8), and all we need is to compute the values of $b(i,j)$ using the backward algorithm. The initialization conditions are as follows:

$$b^M(5,3) = b^X(5,3) = b^Y(5,3) = \tau, \quad \text{all } b(i,4), b(6,j) = 0.$$

The recurrence equations have the following form:

$$b^M(i,j) = (1 - 2\delta - \tau)P_{x_{i+1}y_{j+1}}b^M(i+1,j+1)$$
$$+ \delta(q_{x_{i+1}}b^X(i+1,j) + q_{y_{j+1}}b^Y(i,j+1));$$
$$b^X(i,j) = (1 - \varepsilon - \tau)P_{x_{i+1}y_{j+1}}b^M(i+1,j+1) + \varepsilon q_{x_{i+1}}b^X(i+1,j);$$
$$b^Y(i,j) = (1 - \varepsilon - \tau)P_{x_{i+1}y_{j+1}}b^M(i+1,j+1) + \varepsilon q_{y_{j+1}}b^Y(i,j+1).$$

We calculate the variables $b(i,j)$, $i = 5, \ldots, 1$, $j = 3, 2, 1$, step by step and fill in the cells of the matrix of backward variables (Table 4.3).

Remark The implementations of the backward algorithm in PERL and C++ are available as a part of the Web Supplemental Materials at opal.biology.gatech.edu/PSBSA.

The optimal alignment π^* for x and y (determined in Problem 4.7) is

$$
\begin{array}{ccccc}
T & T & A & C & G \\
T & - & A & - & G
\end{array}
$$

To compute the posterior probabilities for all aligned pairs of π^*, we need the value of the full probability $P(x, y)$ (computed by the forward algorithm in Problem 4.8). Thus, we have

$$
P(x_1 \diamond y_1 | x, y) = \frac{P(x_1 \diamond y_1, x, y)}{P(x, y)} = \frac{f^M(1, 1)b^M(1, 1)}{P(x, y)}
$$

$$
= \frac{0.25 \times 7.574 \times 10^{-5}}{3.632 \times 10^{-5}} = 0.521;
$$

$$
P(x_3 \diamond y_2 | x, y) = \frac{P(x_3 \diamond y_2, x, y)}{P(x, y)} = \frac{f^M(3, 2)b^M(3, 2)}{P(x, y)}
$$

$$
= \frac{0.01 \times 0.00275}{3.632 \times 10^{-5}} = 0.757;
$$

$$
P(x_5 \diamond y_3 | x, y) = \frac{P(x_5 \diamond y_3, x, y)}{P(x, y)} = \frac{f^M(5, 3)b^M(5, 3)}{P(x, y)}
$$

$$
= \frac{2.639 \times 10^{-4} \times 0.1}{3.632 \times 10^{-5}} = 0.727.
$$

(b) The *expected accuracy* of the alignment π^* is given by the following formula (Durbin *et al.*, 19998, p 96):

$$
A(\pi^*) = \sum_{(i,j) \in \pi^*} P(x_i \diamond y_j | x, y),
$$

where the sum is taken over all aligned pairs in π^*. Then

$$
A(\pi^*) = P(x_1 \diamond y_1 | x, y) + P(x_3 \diamond y_2 | x, y) + P(x_5 \diamond y_3 | x, y) = 2.005,
$$

which yields, on average, 0.668 per aligned pair. Let us compare this result with the one obtained for the far from optimal alignment π' considered in Problem 4.9:

$$
\begin{array}{cccccc}
T & T & A & C & G & - \\
- & - & T & - & A & G
\end{array}
$$

The expected accuracy of π' is given by

$$
A(\pi') = \frac{f^M(3, 1)b^M(3, 1) + f^M(5, 2)b^M(5, 2)}{P(x, y)} = 0.016
$$

with 0.008 on average per aligned pair. □

Problem 4.11 For the pair HMM (model M) shown in Figure 4.3 the probabilistic functions v^M, v^X, v^Y of the Viterbi algorithm are defined by the recurrence Equations (4.5)–(4.7). By addition of the independence pair-sequence model R shown in Figure 4.1, the derive analogs of these equations in log-odds terms. Show the relations between the parameters of the probabilistic equations and the log-odds equations of the Viterbi algorithm.

Solution We assume that states X and Y of models M and R emit a symbol a with the same probability q_a. The full independence pair-sequence model R generates a pair of sequences $x = x_1, \ldots, x_n$ and $y = y_1, \ldots, y_m$ with probability

$$P(x, y | R) = \eta^2 (1 - \eta)^{n+m} \prod_{i=1}^{n} q_{x_i} \prod_{j=1}^{m} q_{y_j}.$$

The log-odds scores $\hat{V}^{\bullet}(i, j), i = 1, \ldots, n, j = 1, \ldots, m$, are defined by the following equations:

$$\hat{V}^{\bullet}(i, j) = \log \frac{v^{\bullet}(i, j)}{w(i, j)},$$

where $v^{\bullet}(i, j)$ is the probability of aligned subsequences $x_1, \ldots, x_i$ and $y_1, \ldots, y_j$ from either Equation (4.5), Equation (4.6), or Equation (4.7), and $w(i, j)$ is the probability of these subsequences having being generated by model R. We rewrite the Viterbi equations in terms of the log-odds scores $\hat{V}^{\bullet}(i, j)$ as follows. The initialization step takes the form:

$$\hat{V}^M(0, 0) = \log \frac{1}{\eta^2} = -2 \log \eta, \quad \text{all others } \hat{V}(i, 0), \hat{V}(0, j) = -\infty.$$

and the recursion $i = 1, \ldots, n$ and $j = 1, \ldots, m$:

$\hat{V}^M(i, j)$

$$= \log \frac{P_{x_i y_j}}{q_{x_i} q_{y_j}} + \max \begin{cases} \log \dfrac{1 - 2\delta - \tau}{(1 - \eta)^2} + \log \dfrac{v^M(i - 1, j - 1)}{w(i - 1, j - 1)}, \\[2ex] \log \dfrac{1 - \varepsilon - \tau}{(1 - \eta)^2} + \log \dfrac{v^X(i - 1, j - 1)}{w(i - 1, j - 1)}, \\[2ex] \log \dfrac{1 - \varepsilon - \tau}{(1 - \eta)^2} + \log \dfrac{v^Y(i - 1, j - 1)}{w(i - 1, j - 1)}; \end{cases}$$

$$= \log \frac{P_{x_i y_j}}{q_{x_i} q_{y_j}} + \log \frac{1 - 2\delta - \tau}{(1 - \eta)^2} + \max \begin{cases} \hat{V}^M(i - 1, j - 1), \\[2ex] \log \dfrac{1 - \varepsilon - \tau}{1 - 2\delta - \tau} + \hat{V}^X(i - 1, j - 1), \\[2ex] \log \dfrac{1 - \varepsilon - \tau}{1 - 2\delta - \tau} + \hat{V}^Y(i - 1, j - 1); \end{cases}$$

$$\hat{V}^X(i,j) = \max \begin{cases} \log \dfrac{\delta}{1-\eta}\dfrac{v^M(i-1,j)}{w(i-1,j)}, \\[2mm] \log \dfrac{\varepsilon}{1-\eta}\dfrac{v^X(i-1,j)}{w(i-1,j)}; \end{cases}$$

$$= \max \begin{cases} \log \dfrac{\delta}{1-\eta} + \hat{V}^M(i-1,j), \\[2mm] \log \dfrac{\varepsilon}{1-\eta} + \hat{V}^X(i-1,j); \end{cases}$$

$$\hat{V}^Y(i,j) = \max \begin{cases} \log \dfrac{\delta}{1-\eta}\dfrac{v^M(i,j-1)}{w(i,j-1)}, \\[2mm] \log \dfrac{\varepsilon}{1-\eta}\dfrac{v^Y(i,j-1)}{w(i,j-1)}; \end{cases}$$

$$= \max \begin{cases} \log \dfrac{\delta}{1-\eta} + \hat{V}^M(i,j-1), \\[2mm] \log \dfrac{\varepsilon}{1-\eta} + \hat{V}^Y(i,j-1). \end{cases}$$

The termination step is as follows:

$$\hat{V} = \max(\hat{V}^M(n,m), \hat{V}^X(n,m), \hat{V}^Y(n,m)).$$

The above equations could be converted into the more conventional DP form if we define the new set of log-odds scores $V^\bullet(i,j)$ by the following formulas:

$$V^M(i,j) = \hat{V}^M(i,j) = \log \frac{v^M(i,j)}{w(i,j)},$$

$$V^X(i,j) = \log \frac{1-\varepsilon-\tau}{1-2\delta-\tau} + \hat{V}^X(i,j) = \log \frac{1-\varepsilon-\tau}{1-2\delta-\tau} + \log \frac{v^X(i,j)}{w(i,j)},$$

$$V^Y(i,j) = \log \frac{1-\varepsilon-\tau}{1-2\delta-\tau} + \hat{V}^Y(i,j) = \log \frac{1-\varepsilon-\tau}{1-2\delta-\tau} + \log \frac{v^Y(i,j)}{w(i,j)}.$$

Next, we define the substitution scores, gap-open and gap-extension penalties as follows:

$$s(a,b) = \log \frac{P_{ab}}{q_a q_b} + \log \frac{1-2\delta-\tau}{(1-\eta)^2},$$

$$d = -\log \frac{\delta(1-\varepsilon-\tau)}{(1-\eta)(1-2\delta-\tau)}, \qquad (4.9)$$

$$e = -\log \frac{\varepsilon}{1-\eta}.$$

Then, the recurrence equations of the Viterbi algorithm in the log-odds form become

$$V^M(i,j) = s(x_i, y_j) + \max \begin{cases} V^M(i-1,j-1), \\ V^X(i-1,j-1), \\ V^Y(i-1,j-1); \end{cases}$$

$$V^X(i,j) = \max \begin{cases} V^M(i-1,j) - d, \\ V^X(i-1,j) - e; \end{cases}$$

$$V^Y(i,j) = \max \begin{cases} V^M(i,j-1) - d, \\ V^Y(i,j-1) - e. \end{cases}$$

The termination is given by:

$$V = \max(V^M(n,m), V^X(n,m) + c, V^Y(n,m) + c),$$

with $c = \log(1 - 2\delta - \tau)/(1 - \varepsilon - \tau)$. The value V is the log-odds score of the optimal alignment of sequences x and y associated with the Viterbi path. The Viterbi path and the optimal alignment can be found by the traceback procedure.

Remark It is easy to give a probabilistic interpretation of the parameters of the pairwise alignment scoring system in Equation (4.9) if in a pair HMM (Figure 4.3) transition probabilities to the match state M are equal to each other $(1 - 2\delta - \tau = 1 - \varepsilon - \tau)$.

The term $\log P_{ab}/q_a q_b$ in the substitution score $s(a,b)$ is the log-likelihood ratio of the emission of pair (a,b) under the pair HMM to the emission of this pair under the independence pair-sequence model R. The term $\log(1 - 2\delta - \tau)/(1 - \eta)^2$ is the log-likelihood ratio of the transition to state M in the pair HMM to two transitions in the independence pair-sequence model R that increase the lengths of both sequences x and y by one symbol.

For the gap-open penalty d and the gap-extension penalty e the log-likelihood ratio of emissions is equal to zero, since emission probabilities from X and Y states of the HMM and the independence pair-sequence model R are equal. Therefore, the gap-open penalty $d = -\log \delta/(1 - \eta)$ equals the negative log-odds ratio of transition from match state M to state X or state Y in the pair HMM (which corresponds to the beginning of the gap), to a transition in the independence pair-sequence model which increases the length of one of sequences x or y by one symbol. Similarly, the gap-extension penalty $e = -\log \varepsilon/(1 - \eta)$ becomes the negative log-odds ratio of a loop from either X or Y to itself in the pair HMM, corresponding to the continuation of the gap, to a transition in R which increases the length of one of the sequences x or y by one symbol. Note that in this case parameter $c = 0$ and the optimal score of aligned sequences x and y generated by the pair HMM becomes

$$V = \max(V^M(n,m), V^X(n,m), V^Y(n,m)).$$

Notably, the scoring system (4.9) with parameters $s(a, b)$, d, and e suitable for aligning sequences with gaps is the generalization of the scoring system for ungapped alignments, which is defined solely by the scoring matrix with elements $s(a, b)$. In that case each substitution score $s(a, b)$ is the log-likelihood ratio of the observing pair (a, b) under some model of substitutions M to observing the pair (a, b) under the independence pair-sequence model R (see Section 2.2.1). $\quad\quad\square$

4.3 Further reading

Traditionally, pairwise sequence alignments have been constructed with DP algorithms (Needleman–Wunsch, Smith–Waterman, etc.) without direct reference to pair HMM. However, recent developments in applications of pair HMMs for gene prediction that use alignments of DNA syntenic regions have been described by (Meyer and Durbin 2002, 2004), as well as by Pachter, Alexandersson, and Cawley (2002). The program DOUBLESCAN (Meyer and Durbin (2002)) predicts genes in two syntenic DNA sequences from closely related species by retrieving conserved subsequences within protein-coding and non-coding regions. The Projector program (Meyer and Durbin (2004)) predicts the genes in a DNA sequence with an unknown gene content by building an alignment of this sequence with an annotated DNA sequence of related species.

A novel pairwise statistical alignment method using a pair HMM and the model of sequence evolution with insertions and deletions was proposed by Knudsen and Miyamoto (2003).

5

Profile HMMs for sequence families

Classifying biological sequences into families is one of the major challenges of bioinformatics. In fact, to provide the exact definition of the family (for example, the protein family) is difficult. Even with the introduction of the notion of a conserved in evolution protein domain as a structural determinant of a protein family, to classify multidomain proteins consistently is not a simple task. For practical purposes, nevertheless, it is important to develop efficient computational tools able to assign a protein translated from a newly predicted gene to one of already established families, thus characterizing the protein based on its amino acid sequence alone.

Computational tools of protein characterization have to recognize the family-specific features in a new protein sequence. Frequently, these detectable common features are manifested as statistically significant structural conservations. The computational tools that are required to solve the classification problem should be able to (i) make use of known structural patterns specific for a given family, (ii) detect the family patterns in the new protein sequence by alignment of the new protein to the family model, and (iii) assess the statistical significance of the detected similarity in order to help correctly identify the true family members.

These three properties of the protein characterization algorithm are similar to the properties of the pairwise sequence alignment algorithm, but there are significant differences. First, the availability of several sequences makes the differential scoring of amino acid matches feasible; with matches (mismatches) in conserved positions receiving higher (lower) scores than scores of similar events in non-conserved positions. Secondly, the algorithm of alignment of a sequence to a family model has to reflect the natural difference of the alignment partners, contrary to the pairwise alignment where two aligned sequences have equal status. Thirdly, the method of assessment of the statistical significance has to take into account the details of the alignment score computation. As this score generating scheme becomes more complicated compared with the generation of scores of pairwise alignment, computational rather than analytical approximations to the score distributions become necessary.

Position-specific scoring matrices (PSSMs) and non-probabilistic *profiles* have been the pioneering, early models of the protein domains, the signatures of protein families. The former technique has been further developed and implemented in PSI-BLAST (Altschul *et al.*, 1997) which, during the run, iteratively extracts sets of similar sequences and updates the current PSSM in semiautomatic fashion. The motifs of the latter technique could be traced in the frequently used heuristic algorithm for multiple sequence alignment called CLUSTAL. Again, since the real biological sequences cannot be viewed as precise realizations of a particular probabilistic model, there is no theoretically established superiority of a particular approach to modeling DNA and protein sequence alignments. Parameters of any initial theoretical model may require additional corrections due to the world realities that have not been taken into account. For instance, estimation of the profile HMM parameters from a given training set (a multiple alignment of several sequences) frequently needs correction to compensate both for a small sample effects (introducing pseudocounts) and for a possible bias of the training set towards a subset of the family (introducing sequence weights).

The problems included in Chapter 5 of *BSA* illustrate the relationships between the algorithm of local pairwise alignment and the algorithm of local alignment of a sequence to a profile HMM. Since the analysis of frequencies of amino acid substitutions in aligned sequences of proteins naturally leads to the analysis of properties of the substitution log-odds scores, several problems deal with the analytical properties of the amino acid substitution score matrix and its scaling factor λ. Also covered is the important topic of assignment of the weights to sequences in the training set.

The additional problems provide practice of the construction of a sequence to a profile HMM alignment and of the estimation of parameters of the profile HMM and the PSSM, with and without use of pseudocounts. Approaches to sequence weights assignment seem to vary significantly; the additional problems extend further the discussion of the existing weights assignment methods.

5.1 Original problems

Problem 5.1 The profile HMM for local alignment is shown in Figure 5.1. Show that this profile HMM yields update equations similar to those of the local pairwise sequence alignment (introduced by Smith and Waterman, 1981).

Solution First, we have to remind ourselves of the equations of the Smith–Waterman algorithm, constructing the optimal local alignment with

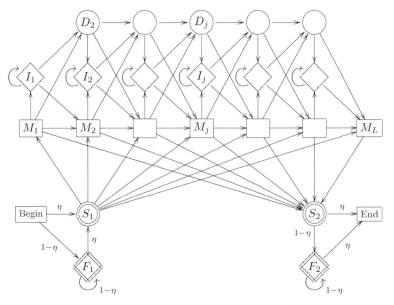

Figure 5.1. The profile HMM for local alignment. Two flanking model states F_1 and F_2 emit symbols with the same probabilities q_a as states X and Y of the independence pair-sequence model R shown in Figure 4.1. The looping probability $1 - \eta$ to the flanking states is close to one and is the same as the looping probabilities in R. Two silent states S_1 and S_2 are shown by double circles. All transition probabilities from the silent state S_1 to the match states $M_j, j = 1, \ldots, L$, and from match states $M_j, j = 1, \ldots, L$, to the silent state S_2 are equal to η/L.

affine gap score for two sequences $x = x_1, \ldots, x_n$ and $y = y_1, \ldots, y_m$. The algorithm finds the three dynamic programming matrices V^M, V^X, and V^Y with elements defined as the best scores of local alignment up to symbols x_i, y_i given that (a) x_i is aligned to y_j ($V^M(i,j)$), (b) x_i is aligned to a gap ($V^X(i,j)$), (c) y_j is aligned to a gap ($V^Y(i,j)$). The update equations for the scores are as follows:

$$V^M(i,j) = \max \begin{cases} 0, \\ V^M(i-1, j-1) + s(x_i, y_j), \\ V^X(i-1, j-1) + s(x_i, y_j), \\ V^Y(i-1, j-1) + s(x_i, y_j); \end{cases}$$

$$V^X(i,j) = \max \begin{cases} 0, \\ V^M(i-1, j) - d, \\ V^X(i-1, j) - e; \end{cases} \tag{5.1}$$

$$V^Y(i,j) = \max \begin{cases} 0, \\ V^M(i, j-1) - d, \\ V^Y(i, j-1) - e. \end{cases}$$

Here $s(x_i, y_j)$ is the substitution score, d is the gap-open penalty, and e is the gap-extension penalty. As soon as all elements of matrices V^M, V^X, and V^Y are determined, the algorithm yields

$$V^M(i^*, j^*) = \max_{i,j} V^M(i,j),$$

the score associated with the optimal local alignments of sequences x and y. The traceback from cell (i^*, j^*) to cell (i', j') with $V^M(i', j') = 0$ defines the alignment of subsequences $x_i, i = i' \ldots i^*$, and $y_j, j = j' \ldots j^*$, the best local alignment of x and y.

To draw an analogy between Equations (5.1) and the equations of the algorithm constructing the optimal local alignment of sequence $x_i, i = 1, \ldots, N$, to the profile HMM (Figure 5.1), we have to show explicitly the equations that determine the Viterbi path for sequence x through the profile HMM. For subsequence $x_1, \ldots, x_i$ let $V_j^M(i)$ denote the log-odds score of the highest probability path through the profile HMM, ending with x_i emitted by state M_j. Similarly, $V_j^I(i)$ designates the score of the highest probability path ending with x_i emitted by state I_j, $V_j^D(i)$ is the score of the highest probability path ending in state D_j, and $V^F(i)$ is the score of the highest probability path ending with x_i emitted by state F_2. The initialization conditions are as follows:

$$V_j^I(1) = -\infty,$$

$$V_j^D(1) = -\infty,$$

$$V^F(1) = -\infty,$$

$$V_j^M(1) = \log \frac{P(x_1|M)}{P(x_1|R)} = \log \frac{(\eta^2/L)e_{M_j}(x_1)}{(1-\eta)q_{x_1}} = \log \frac{e_{M_j}(x_1)\eta^2}{q_{x_1}L(1-\eta)},$$

for any $j : j = 1, \ldots, L$. The update equations are as follows:

$$V_j^M(i) = \log \frac{e_{M_j}(x_i)}{q_{x_i}} + \max \begin{cases} V_{j-1}^M(i-1) + \log \dfrac{a_{M_{j-1}M_j}}{1-\eta}, \\[2mm] V_{j-1}^I(i-1) + \log \dfrac{a_{I_{j-1}M_j}}{1-\eta}, \\[2mm] V_{j-1}^D(i-1) + \log \dfrac{a_{D_{j-1}M_j}}{1-\eta}, \\[2mm] \log \dfrac{\eta^2}{(1-\eta)L}; \end{cases}$$

$$V_j^I(i) = \max \begin{cases} V_j^M(i-1) + \log \dfrac{a_{M_j I_j}}{1-\eta}, \\[2mm] V_j^I(i-1) + \log \dfrac{a_{I_j I_j}}{1-\eta}, \\[2mm] V_j^D(i-1) + \log \dfrac{a_{D_j I_j}}{1-\eta}; \end{cases} \tag{5.2}$$

$$V_j^D(i) = \max \begin{cases} V_{j-1}^M(i) + \log \dfrac{a_{M_{j-1}D_j}}{1-\eta}, \\ V_{j-1}^I(i) + \log \dfrac{a_{I_{j-1}D_j}}{1-\eta}, \\ V_{j-1}^D(i) + \log \dfrac{a_{D_{j-1}D_j}}{1-\eta}; \end{cases}$$

$$V^F(i) = \max \begin{cases} \max_j V_j^M(i-1) + \log \dfrac{\eta}{L}, \\ V^F(i-1). \end{cases}$$

Note that the term $\log \eta^2/(1-\eta)L$ in the equation for $V_j^M(i)$ corresponds to the direct transition from silent state S_1 to match state M_j, the case when the first $i-1$ residues of the sequence x are emitted from the flanking state F_1. Then the log-odds ratio becomes

$$\log \frac{P(x_1,\ldots,x_i|M)}{P(x_1,\ldots,x_i|R)} = \frac{(1-\eta)^{i-1} \prod_{l=1}^{i-1} q_{x_l} \eta^2 e_{M_j}(x_i)/L}{(1-\eta)^i \prod_{l=1}^{i} q_{x_l}}$$

$$= \log \frac{e_{M_j}(x_i)}{q_{x_i}} + \log \frac{\eta^2}{(1-\eta)L}.$$

At the termination step we have

$$V = \max(\max_j(V_j^M(N) - \log L), V^F(N) - \log \eta).$$

The value V is the total score of the Viterbi path for sequence x through the local alignment profile HMM. The path itself is recovered by applying the traceback procedure. The subsequence of sequence x emitted on the path between the appearances of the first matching state $M_{j'}$ and the last matching state M_{j*}, which precedes the silent state S_2, participates in the best local alignment of the sequence x to the profile HMM. This optimal local alignment is described by the part of the path between matching states $M_{j'}$ and M_{j*}, $j' \le j^*$.

Comparison of update Equations (5.1) and (5.2) shows that the two algorithms have much in common. Emission of x_i from match state M_j of the profile HMM corresponds to the aligned pair (x_i, y_j) in the pairwise alignment; emission of x_i from an insert state corresponds to symbol x_i aligned to a gap in the pairwise alignment; and occurrence of delete state D_j corresponds to symbol y_j aligned to a gap in the pairwise alignment. Note that in the extreme case when the set of aligned sequences, serving as the training set for the profile HMM, shrinks to just one sequence designated as y, finding the optimal path through the profile HMM emitting sequence x becomes equivalent to the problem of finding the optimal pairwise local alignment of sequences x and y. Formally, the optimal current score $V_j^M(i)$ in Equations (5.2) corresponds to $V^M(i,j)$ in Equations (5.1), $V_j^I(i)$ to

$V^X(i,j)$, and $V_j^D(i)$ to $V^Y(i,j)$. Interpretation of the substitution score $s(x_i, y_j)$ in Equations (5.1) presents some difficulty as in Equations (5.2) it is split into three terms: $\log e_{M_j}(x_i)/q_{x_i} + \log a_{M_{j-1}M_j}/(1-\eta)$, $\log e_{M_j}(x_i)/q_{x_i} + \log a_{I_{j-1}M_j}/(1-\eta)$, and $\log e_{M_j}(x_i)/q_{x_i} + \log a_{D_{j-1}M_j}/(1-\eta)$. If all the probabilities of the transitions to match state M_j are the same, so are the values of these three terms. Similarly, the gap-open penalty d in Equations (5.1) corresponds to the following terms in Equation (5.2): $-\log a_{M_j I_j}/(1-\eta), j = 1, \ldots, L-1$, and $-\log a_{M_{j-1}D_j}/(1-\eta)$, $j = 2, \ldots, L-1$. These terms are equal if all permitted transitions from match states to insert and delete states have equal probabilities. Finally, the gap-extension penalty e in Equations (5.1) corresponds to the following terms in Equations (5.3): $-\log a_{I_j I_j}/(1-\eta), j = 1, \ldots, L-1$, and $-\log a_{D_{j-1}D_j}/(1-\eta), j = 3, \ldots, L-1$, which again collapse in just one value if the values of transition probabilities $a_{I_j I_j}$ and $a_{D_{k-1}D_k}$ are the same.

Remark The relationship between the probabilistic parameters of a pair HMM and the parameters of the pairwise alignment algorithm, was discussed in Remark to Problem 4.11. $\qquad\square$

Problem 5.2 Explain the reasons (referring to Problem 5.1) for any differences between the optimal local pairwise alignment algorithm (5.1) and the algorithm (5.2) for optimal local alignment of a sequence to a profile HMM.

Solution To determine either the best local pairwise alignment of two sequences x and y or the best alignment of sequence x to the profile HMM with flanking states, we have to calculate elements of the three dynamic programming matrices by using Equations (5.1) or Equations (5.2), respectively. To construct the best local pairwise alignment, we use dynamic programming Equations (5.1) and find values of $V^M(i,j)$, $V^X(i,j)$, and $V^Y(i,j)$ under the assumption that the substitution scores $s(x_i, y_j)$, the gap-open penalty d, and the gap-extension penalty e do not depend on the positions i, j. Parameters of Equations (5.2) used to find the best alignment of x to profile HMM (Figure 5.1), the emission probabilities $e_{M_j}(x_i)$, and the transition probabilities $(a_{M_{j-1}M_j}, a_{I_{j-1}M_j}, a_{D_{j-1}M_j}, \ldots)$, explicitly depend on j. Thus, use of a larger number of parameters in the profile HMM-based algorithm allows us to incorporate information on position-specific conservation patterns of a biological sequence. For instance, for protein sequences algorithm (5.1) needs 212 parameters: 210 substitution scores $s(a, b)$, the gap-open penalty d, and the gap-extension penalty e. Note that the total number of parameters, 212, does not depend on the lengths of sequences x and y. On the other hand, algorithm (5.2) uses the profile HMM with L match states M_j and two flanking states. This model needs $3(3(L-2)+2) = 9L-12$ transition probabilities between hidden states, parameter η, $20L$ emission probabilities $e_{M_j}(x_i)$ for match states, twenty emission probabilities q_{x_i} for insert states (the set q_{x_i}, $i = 1, \ldots, 20$, is assumed to be the same for all $L - 1$ insert states of the profile HMM and to coincide with the set of emission probabilities of the background independence model used for the definition of the log-odds scores in Equations (5.2)). Thus, the total number of parameters of the profile HMM, $29L + 9$, depends on the number L of its match states, and becomes larger than 212 for $L > 7$.

One more difference between these two algorithms is that transitions $I_{j-1} \to D_j$ and $D_j \to I_j$ are normally permitted in a profile HMM (see Figure 5.1), while in a pairwise alignment, as Equations (5.1) show, an x-gap cannot immediately follow a y-gap and vice versa.   □

Problem 5.3 Use the negative bias condition $\sum_{a,b} q_a q_b s(a, b) < 0$ to show that

$$f(\lambda) = \sum_{a,b} q_a q_b e^{\lambda s(a,b)} - 1$$

is negative for small enough $\lambda > 0$.

Solution For $\lambda = 0$ we have

$$f(0) = \sum_{a,b} q_a q_b - 1 = \left(\sum_a q_a \right) \left(\sum_b q_b \right) - 1 = 0.$$

The first derivative of f is defined by the following expression:

$$f'(\lambda) = \sum_{a,b} q_a q_b e^{\lambda s(a,b)} s(a, b),$$

and, due to the negative bias condition, $f'(0) = \sum_{a,b} q_a q_b s(a, b) < 0$. Near the point zero differentiable function $f(\lambda)$ can be expanded as

$$f(\lambda) = f(0) + f'(0)\lambda + o(\lambda^2).$$

Therefore, for small enough $\lambda > 0$ function $f(\lambda)$ is negative.   □

Problem 5.4 Use the condition that there is at least one positive $s(a, b)$ to show that $f(\lambda)$ becomes positive for large enough λ.

Solution In the expression

$$f(\lambda) = \sum_{a,b} q_a q_b e^{\lambda s(a,b)} - 1$$

the sum can be split into two sums Σ_1 and Σ_2, where Σ_1 corresponds to $s(a, b)$ with positive values (there is at least one such term) and Σ_2 corresponds to $s(a, b)$

with non-positive values. As $\lambda > 0$ grows, Σ_1 increases exponentially because of factor(s) $e^{\lambda s(a,b)}$. The second sum Σ_2 is positive and bounded from below. Thus, for sufficiently large λ, such that $\Sigma_1 > 1$, function $f(\lambda)$ becomes positive and even:

$$\lim_{\lambda \to +\infty} f(\lambda) = +\infty. \qquad \square$$

Problem 5.5 Show that the second derivative of $f(\lambda)$ is positive, and that this and the results of the previous two problems show that there is one and only one positive value of λ satisfying $f(\lambda) = 0$.

Solution It was shown that function $f(\lambda)$ possesses the following properties:

(a) $f(0) = 0$,
(b) $f(\lambda)$ is negative for a small enough $\lambda > 0$,
(c) f becomes positive for a sufficiently large λ and tends to $+\infty$ as λ increases.

Since f is a continuous function, there exists at least one positive λ^* such that $f(\lambda^*) = 0$. To prove that there is only one positive λ^*: $f(\lambda^*) = 0$, we find the second derivative of f:

$$f''(\lambda) = \sum_{a,b} q_a q_b e^{\lambda s(a,b)} (s(a,b))^2.$$

It is easy to see that $f''(\lambda) > 0$; therefore, the first derivative f' monotonically increases on the interval $[0; +\infty)$ starting from some negative value at point 0. Obviously, f' has only one root λ' on this interval with the sign of f' changing at λ' from negative to positive. Thus, at this stationary point λ', the function f has a local minimum with negative value. On the interval $[\lambda', +\infty)$, function f monotonically increases as λ increases, and the graph of function f intercepts the x-axis at only one point λ^*: $\lambda^* > \lambda' > 0, f(\lambda^*) = 0$. $\qquad \square$

Problem 5.6 Compute the weights for the sequence set $x_1 = AGAA$, $x_2 = CCTC$, $x_3 = AGTC$, using the weighting methods by (a) Thompson, Higgins, and Gibson (1994a); (b) Gerstein, Sonnhammer, and Chothia (1994); (c) Altschul, Carroll, and Lipman (1989); (d) Henikoff and Henikoff (1994); (e) Krogh and Mitchison (1995).

Solution Some of the weighting methods mentioned above need a tree relating the sequences in question. Therefore, we will start with a tree construction by the classical clustering procedure, the UPGMA, proposed by Sokal and Michener (1958). We define the distance d_{ij} between sequences x_i and x_j aligned without gaps as a fraction of the aligned nucleotide pairs with mismatches. Thus, we obtain $d_{12} = 4/4 = 1$, $d_{13} = 2/4 = 0.5$, and $d_{23} = 2/4 = 0.5$. At the first step of the

UPGMA algorithm each sequence x_i is assigned to its own cluster C_i and a leaf i for each sequence x_i is placed at height zero (the tree will be drawn with leaves at the bottom level and the root at the top). At'the next step we have to select two clusters separated by the minimal distance. Since $d_{13} = d_{23} = 0.5$, we can select either pair C_1 & C_3 or C_2 & C_3. Say we select clusters C_2 and C_3 and join into a new cluster $C_4 = \{C_2, C_3\} = \{x_2, x_3\}$. The distance $d_{14} = 0.75$ is defined by the following formula:

$$d_{ij} = \frac{1}{|C_i||C_j|} \sum_{p \in C_i,\, q \in C_j} d_{pq}.$$

Then, node 4 is placed at height $t_2 = t_3 = d_{23}/2 = 0.25$ above daughter nodes 2 and 3. With two clusters C_1 and C_4 remaining, the root, node 5, is placed at height $t_1 = d_{14}/2 = 0.375$. As result, we have built the following tree:

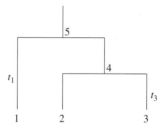

(a) Computation of the sequence weights by the "voltage" method by Thompson *et al.* (1994a).

It is assumed that the frame of the tree above is made of a conducting wire of constant thickness and that a voltage V is applied to the root point. With resistance assigned to be equal to the edge lengths and all the leaf points set to zero potential, we have to find currents in the edges immediately above the leaves. The values of the currents will be the sequence weights we are interested in. If the current and voltage at node n are I_n and V_n, respectively, then from the following diagram:

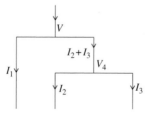

and Ohm's law it follows that $V_4 = 0.25I_2 = 0.25I_3$, $V_5 = 0.375I_1 = 0.125(I_2 + I_3) + 0.25I_2$. Hence, the ratios of the currents (and the weights) come out as:

$$I_1 : I_2 : I_3 = w_1 : w_2 : w_3 = 4 : 3 : 3.$$

(b) Computation of the sequence weights by the method of Gerstein *et al.* (1994).

It is assumed that the initial weights of the tree leaves are defined as the lengths of the leaf edges: $w_1 = 0.375$, $w_2 = w_3 = 0.25$. Then, node 4 with the edge length of 0.125 above it, ancestral to leaves 2 and 3, makes an equal contribution of $0.125/2 = 0.0625$ to each of w_2 and w_3. Thus, $w_2 = w_3 = 0.25+0.0625 = 0.3125$. With $w_1 = 0.375$ we have

$$w_1 : w_2 : w_3 = 6 : 5 : 5.$$

(c) Computation of the sequence weights by the Altschul–Carroll–Lipman method (Altschul *et al.*, 1989).

It is assumed that a continuous variable ξ with a Gaussian distribution is associated with the tree in the following way. The probability of substituting one value, x, of ξ by another value, y, along a tree edge with length t is $\exp -(x - y)^2/2t$. It can be shown that the mean value μ of variable ξ at the tree root depends linearly on the ξ-values x_i at the leaves, so $\mu = \sum w_i x_i$. A coefficient w_i is assigned to be the weight of the leaf sequence x_i, $i = 1, 2, 3$. For the tree above the sequence weights are calculated by the following formulas:

$$w_1 = \frac{t_2 t_3 + t_4(t_2 + t_3)}{t_2 t_3 + (t_1 + t_4)(t_2 + t_3)} = 0.4;$$

$$w_2 = \frac{t_1 t_3}{t_2 t_3 + (t_1 + t_4)(t_2 + t_3)} = 0.3;$$

$$w_3 = \frac{t_1 t_2}{t_2 t_3 + (t_1 + t_4)(t_2 + t_3)} = 0.3.$$

The next two methods do not require a tree for the computation of the sequence weights.

(d) Computation of the sequence weights by the block alignment method of Henikoff and Henikoff (1994).

Given the ungapped "block" alignment of sequences x_1, x_2, and x_3,

$$
\begin{array}{cccc}
A & G & A & A \\
C & C & T & C \\
A & G & T & C
\end{array}
$$

we take counts $k_{i\alpha}$ of letter α in each column (position) i. The first column ($k_{1A} = 2$ and $k_{1C} = 1$) has two different types of symbols ($m_1 = 2$). For sequence x_i the weight w_{1i} associated with the first column weight w_{1i} for sequence x_i is defined by the formula $w_{1i} = 1/m_1 k_{1x_i^k}$. We obtain

$$w_{11} = \frac{1}{2 \times 2} = 0.25, \quad w_{12} = \frac{1}{2 \times 1} = 0.5, \quad w_{13} = \frac{1}{2 \times 2} = 0.25.$$

Here, two sequences x_1 and x_3 sharing the same symbol A are assigned smaller weights than x_2, because together they contribute to the multiple alignment the same amount of information about the possible types of symbols in the first column as does sequence x_2 alone. Similar processing of the remaining columns produces the the following column-specific weights of sequences x_1, x_2, and x_3:

$$w_{21} = 0.25, \quad w_{22} = 0.5, \quad w_{23} = 0.25;$$

$$w_{31} = 0.5, \quad w_{32} = 0.25, \quad w_{33} = 0.25;$$

$$w_{41} = 0.5, \quad w_{12} = 0.25, \quad w_{43} = 0.25.$$

Finally, the column-specific weights for each sequence are summed over all columns and normalized. We obtain $w_1 = 0.375$, $w_2 = 0.375$, and $w_3 = 0.25$.

(e) Computation of the sequence weights by the maximum entropy method of Krogh and Mitchison (1995).

The sequence weights are chosen to maximize the sum of the entropies $H_k(w_1, w_2, w_3) = -\sum_a p_{ka} \log_2 p_{ka}$, defined for each alignment site $k = 1, 2, 3, 4$, given the constraint $\sum_i w_i = 1$. Here p_{ka}, the weighted frequency of symbol a at the kth site, is defined as follows:

$$p_{ka} = \sum_{i=1,2,3} \mathbf{I}(\text{symbol } a \text{ appears at site } k \text{ in sequence } i) w_i,$$

where $\mathbf{I}(A)$ is an indicator of an event A and the sum is taken over all sequences in the multiple alignment. Thus, we have

$$H_1(w_1, w_2, w_3) = -(w_1 + w_3) \log_2(w_1 + w_3) - w_2 \log_2 w_2;$$

$$H_2(w_1, w_2, w_3) = -(w_1 + w_3) \log_2(w_1 + w_3) - w_2 \log_2 w_2;$$

$$H_3(w_1, w_2, w_3) = -(w_2 + w_3) \log_2(w_2 + w_3) - w_1 \log_2 w_1;$$

$$H_4(w_1, w_2, w_3) = -(w_2 + w_3) \log_2(w_2 + w_3) - w_1 \log_2 w_1.$$

To find the maximum of the function $\Sigma = \sum_i H_i(w_1, w_2, w_3)$ with constraint $\sum_i w_i = 1$, we have to solve the system of equations with the Lagrange multiplier λ:

$$\frac{\partial \Sigma}{\partial w_1} + \lambda = -2 \log_2(w_1 + w_3) - 2 \log_2 w_1 - 4 + \lambda = 0;$$

$$\frac{\partial \Sigma}{\partial w_2} + \lambda = -2 \log_2(w_2 + w_3) - 2 \log_2 w_2 - 4 + \lambda = 0;$$

$$\frac{\partial \Sigma}{\partial w_3} + \lambda = -2 \log_2(w_1 + w_3) - 2 \log_2(w_2 + w_3) - 4 + \lambda = 0.$$

The system can be reduced to

$$(w_1 + w_3)w_1 = (w_2 + w_3)w_2 = (w_1 + w_3)(w_2 + w_3),$$

and finally $w_3 = 0$, $w_1 = w_2 = 0.5$. The zero value of weight w_3 seems unexpected. However, it could be argued that sequence x_3 does not bring any new information to the alignment as its prefix AG exists in x_1 and its suffix TC exists in x_2. ☐

5.2 Additional problems and theory

This section illustrates the profile HMM theory by practical applications. A profile HMM can be built from the multiple alignment of DNA fragments either without pseudocounts (Problem 5.7) or with pseudocounts defined by the Laplace rule (Problem 5.8). Similar derivations can be made for the case when the aligned sequences are weighted (Problem 5.10). To construct an alignment of a sequence to a profile HMM with a complete state diagram one can use the modified Viterbi algorithm for a profile HMM (Problem 5.9).

In the theoretical introduction to Problem 5.11, we define the discrimination function of a profile HMM and the maximum discrimination weights of the sequences. The fact that these weighted estimates of emission and transition probabilities maximize the discrimination function as well as the weighted sum of log-likelihood ratios is proved in Problem 5.11. A Bayesian approach to the estimation of the parameters of a profile HMM is illustrated in the context of Problem 5.12: assuming the Dirichlet prior for the emission probabilities, the mean values of the posterior distribution and the maximum of the posterior probability (MAP) estimates are derived.

In Problem 5.13 we find the elements (emission probabilities) of the position-specific scoring matrix (PSSM) given an ungapped alignment of amino acid sequences.

The PSSMs for DNA sequences are considered in Problems 5.14 and 5.15. In Problem 5.14 a position-specific independence model R_1 is used to describe a ribosomal binding site in the *E. coli* genome. In Problem 5.15 the Kullback–Leibler distance between model R_1 and a background independence model Q is compared with the Kullback–Leibler distance between an independence model for a protein-coding region and model Q.

Problem 5.7 Estimate the parameters of a profile HMM for the following multiple alignment of DNA sequences:

$$
\begin{array}{cccc}
G & C & A & G \\
G & - & - & G \\
G & - & A & G \\
G & C & T & G \\
A & - & A & C \\
G & - & A & C \\
G & - & G & G \\
A & - & A & C
\end{array}
\tag{5.3}
$$

Draw the state diagram.

Solution A heuristic rule suggests that an alignment column with a fraction of gap symbols below 0.5 corresponds to a match state of a profile HMM; otherwise, it corresponds to an insert state. Therefore, we assign columns in the multiple alignment (5.3) but the second one to the match states M_1, M_2, and M_3, respectively. An HMM state diagram includes a *begin* state $\mathcal{B}$ (the same as M_0), an *end* state $\mathcal{E}$, along with insert states I_0, I_1, I_2, I_3, and delete states D_1, D_2, D_3. The counts $E_k(x)$ of nucleotide emissions and the counts A_{kl} of the hidden state transition observed in alignment (5.3) are shown in Table 5.1. Then the maximum likelihood estimates of parameters of the profile HMM are given by the following formulas:

$$
a_{kl} = \frac{A_{kl}}{\sum_{l'} A_{kl'}}, \quad e_k(x) = \frac{E_k(x)}{\sum_{x'} E_k(x')}.
\tag{5.4}
$$

Here, indices k and l designate hidden states; a_{kl} and $e_k(x)$ are estimates of the transition and emission probabilities, respectively. Equations (5.4) generate the estimates of the profile HMM parameters that fill Table 5.2.

Since many of the emission and transition probabilities are zeros, several states defined for a generic profile HMM are unreachable and do not appear in the final state diagram (Figure 5.2) showing the rather sparse profile HMM. $\square$

Problem 5.8 Estimate the parameters of the profile HMM derived from the same multiple alignment (5.3) if the emission and transition counts are augmented by the pseudocounts defined by Laplace's rule. Draw the state diagram.

Table 5.1. *Counts of emissions of nucleotides and transitions between hidden states observed in the alignment block (5.3)*

		0	1	2	3
Emission counts for match states	A	–	2	5	0
	C	–	0	0	3
	G	–	6	1	5
	T	–	0	1	0
Emission counts for insert states	A	0	0	0	0
	C	0	2	0	0
	G	0	0	0	0
	T	0	0	0	0
Counts of transitions between hidden states	M − M	8	5	7	8
	M − D	0	1	0	–
	M − I	0	2	0	0
	I − M	0	2	0	0
	I − D	0	0	0	–
	I − I	0	0	0	0
	D − M	–	0	1	0
	D − D	–	0	0	–
	D − I	–	0	0	0

Table 5.2. *Values of the estimated parameters of the profile HMM*

		0	1	2	3
Emission probabilities for match states	A	–	0.25	0.72	0
	C	–	0	0	0.375
	G	–	0.75	0.14	0.625
	T	–	0	0.14	0
Emission probabilities for insert states	A	0	0	0	0
	C	0	1	0	0
	G	0	0	0	0
	T	0	0	0	0
Transition probabilities between hidden states	M − M	1	0.625	1	1
	M − D	0	0.125	0	–
	M − I	0	0.25	0	0
	I − M	0	1	0	0
	I − D	0	0	0	–
	I − I	0	0	0	0
	D − M	–	0	1	0
	D − D	–	0	0	–
	D − I	–	0	0	0

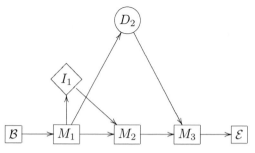

Figure 5.2. The state diagram of the profile HMM derived from the multiple alignment (5.3) with only states and transitions permitted by the alignment data.

Solution To implement Laplace's rule, each count shown in Table 5.1 should be increased by one. Note that the cells corresponding to emissions (or transitions) that are impossible for the generic profile HMM state diagram are not affected by this operation. The updated counts can be used directly to estimate the transition and emission probabilities, a_{kl} and $e_k(x)$, of the profile HMM by Equations (5.4). The same estimates can be obtained using the updated equations

$$a_{kl} = \frac{A_{kl} + 1}{\sum_{l'} A_{kl'} + m_k}, \quad e_k(x) = \frac{E_k(x) + 1}{\sum_{x'} E_k(x') + K}, \tag{5.5}$$

with A_{kl} and $E_k(x)$ values taken from Table 5.1. In Equations (5.5) again k and l designate hidden states and A_{kl} and $E_k(x)$ stand for counts of observed transitions and emissions, respectively. Also, m_k denotes the number of transitions from state k permitted by the generic state diagram of a profile HMM, while K is the size of the alphabet. The resulting profile HMM parameters are given in Table 5.3.

  The state diagram of the defined profile HMM is shown in Figure 5.3. The use of Laplace's rule makes all regular transitions between states of the profile HMM possible, and the graph of connections between hidden states becomes complete as compared with the graph constructed earlier (Figure 5.2). □

Problem 5.9 Use the log-odds score version of the Viterbi algorithm to align sequence *GCCAG* to the profile HMM built in Problem 5.8. To define the log-odds scores, assume that the background model is an independence model with $P(A) = P(T) = 0.3$ and $P(C) = P(G) = 0.2$.

Solution The optimal (most probable) alignment of a sequence to a profile HMM is defined by a path of the hidden states, emitting the symbols of the given sequence, as determined by the Viterbi algorithm. The log-odds version of the algorithm works as follows. Let $V_j^M(i)$ be the log-odds score of the highest scoring alignment of

Table 5.3. *The profile HMM parameters estimated using the pseudo counts defined by Laplace's rule*

		0	1	2	3
Emission probabilities for match states	A	—	0.25	0.55	0.08
	C	—	0.08	0.09	0.33
	G	—	0.58	0.18	0.5
	T	—	0.08	0.18	0.08
Emission probabilities for insert states	A	0.25	0.17	0.25	0.25
	C	0.25	0.5	0.25	0.25
	G	0.25	0.17	0.25	0.25
	T	0.25	0.17	0.25	0.25
Transition probabilities between hidden states	$M - M$	0.82	0.55	0.8	0.9
	$M - D$	0.09	0.18	0.1	—
	$M - I$	0.09	0.27	0.1	0.1
	$I - M$	0.33	0.6	0.33	0.5
	$I - D$	0.33	0.2	0.33	—
	$I - I$	0.33	0.2	0.33	0.5
	$D - M$	—	0.33	0.5	0.5
	$D - D$	—	0.33	0.25	—
	$D - I$	—	0.33	0.25	0.5

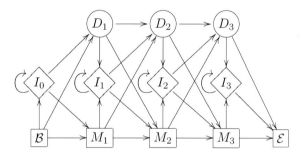

Figure 5.3. The state diagram of the profile HMM derived from the sequence alignment (5.3) with parameters estimated with pseudoconts defined by the Laplace rule.

subsequence $x_1 \ldots x_i$ with symbol x_i emitted by state M_j. Similarly, $V_j^I(i)$ is the score of the highest scoring alignment of the same subsequence with x_i emitted by I_j, and $V_j^D(i)$ is the score of the highest scoring alignment of subsequence $x_1, \ldots, x_i$ ending in state D_j. For a profile HMM with N match states and sequence x of length L, the algorithm goes through the following steps.

The initialization step: $V_0^M(0) = 0$, $V_0^I(0) = -\infty$, $V_0^D(0) = -\infty$.

The update equations for the log-odds scores are as follows:

$$V_j^M(i) = \log \frac{e_{M_j}(x_i)}{q_{x_i}} + \max \begin{cases} V_{j-1}^M(i-1) + \log a_{M_{j-1}M_j}, \\ V_{j-1}^I(i-1) + \log a_{I_{j-1}M_j}, \\ V_{j-1}^D(i-1) + \log a_{D_{j-1}M_j}; \end{cases}$$

$$V_j^I(i) = \log \frac{e_{I_j}(x_i)}{q_{x_i}} + \max \begin{cases} V_j^M(i-1) + \log a_{M_jI_j}, \\ V_j^I(i-1) + \log a_{I_jI_j}, \\ V_j^D(i-1) + \log a_{D_jI_j}; \end{cases}$$

$$V_j^D(i) = \max \begin{cases} V_{j-1}^M(i) + \log a_{M_{j-1}D_j}, \\ V_{j-1}^I(i) + \log a_{I_{j-1}D_j}, \\ V_{j-1}^D(i) + \log a_{D_{j-1}D_j}. \end{cases}$$

The termination step,

$$\mathbf{V} = \max\{V_N^M(L) + \log a_{M_N\varepsilon}; V_N^I(L) + \log a_{I_N\varepsilon}; V_N^D(L) + \log a_{D_N\varepsilon}\},$$

calculates the log-odds score $\mathbf{V}$ of the optimal path. Now we apply the Viterbi algorithm to find the alignment of the given profile HMM (Figure 5.3) and sequence *GCCAG*. The algorithm proceeds as follows (natural logarithms are used):

$$V_0^M(0) = 0,$$

$$V_1^D(0) = V_0^M(0) + \ln a_{M_0D_1} = 0 + \ln 0.09 = -2.408,$$

$$V_2^D(0) = V_1^D(0) + \ln a_{D_1D_2} = -2.408 + \ln 0.33 = -3.5166,$$

$$V_3^D(0) = V_2^D(0) + \ln a_{D_2D_3} = -3.5166 + \ln 0.25 = -4.903;$$

$$V_1^M(1) = \ln \frac{e_{M_1}(x_1)}{q_{x_1}} + V_0^M(0) + \ln a_{M_0M_1} = \ln \frac{0.58}{0.2} + \ln 0.82 = 0.866,$$

$$V_2^M(1) = \ln \frac{e_{M_2}(x_1)}{q_{x_1}} + V_1^D(0) + \ln a_{D_1M_2} = \ln \frac{0.18}{0.2} - 2.408 + \ln 0.33 = -3.622,$$

$$V_3^M(1) = \ln \frac{e_{M_3}(x_1)}{q_{x_1}} + V_2^D(0) + \ln a_{D_2M_3} = \ln \frac{0.5}{0.2} - 3.5166 + \ln 0.5 = -3.293;$$

Table 5.4. *The log-odds scores $V_j^\bullet(i)$ as determined by the Viterbi algorithm*

Entries on the best path are highlighted in bold.

	$x_1 = G$	$x_2 = C$	$x_3 = C$	$x_4 = A$	$x_5 = G$
M_1	**0.866**	−4.210	−5.095	−5.247	−5.291
M_2	−3.622	−0.530	−0.836	**−0.125**	−3.014
M_3	−3.293	−1.041	−0.252	−2.381	**0.569**
I_0	−2.185	−3.070	−3.956	−5.247	−6.132
I_1	−3.679	**0.473**	**−0.220**	−2.397	−4.169
I_2	−4.680	−2.012	−2.299	−3.321	−2.204
I_3	−5.373	−2.705	−2.993	−2.737	−2.897
D_1	−3.293	−4.179	−5.065	−6.355	−7.241
D_2	−0.849	−1.136	−1.829	−4.007	−5.779
D_3	−2.235	−2.523	−3.139	−2.427	−3.313

$$V_0^I(1) = \ln \frac{e_{I_0}(x_1)}{q_{x_1}} + V_0^M(0) + \ln a_{M_0 I_0} = \ln \frac{0.25}{0.2} + \ln 0.09 = -2.185,$$

$$V_1^I(1) = \ln \frac{e_{I_1}(x_1)}{q_{x_1}} + V_1^D(0) + \ln a_{D_1 I_1} = \ln \frac{0.17}{0.2} - 2.408 + \ln 0.33 = -3.679,$$

$$V_2^I(1) = \ln \frac{e_{I_2}(x_1)}{q_{x_1}} + V_2^D(0) + \ln a_{D_2 I_2} = \ln \frac{0.25}{0.2} - 3.5166 + \ln 0.25 = -4.680,$$

$$V_3^I(1) = \ln \frac{e_{I_3}(x_1)}{q_{x_1}} + V_3^D(0) + \ln a_{D_3 I_3} = \ln \frac{0.25}{0.2} - 4.903 + \ln 0.5 = -5.373;$$

$$V_1^D(1) = V_0^I(1) + \ln a_{I_0 D_1} = -2.185 + \ln 0.33 = -3.293,$$

$$V_2^D(1) = V_1^M(1) + \ln a_{M_1 D_2} = 0.866 + \ln 0.18 = -0.849,$$

$$V_3^D(1) = V_2^D(1) + \ln a_{D_2 D_3} = -0.849 + \ln 0.25 = -2.235.$$

The calculations continue until all values $V_j^M(i)$, $V_j^I(i)$, and $V_j^D(i)$ for $i = 1, \ldots, 5$, $j = 1, 2, 3$ are computed (Table 5.4). We find the score of the optimal alignment $V = V_3^M(5) = 0.569$. The traceback procedure starts from $V_3^M(5)$ and reveals the sequence of log-odds scores

$$V_3^M(5) \rightarrow V_2^M(4) \rightarrow V_1^I(3) \rightarrow V_1^I(2) \rightarrow V_1^M(1) \rightarrow V_0^M(0)$$

shown in bold in Table 5.4. The optimal path π associated with this sequence of the log-odds scores defines the optimal alignment of sequence *GCCAG* to the profile HMM. Since the choice of how to put the residues emitted from an insert state in the alignment constructed by the profile HMM is arbitrary (Durbin *et al.*, 1998, p. 151), path π also defines two alignments (with permutation of columns 2 and 3) of sequence *GCCAG* to the multiple alignment (5.3), the training set for the profile HMM:

$$
\begin{array}{ccccc@{\qquad}ccccc}
G & C & - & A & G & G & - & C & A & G \\
G & - & - & - & G & G & - & - & - & G \\
G & - & - & A & G & G & - & - & A & G \\
G & C & - & T & G & G & - & C & T & G \\
A & - & - & A & C & A & - & - & A & C \\
G & - & - & A & C & G & - & - & A & C \\
G & - & - & G & G & G & - & - & G & G \\
A & - & - & A & C & A & - & - & A & C \\
\mathbf{G} & \mathbf{C} & \mathbf{C} & \mathbf{A} & \mathbf{G} & \mathbf{G} & \mathbf{C} & \mathbf{C} & \mathbf{A} & \mathbf{G} \\
\end{array}
$$

Remark Note that the state diagram of the profile HMM derived from the multiple alignment (5.3) is quite different in the absence of pseudocounts (Figure 5.2). That profile HMM could accommodate sequences of lengths 2, 3, and 4, but not of greater length (*GCCAG* included). Even some short sequences of length 2, 3, 4 will not fit the model (Figure 5.2) due to zero values of emission probabilities of some nucleotides (See Table 5.2). The use of pseudocounts for the estimation of the HMM parameters in Equations (5.5) results in the HMM architecture capable of accommodating a wider set of sequences potentially related to the training sequence set. □

Problem 5.10 A multiple alignment of a set of DNA sequences is given as follows:

$$
\begin{array}{ccccccc}
C & C & - & - & A & T & C \\
G & C & A & A & A & G & C \\
G & T & A & T & A & T & C \\
G & C & - & - & C & T & G \\
C & C & - & - & A & T & C \\
\end{array}
\tag{5.6}
$$

Determine the parameters and the state diagram of a profile HMM for the multiple alignment (5.6). Take into account the sequence weights defined by the method suggested in Henikoff and Henikoff (1994). Consider an optional use of pseudocounts defined by Laplace's rule.

Solution First, the multiple alignment (5.6) should be reduced to the ungapped block as follows:

$$
\begin{array}{ccccc}
C & C & A & T & C \\
G & C & A & G & C \\
G & T & A & T & C \\
G & C & C & T & G \\
C & C & A & T & C
\end{array}
\tag{5.7}
$$

Next, the weights of the sequences in the rows of block (5.7) are derived as a linear combination of the column-specific weights. The weight w_{ji} for sequence i per column j is calculated by the formula $w_{ji} = 1/m_j k_{jx_i^j}$, where m_j is the number of letter types in column j, x_i^j is the letter at site (column) j of sequence i, and $k_{jx_i^j}$ is the number of letters of x_i^j in column j. Then, the first column weights w_{1i} are given by

$$w_{11} = 1/4, \quad w_{12} = 1/6, \quad w_{13} = 1/6, \quad w_{14} = 1/6, \quad w_{15} = 1/4.$$

Similarly, the column-specific weights for the other four columns are given by

$$w_{21} = 1/8, \quad w_{22} = 1/8, \quad w_{23} = 1/2, \quad w_{24} = 1/8, \quad w_{25} = 1/8;$$

$$w_{31} = 1/8, \quad w_{32} = 1/8, \quad w_{33} = 1/8, \quad w_{34} = 1/2, \quad w_{35} = 1/8;$$

$$w_{41} = 1/8, \quad w_{42} = 1/2, \quad w_{43} = 1/8, \quad w_{44} = 1/8, \quad w_{45} = 1/8;$$

$$w_{51} = 1/8, \quad w_{52} = 1/8, \quad w_{53} = 1/8, \quad w_{54} = 1/2, \quad w_{55} = 1/8.$$

The summation of the column-specific weights w_{ji} for each sequence i, and subsequent normalization, produce the sequence weights w_i:

$$w_1 = 3/20, \quad w_2 = 5/24, \quad w_3 = 5/24, \quad w_4 = 17/60, \quad w_5 = 3/20.$$

Now, to determine the number of match states of the profile HMM from the sequence alignment, we use the heuristic rule stated earlier: a column with a fraction of gap symbols below 0.5 is marked as a match state; otherwise, it is marked as an insert state. Thus, there are five match states $M_1, \ldots, M_5$ corresponding, respectively, to columns 1, 2, 5, 6, and 7 of the multiple alignment (5.6). The state diagram should also include the *begin* state $\mathcal{B}$, the *end* state $\mathcal{E}$, the insert states $I_0, I_1, \ldots, I_5$, and the delete (silent) states $D_1, D_2, \ldots, D_5$. To estimate the parameters of the profile HMM, we will use weighted emission and transition counts $E_k^w(x)$ and A_{kl}^w, respectively. For a given state k, the weighted emission (transition) count is equal to the sum of the weights of the training sequences where a particular symbol is emitted from state k (transition is observed). For example, for match state M_2, the weighted emission count of nucleotide C is the sum of the weights of sequences 1, 2, 4, and 5:

$$E_2^w(C) = w_1 + w_2 + w_4 + w_5 = \frac{3}{20} + \frac{5}{24} + \frac{17}{60} + \frac{3}{20} = \frac{19}{24}.$$

Table 5.5. *Counts of observed nucleotide emissions and transitions between hidden states modified by the use of the sequence weights*

		0	1	2	3	4	5
Emission counts for match states	A	–	0	0	43/60	0	0
	C	–	3/10	19/24	17/60	0	43/60
	G	–	7/10	0	0	5/24	17/60
	T	–	0	5/24	0	19/24	0
Emission counts for insert states	A	0	0	5/8	0	0	0
	C	0	0	0	0	0	0
	G	0	0	0	0	0	0
	T	0	0	5/24	0	0	0
Counts of transitions between hidden states	$M - M$	1	1	7/12	1	1	1
	$M - D$	0	0	0	0	0	–
	$M - I$	0	0	5/12	0	0	0
	$I - M$	0	0	5/12	0	0	0
	$I - D$	0	0	0	0	0	–
	$I - I$	0	0	5/12	0	0	0
	$D - M$	–	0	0	0	0	0
	$D - D$	–	0	0	0	0	–
	$D - I$	–	0	0	0	0	0

The weighted emission and transition counts are listed in Table 5.5. To find estimates of the parameters of the profile HMM, we could either use the modified counts shown in Table 5.5 and Equations (5.4), or, equivalently, the counts of emissions and transitions determined in the straightforward way (see Problem 5.7) and the generalized variants of Equations (5.4):

$$a_{kl} = \frac{A_{kl}^{w}}{\sum_{l'} A_{kl'}^{w}} = \frac{\sum_{i=1}^{N} \varepsilon_{kl}(i) w_i}{\sum_{l'} \sum_{i=1}^{N} \varepsilon_{kl'}(i) w_i},$$

(5.8)

$$e_k(x) = \frac{E_k^{w}(x)}{\sum_{x'} E_k^{w}(x')} = \frac{\sum_{i=1}^{N} \delta_{kx}(i) w_i}{\sum_{x'} \sum_{i=1}^{N} \delta_{kx'}(i) w_i}.$$

Here w_i, $i = 1, \ldots, N$, is the weight of the ith sequence; $\varepsilon_{kl}(i)$ is the number of transitions from hidden state k to hidden state l in training sequence i; $\delta_{kx}(i)$ is the number of emissions of symbol x from state k of sequence i. The estimates of the emission and transition probabilities calculated by Equations (5.8) are shown in Table 5.6.

The state diagram of this profile HMM (Figure 5.4) shows that all but one insert states and all the delete states are missing, since the transition probabilities to these

Table 5.6. *Parameters of the profile HMM estimated from the alignment of weighted sequences*

		0	1	2	3	4	5
Emission probabilities for match states	A	—	0	0	0.717	0	0
	C	—	0.3	0.792	0.293	0	0.717
	G	—	0.7	0	0	0.218	0.293
	T	—	0	0.218	0	0.792	0
Emission probabilities for insert states	A	0	0	0.75	0	0	0
	C	0	0	0	0	0	0
	G	0	0	0	0	0	0
	T	0	0	0.25	0	0	0
Transition probabilities between hidden states	$M - M$	1	1	0.583	1	1	1
	$M - D$	0	0	0	0	0	—
	$M - I$	0	0	0.417	0	0	0
	$I - M$	0	0	0.5	0	0	0
	$I - D$	0	0	0	0	0	—
	$I - I$	0	0	0.5	0	0	0
	$D - M$	—	0	0	0	0	0
	$D - D$	—	0	0	0	0	—
	$D - I$	—	0	0	0	0	0

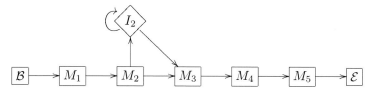

Figure 5.4. The state diagram of the profile HMM derived from the multiple alignment (5.6) with only the states and transitions permitted by the alignment data.

missing states are equal to zero. Note that this "reduced" HMM is limited in its ability to accommodate sequences, though not in terms of restriction on the sequence length (see the Remark to Problem 5.9) but in terms of their nucleotide composition. This is yet an other example of how, in the absence of pseudocounts, parameter estimation from a limited training set may lead to overfitting.

To implement the Laplace rule, one normalized pseudocount $1/5$ (in general, $1/N$ if N is the number of sequences) is added to all weighted counts (Table 5.5) except counts for those states from which emissions (or transitions) are not possible in the generic profile HMM architecture. The results are shown in Table 5.7. Finally, to estimate the profile HMM parameters, we can either plug the counts

Table 5.7. *Counts of observed nucleotide emissions and transitions between hidden states calculated from counts in Table 5.5 using Laplace's pseudocounts*

		0	1	2	3	4	5
Emission counts for	A	–	1/5	1/5	11/12	1/5	1/5
match states	C	–	1/2	119/120	29/60	1/5	11/12
	G	–	9/10	1/5	1/5	49/120	29/60
	T	–	1/5	49/120	1/5	119/120	1/5
Emission counts for	A	1/5	1/5	33/40	1/5	1/5	1/5
insert states	C	1/5	1/5	1/5	1/5	1/5	1/5
	G	1/5	1/5	1/5	1/5	1/5	1/5
	T	1/5	1/5	49/120	1/5	1/5	1/5
Counts of transitions	M − M	6/5	6/5	47/60	6/5	6/5	6/5
between hidden states	M − D	1/5	1/5	1/5	1/5	1/5	–
	M − I	1/5	1/5	37/60	1/5	1/5	1/5
	I − M	1/5	1/5	37/60	1/5	1/5	1/5
	I − D	1/5	1/5	1/5	1/5	1/5	–
	I − I	1/5	1/5	37/60	1/5	1/5	1/5
	D − M	–	1/5	1/5	1/5	1/5	1/5
	D − D	–	1/5	1/5	1/5	1/5	–
	D − I	–	1/5	1/5	1/5	1/5	1/5

shown in Table 5.5 into Equations (5.8) modified by the addition of the normalized pseudocounts:

$$a_{kl} = \frac{A_{kl}^{w} + 1/N}{\sum_{l'} A_{kl'}^{w} + m_k/N} = \frac{\sum_{i=1}^{N}(\varepsilon_{kl}(i) + 1/N)w_i}{\sum_{l'} \sum_{i=1}^{N}(\varepsilon_{kl'}(i) + 1/N)w_i},$$

$$e_k(x) = \frac{E_k^{w}(x) + 1/N}{\sum_{x'} E_k^{w}(x') + K/N} = \frac{\sum_{i=1}^{N}(\delta_{kx}(i) + 1/N)w_i}{\sum_{x'} \sum_{i=1}^{N}(\delta_{kx'}(i) + 1/N)w_i},$$

(5.9)

or directly use the counts shown in Table 5.7 to plug into Equations (5.8). In both cases, the result of computations of the emission and transition probabilities is the same and shown in Table 5.8.

We see from Table 5.8 that the addition of pseudocounts makes the probabilities of all generic transitions positive and establishes all possible (permitted) connections between hidden states. Now, the profile HMM shown in Figure 5.5 is capable to accommodate any nucleotide sequence. The higher alignment scores, however, will be given to those sequences which carry more affinity to the initial training set.

Table 5.8. *The final estimates of the parameters of the profile HMM*

		0	1	2	3	4	5
Emission probabilities for match states	A	–	0.111	0.111	0.509	0.111	0.112
	C	–	0.278	0.551	0.267	0.111	0.509
	G	–	0.5	0.111	0.112	0.227	0.267
	T	–	0.111	0.227	0.112	0.551	0.112
Emission probabilities for insert states	A	0.25	0.25	0.505	0.25	0.25	0.25
	C	0.25	0.25	0.122	0.25	0.25	0.25
	G	0.25	0.25	0.122	0.25	0.25	0.25
	T	0.25	0.25	0.25	0.25	0.25	0.25
Transition probabilities between hidden states	$M-M$	0.75	0.75	0.49	0.75	0.75	0.857
	$M-D$	0.125	0.125	0.125	0.125	0.125	–
	$M-I$	0.125	0.125	0.385	0.125	0.125	0.143
	$I-M$	0.33	0.33	0.43	0.33	0.33	0.5
	$I-D$	0.33	0.33	0.14	0.33	0.33	–
	$I-I$	0.33	0.33	0.43	0.33	0.33	0.5
	$D-M$	–	0.33	0.33	0.33	0.33	0.5
	$D-D$	–	0.33	0.33	0.33	0.33	–
	$D-I$	–	0.33	0.33	0.33	0.33	0.5

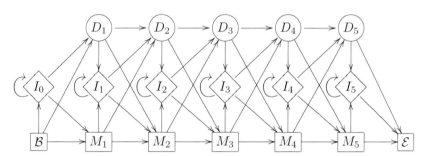

Figure 5.5. The state diagram of the profile HMM derived from multiple alignment (5.6) of weighted sequences with parameters estimated using the pseudocounts defined by Laplace's rule.

Remark From Equations (5.4) and (5.8) it is easy to see that the estimation of the parameters of the profile HMM without sequence weights is equivalent to the case when all sequence weights are equal to $1/N$, where N is the number of sequences in multiple alignment. This type of extrapolation also holds true for the weighted pseudocounts, as follows from a comparison of Equations (5.5) and (5.9). □

Profile HMMs for sequence families

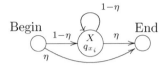

Figure 5.6. The full independence model R generating a single sequence.

5.2.1 Discrimination function and maximum discrimination weights

5.2.1.1 Theoretical introduction to Problem 5.11

The estimation of the profile HMM parameters could be biased if some sequence families are over-represented in the training set. Several methods designed to eliminate the bias by weighting training sequences have already been considered (Problems 5.6 and 5.10). Yet another weighting scheme was proposed by Eddy, Mitchison, and Durbin (1995). In this scheme the sequence weights $w_1, w_2, \ldots, w_N$, to be used in Equations (5.8), are defined as follows:

$$w_i = \lambda(1 - P(M(\Theta)|x_i)), \quad \sum_i w_i = 1. \tag{5.10}$$

Here $M(\Theta)$ is a probabilistic model with a set of parameters Θ. In the case when M is the profile HMM, Θ is a set of the emission probabilities $e_k(x)$ and transition probabilities a_{kl}. Model M is considered along with a background independence model R (the full probabilistic version of R is shown in Figure 5.6). The motivation behind the definition given in Equations (5.10) is as follows. Suppose that the set of parameters Θ maximizes the *discrimination function* D of model $M(\Theta)$:

$$D = \prod_{i=1}^{N} P(M(\Theta)|x_i) = \prod_{i=1}^{N} \frac{P(x_i|M(\Theta))P(M(\Theta))}{P(x_i|M(\Theta))P(M(\Theta)) + P(x_i|R)P(R)}.$$

The discrimination function D is the product of the posterior probabilities of model M given the sequences $x_1, \ldots, x_N$. The set of parameters Θ maximizing the discrimination function D can be found from an algorithm that uses Equations (5.8) and (5.10) for the iterative re-estimation of the parameters and the sequence weights (Eddy, Mitchison, and Durbin, 1995). Starting with the initial set of weights, parameters $\Theta = \{e_k(x), a_{kl}\}$ are determined by Equations (5.8). The values of the weights are then updated by Equations (5.10). These two steps are repeated until a convergence to a (possibly local) maximum of D is reached. At this point, Equations (5.10) for the sequence weights determine the *maximum discrimination weights*.

Re-estimation equations with pseudocounts could be defined by Equations (5.9) and (5.10), with the normalized simple counts $1/N$ in Equations (5.9) replaced by the normalized Dirichlet pseudocounts α_i/N (see also Problem 5.13).

Problem 5.11 (a) Show that estimates of the parameters which deliver the maximum of the discrimination function D satisfy Equations (5.8) with the sequence weights $w_1, \ldots, w_N$ defined by Equations (5.10).

(b) Prove that if the set of parameters Θ^* maximizes D then it also maximizes the weighted sum of the log-odds ratios L^w, where

$$L^w = \sum_{i=1}^{N} w_i \log \frac{P(x_i|M)}{P(x_i|R)}$$

with weights $w_i = \lambda(1 - P(M(\Theta^*)|x_i))$, $i = 1, \ldots, N$.

Solution (a) We consider two competing models: a profile HMM $M(\Theta)$ with state diagram and parameters derived from a given multiple alignment of sequences x_1, $x_2, \ldots, x_N$; and the independence model R.

Let K be the set of all hidden states (including silent ones) and let K_1 be the set of states emitting observed symbols (match states and insert states), $K_1 \subset K$. We want to find a set Θ of parameters of the profile HMM, $\Theta = (\theta_1, \theta_2, \ldots, \theta_L) = \{e_{k'}(x), a_{kl}|k' \in K_1, k, l \in K\}$, maximizing the discrimination function D (and its logarithm) under the following conditions:

$$\begin{aligned} \sum_l a_{kl} &= 1, & k \in K, \\ \sum_{x \in \mathcal{A}} e_k(x) &= 1, & k \in K_1. \end{aligned} \tag{5.11}$$

We assume that the ratio of priors $P(R)/P(M) = m$ does not depend on Θ, and we designate $\mu_i = P(x_i|R)$, $y_i = P(x_i|M)$, for $i = 1, \ldots, N$. Then we have

$$\log D = \sum_{i=1}^{N} \log P(M|x_i) = \sum_{i=1}^{N} \log \frac{P(x_i|M)P(M)}{P(x_i|M)P(M) + P(x_i|R)P(R)}$$

$$= \sum_{i=1}^{N} (\log y_i - \log(y_i + m\mu_i)).$$

Next, for the partial derivatives of $\log D$ we have the following expressions:

$$\frac{\partial \log D}{\partial \theta_j} = \sum_{i=1}^{N} \left(\frac{1}{y_i} \frac{\partial y_i}{\partial \theta_j} - \frac{1}{y_i + m\mu_i} \frac{\partial y_i}{\partial \theta_j} \right)$$

$$= m \sum_{i=1}^{N} \frac{\mu_i}{y_i(y_i + m\mu_i)} \frac{\partial y_i}{\partial \theta_j} = m \sum_{i=1}^{N} \frac{\mu_i}{y_i + m\mu_i} \frac{\partial \log y_i}{\partial \theta_j}. \tag{5.12}$$

The analytical expression for y_i in terms of the parameters of model M is as follows:

$$y_i = P(x_i|M) = \prod_{k \in K_1} \prod_{x \in \mathcal{A}} (e_k(x))^{\delta_{kx}(i)} \prod_{k,l \in K} (a_{kl})^{\varepsilon_{kl}(i)}.$$

Here $\varepsilon_{kl}(i)$ is the number of transitions from the hidden state k to the hidden state l in sequence x_i, and $\delta_{kx}(i)$ is the number of emissions of symbol x from state k in sequence x_i. Thus, the partial derivative $\partial \log y_i / \partial \theta_j$ becomes

$$\frac{\partial \log y_i}{\partial \theta_j} = \begin{cases} \dfrac{\delta_{kx}(i)}{e_k(x)}, & \text{if } \theta_j = e_k(x); \\[2mm] \dfrac{\varepsilon_{kl}(i)}{a_{kl}}, & \text{if } \theta_j = a_{kl}. \end{cases} \tag{5.13}$$

The conditional optimization problem for $\log D$ under the constraints given in Equations (5.11) can be solved with the use of the Lagrange multipliers $\lambda_j, j = 1, \ldots, L$. Then Equations (5.12) and (5.13) lead to the following system of equations:

$$\frac{\partial \log D}{\partial \theta_j} + \lambda_j = m \sum_{i=1}^{N} \frac{\mu_i}{y_i + m\mu_i} \frac{\delta_{kx}(i)}{e_k(x)} + \lambda_j = 0, \quad \text{if } \theta_j = e_k(x);$$

$$\frac{\partial \log D}{\partial \theta_j} + \lambda_j = m \sum_{i=1}^{N} \frac{\mu_i}{y_i + m\mu_i} \frac{\varepsilon_{kl}(i)}{a_{kl}} + \lambda_j = 0, \quad \text{if } \theta_j = a_{kl}. \tag{5.14}$$

Note that the equations for parameters θ_j, which are the values of probabilities from the same probability distribution, contain the same λ_j. From Equations (5.14) we find

$$e_k(x) = -\frac{m \sum_{i=1}^{N}(\mu_i/y_i + m\mu_i)\delta_{kx}(i)}{\lambda_j},$$

$$a_{kl} = -\frac{m \sum_{i=1}^{N}(\mu_i/y_i + m\mu_i)\varepsilon_{kl}(i)}{\lambda_j}. \tag{5.15}$$

Values of all λ_j, λ_j' can be determined from Equations (5.11) as follows:

$$\sum_{l'} a_{kl'} = -\frac{m}{\lambda_j'} \sum_{l'} \sum_{i=1}^{N} \frac{\mu_i}{y_i + m\mu_i} \varepsilon_{kl'}(i) = 1,$$

$$\sum_{x' \in A} e_k(x) = -\frac{m}{\lambda_j} \sum_{x' \in A} \sum_{i=1}^{N} \frac{\mu_i}{y_i + m\mu_i} \delta_{kx'}(i) = 1.$$

Substitution of values λ_j, λ_j' into Equations (5.15) and use of the definitions of μ_i and y_i lead to the equations we are looking for:

$$e_k(x) = \frac{\sum_{i=1}^{N} \delta_{kx}(i)(1 - P(M|x_i))}{\sum_{x'} \sum_{i=1}^{N} \delta_{kx'}(i)(1 - P(M|x_i))},$$

$$a_{kl} = \frac{\sum_{i=1}^{N} \varepsilon_{kl}(i)(1 - P(M|x_i))}{\sum_{l'} \sum_{i=1}^{N} \varepsilon_{kl'}(i)(1 - P(M|x_i))}. \tag{5.16}$$

(b) To solve the conditional maximization problem for the function L^w, where

$$L^w = \sum_{i=1}^{N} w_i \log \frac{P(x_i|M)}{P(x_i|R)} = \sum_{i=1}^{N} w_i \log \frac{y_i}{\mu_i},$$

with constraints as in Equations (5.11), we need the expressions for the partial derivatives of L^w. These are as follows:

$$
\begin{aligned}
\frac{\partial L^w}{\partial \theta_j} &= \sum_{i=1}^{N} w_i \frac{\mu_i}{y_i \mu_i} \frac{\partial y_i}{\partial \theta_j} = \sum_{i=1}^{N} w_i \frac{\partial \log y_i}{\partial \theta_j} \\
&= \begin{cases} \sum_{i=1}^{N} w_i \dfrac{\delta_{kx}(i)}{e_k(x)}, & \text{if } \theta_j = e_k(x); \\ \sum_{i=1}^{N} w_i \dfrac{\varepsilon_{kl}(i)}{a_{kl}}, & \text{if } \theta_j = a_{kl}. \end{cases}
\end{aligned}
\tag{5.17}
$$

From Equations (5.17) we obtain the system of equations with the Lagrange multipliers $\lambda_j, j = 1, \ldots, L$ as follows:

$$\frac{\partial L^w}{\partial \theta_j} + \lambda_j = \sum_{i=1}^{N} \frac{w_i \delta_{kx}(i)}{e_k(x)} + \lambda_j = 0, \qquad \text{if } \theta_j = e_k(x);$$

$$\frac{\partial \log D}{\partial \theta_j} + \lambda_j = \sum_{i=1}^{N} \frac{w_i \varepsilon_{kl}(i)}{a_{kl}} + \lambda_j = 0, \qquad \text{if } \theta_j = a_{kl}.$$

Then

$$e_k(x) = - \frac{\sum_{i=1}^{N} w_i \delta_{kx}(i)}{\lambda_j},$$

$$a_{kl} = - \frac{\sum_{i=1}^{N} w_i \varepsilon_{kl}(i)}{\lambda_j'}.$$

With values of λ_j determined from Equations (5.11):

$$\sum_{l'} a_{kl'} = -\frac{1}{\lambda_j} \sum_{l'} \sum_{i=1}^{N} w_i \varepsilon_{kl}(i) = 1,$$

$$\sum_{x \in A} e_k(x) = -\frac{1}{\lambda_j} \sum_{x \in A} \sum_{i=1}^{N} w_i \delta_{kx}(i) = 1,$$

we finally arrive at the equations for the parameters of the HMM which maximize the weighted sum of the log-odds ratios L^w:

$$e_k(x) = \frac{\sum_{i=1}^N \delta_{kx}(i) w_i}{\sum_{x'} \sum_{i=1}^N \delta_{kx'}(i) w_i},$$

$$a_{kl} = \frac{\sum_{i=1}^N \varepsilon_{kl}(i) w_i}{\sum_{l'} \sum_{i=1}^N \varepsilon_{kl'}(i) w_i}.$$

(5.18)

One can see that the set of parameters satisfying Equations (5.16) will also satisfy Equations (5.18) with weights w_i, $i = 1, \ldots, N$, defined by Equations (5.10). Therefore, the set of parameters Θ^* of the profile HMM which maximizes the discrimination function D will also maximize the weighted sum of the log-odds ratios L^w if $w_i = \lambda(1 - P(M(\Theta^*)|x_i))$, $i = 1, \ldots, N$. □

Problem 5.12 A basic step in the construction of a profile HMM from a multiple alignment of N protein sequences is an estimation of the emission probabilities for a match state associated with a given column $x = x_1, \ldots, x_N$ of the multiple alignment. Use the Dirichlet priors to find the estimates of emission probabilities as (a) the mean values of posterior distribution and (b) the maximum *a posteriori* probability (MAP) estimates.

Solution First, we introduce some notation and formulas that we need for both (a) and (b). We designate the emission probabilities from the hidden match state as $e(a_i) = p_i$, $i = 1, \ldots, K$, where K is the size of the alphabet. Next, we assume that the independence model M with probabilities p_i determines the probability distribution of emissions from the hidden state. Then, the likelihood of model M for the observed column $x = x_1, \ldots, x_N$ is given by

$$P(x|M) = \prod_{i=1}^K p_i^{n_i}.$$

Here n_i is the number of symbols a_i among $x_1, \ldots, x_N$. We assume that the prior probability for parameters p_i, $i = 1, \ldots, K$, of model M is given by the Dirichlet distribution with parameters $\alpha_1, \ldots, \alpha_K$, $\alpha_i > 0$:

$$P(M) = P_{\mathcal{D}}(p_1, \ldots, p_K) = Z^{-1} \prod_{i=1}^K p_i^{\alpha_i - 1},$$

where $Z = \prod_i \Gamma(\alpha_i)/\Gamma(\sum_i \alpha_i)$ is the normalization constant of the Dirichlet distribution. Therefore, a posterior probability $P(M|x)$ of model M becomes

$$P(M|x) = \frac{P(x|M)P(M)}{P(x)} = (ZP(x))^{-1} \prod_{i=1}^{K} p_i^{n_i+\alpha_i-1}. \qquad (5.19)$$

(a) Equation (5.19) implies that the posterior distribution of parameters of M is the Dirichlet distribution with parameters $n_1 + \alpha_1, \ldots, n_K + \alpha_K$. Thus, the mean value estimates of the emission probabilities are given by

$$\hat{e}(a_i) = \hat{p}_i = \frac{n_i + \alpha_i}{\sum_{j=1}^{K}(n_j + \alpha_j)} = \frac{n_i + \alpha_i}{N + \sum_{j=1}^{K} \alpha_j},$$

$i = 1, 2, \ldots, K$. It is easy to see that the mean value estimates coincide with the maximum likelihood estimates with the use of pseudocounts α_i, $i = 1, \ldots, K$. For example, if $\alpha_i = 1$, $i = 1, \ldots, K$, $\hat{p}_i$ will coincide with the maximum likelihood estimates derived using Laplace's pseudocounts.

(b) To determine the MAP estimates of the emission probabilities, we have to find the point of maximum of the posterior distribution $P(M|x)$. In logarithmic form (base 2) we have:

$$\log_2 P(M|x) = \sum_{i=1}^{K}(n_i + \alpha_i - 1) \log_2 p_i - \log_2 Z - \log_2 P(x),$$

where only the first sum of K terms depends on the parameters p_i. For two discrete distributions P and Q, the relative entropy

$$H(P|Q) = \sum_i P(y_i) \log_2 \frac{P(y_i)}{Q(y_i)},$$

is always non-negative and attains its minimum zero value if and only if $P = Q$. Therefore,

$$\sum_i P(y_i) \log_2 Q(y_i) \le \sum_i P(y_i) \log_2 P(y_i),$$

with equality if and only if $P = Q$. If distribution Q is given by the set of probabilities $p_1, \ldots, p_K$, and distribution P is given by the set of probabilities

$$\frac{n_1 + \alpha_1 - 1}{N + \sum_i \alpha_i - K}, \ldots, \frac{n_K + \alpha_K - 1}{N + \sum_i \alpha_i - K},$$

the sum $\sum_i P(y_i) \log Q(y_i)$ along with $\log_2 P(M|x)$ becomes maximal when $p_j = (n_j + \alpha_j - 1)/(N + \sum_i \alpha_i - K)$ for $j = 1, \ldots, K$. Then, the MAP estimates are defined by the following formulas:

$$e^*(a_i) = p_i^* = \frac{n_i + \alpha_i - 1}{N + \sum_j \alpha_j - K},$$

$i = 1, 2, \ldots, K$. If $\alpha_i = 1$, $i = 1, \ldots, K$, the MAP estimates coincide with the maximum likelihood estimates. $\qquad\square$

Problem 5.13 An alignment of sequence fragments of cytochrome c from *Rickettsia conorii, Rickettsia prowazekii, Bradyrhizobium japonicum,* and *Agrobacterium tumefaciens* (in top down order) is as follows:

```
NIPELMKTANADNGREIAKK
NIQELMKTANANHGREIAKK
PIEKLLQTASVEKGAAAAKK
PIAKLLASADAAKGEAVFKK
```

A position-specific scoring matrix (PSSM) is derived from this alignment block. Determine the parameters of the (3×3) section of the PSSM containing the scores for amino acids A, R, and N in columns 4, 5, and 6. Use pseudocounts defined by the substitution matrix mixtures with parameter $A_i = 5R_i$, where R_i is the number of different residue types observed in the ith column. The BLOSUM50 substitution matrix is given in Table 2.1. For simplicity, the background frequencies of the twenty amino acids are assumed to be equal to $1/20$.

Solution The score S^i_α of amino acid α in column i is defined by the following formula:

$$S^i_\alpha = \log \frac{e_i(\alpha)}{q_\alpha},$$

where q_α is the background frequency of amino acid α and $e_i(\alpha)$ is the emission probability of amino acid α in position i. The estimates of the emission probabilities are given by the substitution matrix mixtures (Henikoff and Henikoff, 1996):

$$e_i(\alpha) = \frac{C_{i\alpha} + g_{i\alpha}}{\sum_{\alpha'}(C_{i\alpha'} + g_{i\alpha'})}, \tag{5.20}$$

where $C_{i\alpha}$ is the number of occurrences of amino acid α in column i, and the pseudocount $g_{i\alpha}$ is defined by

$$g_{i\alpha} = A_i q_\alpha \sum_\beta \frac{C_{i\beta}}{N} e^{S(\alpha, \beta)}.$$

Here N is the number of aligned sequences and $S(\alpha, \beta)$ is the log-odds score of α-to-β substitution taken from the BLOSUM50 substitution matrix. For the given block of sequences, $N = 4$, $A_4 = 5 \times 2 = 10$. Then, for instance, the pseudocount for alanine in column 4 is calculated as follows:

$$g_{4A} = \frac{10}{20}\left(\frac{2}{4}e^{S(A,E)} + \frac{2}{4}e^{S(A,K)}\right) = \frac{1}{4}(e^{-1} + e^{-1}) = 0.184.$$

To determine the estimates of the emission probabilities $e_4(A)$, $e_4(R)$, and $e_4(N)$, we have to use similar formulas to obtain pseudocounts for all twenty amino acids (calculations are not shown). Then we find by using Equation (5.20):

$$e_4(A) = 0.00084, \quad e_4(R) = 0.02394, \quad e_4(N) = 0.00227.$$

Subsequently, the scores for amino acids A, R, and N in column 4 of the PSSM are as follows (natural logarithms are used):

$$S_A^4 = \log \frac{e_4(A)}{q_A} = \log \frac{0.00084}{0.05} = -4.086,$$

$$S_R^4 = \log \frac{e_4(R)}{q_R} = \log \frac{0.02394}{0.05} = -0.736,$$

$$S_N^4 = \log \frac{e_4(N)}{q_N} = \log \frac{0.00227}{0.05} = -3.092.$$

Similar computations yield the score parameters of the PSSM in columns 5 and 6:

$$S_A^5 = \log \frac{e_5(A)}{q_A} = \log \frac{0.00068}{0.05} = -4.298,$$

$$S_A^6 = \log \frac{e_6(A)}{q_A} = \log \frac{0.00038}{0.05} = -4.880,$$

$$S_R^5 = \log \frac{e_5(R)}{q_R} = \log \frac{0.00025}{0.05} = -5.298,$$

$$S_R^6 = \log \frac{e_6(R)}{q_R} = \log \frac{0.00014}{0.05} = -5.878,$$

$$S_N^5 = \log \frac{e_5(N)}{q_N} = \log \frac{0.00009}{0.05} = -6.320;$$

$$S_N^6 = \log \frac{e_6(N)}{q_N} = \log \frac{0.00012}{0.05} = -6.032.$$

The substitution scores for amino acids A, R, and N rounded to the nearest integer are shown in the PSSM submatrix of order 3×3 as follows:

	4	5	6
A	−4	−4	−5
R	−1	−5	−6
N	−3	−6	−6

The negative scores could be expected as amino acids A, R, and N possess different physico-chemical properties in comparison with amino acids present in columns 4, 5, and 6 of the alignment, except for the case of R, which, is positively charged, similar to K (situated in column 4).   □

Problem 5.14 Parameters of the positional independence model of the *E. coli* ribosomal binding site (RBS) were estimated by the experimental positional nucleotide frequencies shown in Table 5.9.

Table 5.9

	1	2	3	4	5	6
T	0.16	0.05	0.01	0.07	0.12	0.24
C	0.08	0.04	0.01	0.03	0.05	0.11
A	0.68	0.11	0.02	0.86	0.16	0.41
G	0.08	0.80	0.96	0.04	0.67	0.24

Determine the parameters of the "logo" graph introduced by Schneider, Stormo, and Ehrenfeucht (1986). Use the bit units for the entropy and information content values.

Solution The rules for drawing the "logo" graph suggest that at position j the sum of heights of four letters is equal to the information content I_j of the position, and height H_α^j of a letter α is proportional to its probability in this position:

$$H_\alpha^j = p_\alpha^j I_j,$$
$$I_j = H_{max} - H_j.$$

Thus, the information content in position j is the difference of the maximum entropy, corresponding to the uniform discrete distribution and the entropy for the nucleotide frequencies of the RBS model at position j. First, we calculate H_{max} and H_1:

$$H_{max} = -\sum_\alpha \frac{1}{4} \log_2 \frac{1}{4} = 2,$$

$$H_1 = -(p_T^1 \log_2 p_T^1 + p_C^1 \log_2 p_C^1 + p_A^1 \log_2 p_A^1 + p_G^1 \log_2 p_G^1)$$
$$= -(-0.423 - 0.292 - 0.378 - 0.292) = 1.385.$$

Similarly, we determine the entropy $H_j, j = 2, \ldots, 6$:

$$H_2 = 1.01, \quad H_3 = 0.303, \quad H_4 = 0.794, \quad H_5 = 1.393, \quad H_6 = 1.866.$$

Now we can find the information content I_j and the heights of letters $H_\alpha^j, j = 1, \ldots, 6$. Here we give the value of I_j and the maximal height of the letter in position

Figure 5.7. The "logo" graph for the positional independence model of the nucleotide composition of the *E. coli* ribosomal binding site (Table 5.9). The vertical
bar on the left corresponds to the maximal possible information content *I*, 2 bits,
observed in a position where one of the four symbols has probability one.

$j, j = 1, \ldots, 6$. Other values are omitted (however, they are used to draw the "logo"
graph):

$I_1 = H_{\max} - H_1 = 2 - 1.385 = 0.615,$ $I_4 = H_{\max} - H_4 = 2 - 0.794 = 1.21,$
$H_A^1 = p_A^1 I_1 = 0.68 \ldots 0.615 = 0.4182;$ $H_A^4 = p_A^4 I_4 = 0.86 \ldots 1.21 = 1.04;$
$I_2 = H_{\max} - H_2 = 2 - 1.01 = 0.99,$ $I_5 = H_{\max} - H_5 = 2 - 1.393 = 0.607,$
$H_G^2 = p_G^2 I_2 = 0.8 \ldots 0.99 = 0.792;$ $H_G^5 = p_G^5 I_5 = 0.67 \ldots 0.607 = 0.4067;$
$I_3 = H_{\max} - H_3 = 2 - 0.303 = 1.697,$ $I_6 = H_{\max} - H_6 = 2 - 1.866 = 0.134,$
$H_G^3 = p_G^3 I_3 = 0.96 \ldots 1.697 = 1.63;$ $H_A^6 = p_A^6 I_6 = 0.41 \ldots 0.134 = 0.055.$

The "logo" graph is shown in Figure 5.7. $\square$

Problem 5.15 Determine the value $H(P_1 || Q)$ of the relative entropy (Kullback–
Leibler distance) for the RBS positional independence model described in
Problem 5.14 (model P_1) assuming that the background model (model Q) for
the *E. coli* non-coding DNA sequence is the independence model with parameters $q_T = 0.26$, $q_C = 0.23$, $q_A = 0.26$, $q_G = 0.25$. Determine the length of
the *E. coli* from protein-coding DNA that would be sufficient to make the same
Kullback–Leibler distance with the same background distribution Q. As a model
of the *E. coli* protein-coding region use the independence model (model P_2) with
parameters $p_T = 0.23$, $p_C = 0.25$, $p_A = 0.25$, $p_G = 0.27$. Note that model P_2 is
not the same as model Q.

Solution The relative entropy of two distributions P_1 and Q is defined as the sum of the entropies over six positions (note that distribution P_1 is position-dependent and Q is position-independent)

$$H(P_1||Q) = \sum_{j=1}^{6} \sum_{\alpha=A,T,C,G} p_\alpha^j \log_2 \frac{p_\alpha^j}{q_\alpha}.$$

As an approximation of probabilities p_α^j, $j = 1,\ldots,6$, we take their maximum likelihood estimates given in the positional frequency matrix (Table 5.9) to find

$$H(P_1||Q) = \left(0.16 \log_2 \frac{0.16}{0.26} + 0.08 \log_2 \frac{0.08}{0.23} + 0.68 \log_2 \frac{0.68}{0.26} + 0.08 \log_2 \frac{0.08}{0.23}\right)$$

$$+ \left(0.05 \log_2 \frac{0.05}{0.26} + 0.04 \log_2 \frac{0.04}{0.23} + 0.11 \log_2 \frac{0.11}{0.26} + 0.8 \log_2 \frac{0.8}{0.23}\right)$$

$$+ \left(0.01 \log_2 \frac{0.01}{0.26} + 0.01 \log_2 \frac{0.01}{0.23} + 0.02 \log_2 \frac{0.02}{0.26} + 0.96 \log_2 \frac{0.96}{0.23}\right)$$

$$+ \left(0.07 \log_2 \frac{0.07}{0.26} + 0.03 \log_2 \frac{0.03}{0.23} + 0.86 \log_2 \frac{0.86}{0.26} + 0.04 \log_2 \frac{0.04}{0.23}\right)$$

$$+ \left(0.12 \log_2 \frac{0.12}{0.26} + 0.05 \log_2 \frac{0.05}{0.23} + 0.16 \log_2 \frac{0.16}{0.26} + 0.67 \log_2 \frac{0.67}{0.23}\right)$$

$$+ \left(0.24 \log_2 \frac{0.24}{0.26} + 0.11 \log_2 \frac{0.11}{0.23} + 0.41 \log_2 \frac{0.41}{0.26} + 0.24 \log_2 \frac{0.24}{0.23}\right)$$

$$= 5.125.$$

Since both distributions P_2 and Q are position-independent, the relative entropy $H(P_2||Q)$ for n positions is just n times the relative entropy at one position:

$$H(P_2||Q) = n \sum_{\alpha=A,T,C,G} p_\alpha \log \frac{p_\alpha}{q_\alpha}$$

$$= n \left(0.23 \log_2 \frac{0.23}{0.26} + 0.25 \log_2 \frac{0.25}{0.23} + 0.25 \log_2 \frac{0.25}{0.26} + 0.27 \log_2 \frac{0.27}{0.23}\right)$$

$$= 0.0066n.$$

Hence, the relative entropy (Kullback–Leibler distance) between distributions P_2 and Q for six positions will be equal to the Kullback–Leibler distance between distributions P_1 and Q for n sites when

$$H(P_2||Q) = 0.0066n = 5.125 = H(P_1||Q).$$

So, we need $n = 777$ positions, which is much larger than six positions of the RBS, since the per position value of $H(P_2||Q)$ is much less than the average per position value of $H(P_1||Q)$. $\square$

.

5.3 Further reading

The PROSITE database has been an important resource of data on evolutionary conserved protein domains (Bairoch, Bucher, and Hofmann, 1997; Hulo *et al.*, 2004). PROSITE is a derivative of the SWISS-PROT protein sequence data bank (Bairoch and Apweiler, 1999) respected for the high standards of functional annotation. Sequence patterns discovered in the conserved protein domains have been modeled in PROSITE with the aid of regular expressions.

PROSITE gave rise to Pfam, a collection of profile HMMs and multiple alignments of protein domain sequences (Sonnhammer *et al.*, 1998). A new version of Pfam (Bateman *et al.*, 2002) utilizes structural data to improve domain-based annotation of proteins.

Further development of the BLAST program, Position-Specific Iterated BLAST (Altschul *et al.*, 1997) became a frequently used tool for detecting homologous protein sequences. The use of PSI-BLAST requires special care to avoid false positive matches (Altschul and Koonin, 1998; Schäffer *et al.*, 2001).

The development and improvement of methods, including the profile HMM based ones, for detecting remote homology is an issue that continue to attract much attention (Karplus, Barrett, and Hughey, 1998; Park *et al.*, 1997, 1998). Lyngsø, Pedersen, and Nielsen (1999) proposed several similarity measures between profile HMMs as well as an algorithm that computes the measures. This approach allows the comparison of sequence families by comparing profiles of the families instead of individual members of the families.

The profile HMM technique has not only been used in modeling of protein sequence domains. A novel profile HMM based algorithm for separating pseudogenes from functional genes was introduced by Coin and Durbin (2004) and was implemented in the computer program PSILC.

Rychlewski *et al.* (2000) used sequence profiles for the prediction of the protein three-dimensional structure. Kelley, MacCallum, and Sternberg (2000) proposed a new method for detecting protein homologues, the 3D-PSSM (three-dimensional position-specific scoring matrix), which combines multiple-sequence alignments produced by PSI-BLAST with structure based profiles.

The literature on profile HMM based algorithms and software was reviewed by Eddy (1998).

6

Multiple sequence alignment methods

The theory described in Chapter 5 of *BSA* suggests that constructing the multiple alignment of several biological sequences should be a part of the algorithm of the profile HMM training. Such an iterative expectation maximization method is supposed to estimate parameters of the profile HMM from unaligned sequences by means of the construction of the multiple alignment in parallel with the HMM parameter estimation. The resulting alignment can be evoked at the last step of the algorithm via an optimal alignment of each individual sequence to the just built profile HMM. Nevertheless, since this impressive theoretical design meets many practical difficulties, discussed in great detail in *BSA*, it has not yet been implemented in its pure form as an efficient tool for multiple sequence alignment.

One of the major difficulties on the road to a universal and efficient multiple sequence alignment algorithm is as follows. Establishing a gold standard for a multiple sequence alignment that would help to distinguish a good alignment from a better one is difficult. Since both sequence and structure are evolving and the ancestral sequences and structures can be reconstructed only by theoretical means, it is impossible to verify experimentally either alignments or phylogenies. Nevertheless, a formal assignment of the alignment score immediately leads to the notion of the best alignment for a given set of sequences; however, the implications of a so defined optimal alignment have to be taken cautiously. There are several biologically motivated options for the score assignment. For instance, the sum-of-pairs score is computationally convenient and frequently used, but it has well known theoretical drawbacks (Durbin *et al.* (1998), p. 141).

Besides the important topic of alignment score definition, we are interested in a rigorous algorithm that would deliver the multiple alignment with the best score. Such an algorithm, the standard dynamic programming (DP) algorithm, is available in theory, but it is impractical, as it needs time comparable with the age of the Earth to align only ten protein sequences of average length (Problem 6.1). Therefore, the prohibitive computational complexity calls for creative algorithmic solutions that would be able to find an approximate optimal alignment in a reasonable time.

The burden of significant computational efforts associated with solving multiple sequence alignment problems limits the number of theoretical problems that could be offered for an exercise without much computer programming. This consideration might explain why only one problem was offered in this *BSA* chapter.

Additional problems provide "toy examples" to practice using the intricate yet rigorous DP algorithm invented by Carrillo and Lipman (1988) and with the approximate method of multiple alignment, the progressive alignment algorithm.

6.1 Original problem

Problem 6.1 Assume that we have a number of sequences that are fifty residues long, and that a pairwise comparison of two such sequences takes one second of CPU time on our computer. An alignment of four sequences ($N = 4$) takes $(2L)^{N-2} = 10^{2N-4} = 10^4$ seconds (a few hours). If we had unlimited memory and we were willing to wait for the answer until just before the Sun burns out in five billion years, how many sequences could our computer align?

Solution The time T needed to align N sequences of fifty residues long is $T = (2L)^{N-2} = 10^{2N-4}$ seconds. For $T = 5$ billion years, we have (in seconds): $10^{2N-4} = 5 \times 10^9 \times 365 \times 24 \times 60 \times 60 = 1.57 \times 10^{17}$, thus $N = 10.6$. Therefore, the full dynamic programming algorithm implemented on that computer will have enough time to construct the multiple alignment of ten sequences of fifty residues long, before the Sun burns out in five billion years. □

6.2 Additional problems and theory

In this section, we show how to apply the Carrillo–Lipman algorithm for finding the optimal (minimum cost) multiple alignment for three nucleotide sequences (Problem 6.2). We also give an illustrative example of the application of the heuristic progressive multiple alignment algorithm by Feng and Doolittle for the construction of the multiple alignment of four short protein sequences (Problem 6.3). The construction of an ungapped multiple alignment in parallel with the derivation of the probabilistic model can be done using the maximum likelihood approach. We show that such a maximum likelihood alignment will possess minimum entropy (Problem 6.4).

6.2.1 Carrillo–Lipman multiple alignment algorithm

Theoretical introduction to Problem 6.2

An overall score of a multiple alignment is often defined as the sum of scores of all the pairwise alignments unambiguously defined by the multiple alignment. This type of score is called the sum of pairs (SP) score. The maximum SP-score (or the minimum SP-cost) of an alignment of N sequences of lengths $k_1, \ldots, k_N$ can be determined by a DP algorithm on an N-dimensional lattice L of size $k_1 \times \cdots \times k_N$. The optimal multiple alignment (the optimal path through lattice L) is then determined by the traceback procedure. The running time of such an algorithm is proportional to the product $k_1 \times \cdots \times k_N$, which is practically prohibitive for more than ten protein sequences of average length (Problem 6.1).

Carrillo and Lipman (1988) proposed an algorithm that reduces the lattice L to a smaller N-dimensional region which still contains the optimal path with the minimal cost. Thus, the DP algorithm will need much less computational time to find the optimal solution.

The algorithm works as follows. For N sequences $x_1, \ldots, x_N$ we define the N-dimensional lattice L with the *original corner* corresponding to the first symbols and the *end corner* corresponding to the last symbols of all sequences. Lattice L consists of $k_1 \times \cdots \times k_N$ N-dimensional cubes. A permitted path γ is defined as a continuous segmented line through lattice L connecting the original corner and the end corner, with each segment of the line joining vertices of a single cube in L. The end point of each segment must be closer to the end corner than the start point of the segment. It can be shown that any given path γ corresponds to a unique multiple alignment of $x_1, \ldots, x_N$ and vice versa. A projection $p_{ij}(\gamma)$ of γ into plane (i,j) associated with sequences x_i, x_j defines the pairwise alignment of x_i and x_j. The SP-cost $S(\gamma)$ of path γ is defined as the sum of costs of path projections $p_{ij}(\gamma)$:

$$S(\gamma) = \sum_{1 \leq i < j \leq N} S(p_{ij}(\gamma)).$$

Obviously, this definition implies that the scoring of a pairwise alignment induced by a multiple alignment ignores the sites where the multiple alignment contains two aligned gap characters in both sequences x_i and x_j. This happens because the corresponding segment of path γ is orthogonal to plane (i,j) and, therefore, does not appear in the path projection $p_{ij}(\gamma)$.

Let S_{kl}^* designate the cost of the optimal pairwise alignment of x_k and x_l. It was proved (Carrillo and Lipman, 1988) that if γ^e is a permitted path through L, then for any i,j, $1 \leq i < j \leq N$, value U_{ij} defined by

$$U_{ij} = S(\gamma^e) - \sum_{\substack{1 \leq k < l \leq N \\ (k,l) \neq (i,j)}} S_{kl}^* \tag{6.1}$$

gives the upper bound for the cost of $p_{ij}(\gamma^*)$, the projection of the optimal path γ^* into plane (i,j).

Let X be a set of paths γ' through lattice L satisfying the following condition: $S(p_{ij}(\gamma')) \leq U_{ij}$, $1 \leq i < j \leq N$. We need a constructive definition of X to use in the optimization algorithm. For a two-dimensional lattice $L(i,j)$ we consider a set y_{ij} of cells in $L(i,j)$, which are traversed by two-dimensional paths, each path with cost smaller than U_{ij}, and then single out a set of cubes Y_{ij} of the N-dimensional lattice L satisfying the condition $p_{ij}(Y_{ij}) = y_{ij}$. If a set Y is defined as

$$Y = \bigcap_{1 \leq i < j \leq N} Y_{ij},$$

then $X \subset Y$ due to Equation (6.1). Therefore, it is sufficient to apply the N-dimensional DP algorithm just to subregion Y of lattice L to find the optimal path among all paths in L. Note that the "closer" the selected path γ^e to the yet unknown optimal path γ^*, the smaller the volume of region Y and the less computationally expensive the task of dynamic programming.

Problem 6.2 Construct the optimal multiple alignment of sequences $x_1 = CTCACA$, $x_2 = CAC$, and $x_3 = GTAC$ using the Carrillo–Lipman algorithm with the nucleotide substitution costs

$$S(A,A) = S(C,C) = S(G,G) = S(T,T) = 0,$$
$$S(C,T) = S(T,C) = S(A,G) = S(G,A) = 1,$$
$$S(A,T) = S(T,A) = S(A,C) = S(C,A)$$
$$= S(G,C) = S(C,G) = S(G,T) = S(T,G) = 2,$$

the gap-open cost $d = 3$, and the gap-extension cost $e = 2$.

Solution We define a path γ^e as the path through L corresponding to the following multiple alignment:

C	T	C	A	C	A
C	–	–	A	C	–
G	–	T	A	C	–

The cost of γ^e is given by

$$S(\gamma^e) = \sum_{1 \leq i < j \leq N} S(p_{ij}(\gamma^e))$$

$$= S\begin{pmatrix} C & T & C & A & C & A \\ C & - & - & A & C & - \end{pmatrix} + S\begin{pmatrix} C & T & C & A & C & A \\ G & - & T & A & C & - \end{pmatrix} + S\begin{pmatrix} C & - & A & C \\ G & T & A & C \end{pmatrix}$$

$$= S(C,C) + d + e + S(A,A) + S(C,C) + d + S(G,C) + d + S(C,T)$$

$$+ S(A,A) + S(C,C) + d + S(C,G) + d + S(A,A) + S(C,C) = 22.$$

Optimal (global) pairwise alignment of sequences x_i and x_j, $i,j = 1,2,3$, along with the minimal cost S^*_{ij} can be found by the DP algorithm. For a given pair (i,j) we consider three types of alignments of sequences x_i and x_j up to elements $x_i(k), x_j(l)$: an alignment with match $x_i(k)$ to $x_j(l)$ (with the minimal cost $S^M(k,l)$); an alignment with $x_i(k)$ aligned to a gap (with the minimal cost $S^I(k,l)$); and an alignment with $x_j(l)$ aligned to a gap (with the minimal cost $S^J(k,l)$).

The recurrence equations are as follows:

$$S^M(k,l) = S(x_i(k), x_j(l)) + \min \begin{cases} S^M(k-1,l-1), \\ S^I(k-1,l-1), \\ S^J(k-1,l-1); \end{cases} \tag{6.2}$$

$$S^I(k,l) = \min \begin{cases} S^M(k-1,l) + d, \\ S^I(k-1,l) + e; \end{cases} \tag{6.3}$$

$$S^J(k,l) = \min \begin{cases} S^M(k,l-1) + d, \\ S^J(k,l-1) + e. \end{cases} \tag{6.4}$$

For $x_1 = CTCACA$ and $x_2 = CAC$, the matrix of costs $S^\bullet(k,l)$ is determined by Equations (6.2)–(6.4) and is shown in Figure 6.1. The entry with minimal value in the right bottom cell of Figure 6.1 ($k = 6$, $l = 3$) defines the minimum cost $S^*_{12} = 8$. The traceback through the DP matrix recovers two paths with the same cost. These paths correspond to the two best alignments of sequences x_1 and x_2 with cost $S^*_{12} = 8$:

$$
\begin{array}{llcccccc}
(1) & C & T & C & A & C & A \\
    & C & - & - & A & C & - \\
\end{array}
\qquad
\begin{array}{llcccccc}
(2) & C & T & C & A & C & A \\
    & - & - & - & C & A & C & - \\
\end{array}
$$

Similarly, for sequences $x_1 = CTCACA$ and $x_3 = GTAC$, Equations (6.2)–(6.4) produce the optimal alignment with cost $S^*_{13} = 8$:

$$
\begin{array}{llcccccc}
(3) & C & T & C & A & C & A \\
    & G & T & - & A & C & - \\
\end{array}
$$

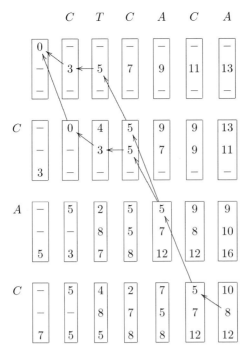

Figure 6.1. The matrix of costs $S^\bullet(k,l)$ for x_1 and x_2 as computed by the DP algorithm in Equations (6.2)–(6.4). Entries $S^M(k,l)$, $S^I(k,l)$, and $S^J(k,l)$ of each cell (k,l) are listed in top down order. Two minimal cost paths through the matrix are shown by arrows.

Finally, for sequences x_2 and x_3, the DP algorithm yields the best alignment with cost $S_{23}^* = 4$:

$$(4) \quad \begin{matrix} - & C & A & C \\ G & T & A & C \end{matrix}$$

Now we can find, by applying Equation (6.1), the upper bounds U_{ij} for the costs of projections of the yet unknown optimal multiple alignment γ^* into planes (i,j):

$$U_{12} = S(\gamma^e) - (S_{13}^* + S_{23}^*) = 22 - 12 = 10,$$

$$U_{13} = S(\gamma^e) - (S_{12}^* + S_{23}^*) = 22 - 12 = 10,$$

$$U_{23} = S(\gamma^e) - (S_{12}^* + S_{13}^*) = 22 - 16 = 6.$$

The next goal is to find sets y_{ij}, $i < j$. For each pair of sequences x_i, x_j, we apply the DP algorithm in the backward direction. As a result, for each cell (k,l) of the two-dimensional lattice $L(i,j)$ we know both the optimal path from the top corner down to this cell (from the initial forward run of the algorithm) and the optimal path

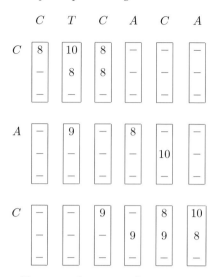

Figure 6.2. The costs $S_{total}^M(k,l)$, $S_{total}^I(k,l)$, $S_{total}^J(k,l)$ of optimal paths traversing through cells (k,l) of the lattice $L(1,2)$ are shown inside (k,l) cells if their values satisfy $S^\bullet(k,l) \leq U_{12} = 10$. Costs $S_{total}^\bullet(k,l)$ greater than 10 are omitted.

from the bottom corner up to this cell (from the backward run of the algorithm). Note that the latter path coincides with the optimal path from the cell to the bottom corner of lattice $L(i,j)$. Two optimal cost values are associated with these two paths, which meet in cell (k,l). Therefore, for each cell (k,l) in lattice $L(i,j)$, we determine the optimal path traversing through this cell as a concatenation of these two optimal paths. The cost $S_{total}^\bullet(k,l)$ of the whole path is the sum of costs of its parts. If $S_{total}^\bullet(k,l) \leq U_{ij}$, then cell (k,l) belongs to set y_{ij}. The sets y_{12}, y_{13}, y_{23} are shown in Figures 6.2, 6.3, and 6.4 as sets of cells with non-empty entries $S_{total}^\bullet(k,l)$ on lattices $L(1,2)$, $L(1,3)$, $L(2,3)$, respectively.

Now, from Figure 6.2 we derive a set of pairwise alignments of sequences x_1 and x_2 corresponding to the possible projections p_{12} of the optimal path γ^*: the above optimal alignment (3) with cost 8; alignment (5) with cost 9; and alignment (6), (7), and (8) with cost 10:

$$
\begin{array}{llllllll}
(5) & C & T & C & A & C & A \\
    & C & A & C & - & - & -
\end{array}
$$

$$
\begin{array}{llllllll}
(6) & C & T & C & A & C & A & \qquad (7) & C & T & C & A & C & A \\
    & - & C & - & A & C & - & \qquad   & - & - & C & A & - & C
\end{array}
$$

$$
\begin{array}{llllllll}
(8) & C & T & C & A & C & A \\
    & C & - & - & A & - & C
\end{array}
$$

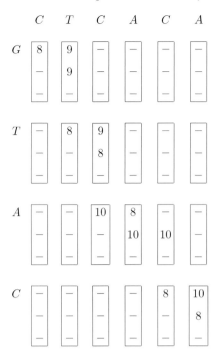

Figure 6.3. The costs (as in Figure 6.2) of optimal paths traversing through cells (k, l) of the lattice $L(1, 3)$.

Similarly, Figure 6.3 allows us to select pairwise alignments of sequences x_1 and x_3 corresponding to the possible projections p_{13} of the best path γ^*: the above optimal alignment (3) with cost 8; alignments (9) and (10) with cost 9,

$$
\begin{array}{lcccccc}
(9) & C & T & C & A & C & A \\
   & - & G & T & A & C & - \\
\end{array}
\qquad
\begin{array}{lcccccc}
(10) & C & T & C & A & C & A \\
    & G & - & T & A & C & - \\
\end{array}
$$

and alignments (11) and (12) with cost 10,

$$
\begin{array}{lcccccc}
(11) & C & T & C & A & C & A \\
    & G & T & A & - & C & - \\
\end{array}
\qquad
\begin{array}{lcccccc}
(12) & C & T & C & A & C & A \\
    & G & T & - & A & - & C \\
\end{array}
$$

Finally, from Figure 6.4 we derive two pairwise alignments of sequences x_2 and x_3 corresponding to possible projections p_{23} of the optimal path γ^*: the above alignment (4) with cost 4 and alignment (13) with cost 5,

$$
\begin{array}{lcccc}
(13) & G & - & A & C \\
    & C & T & A & C \\
\end{array}
$$

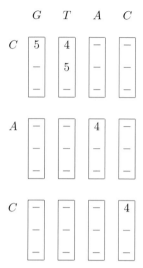

Figure 6.4. The costs (as in Figure 6.2) of optimal paths traversing through cells (k, l) of the lattice $L(2, 3)$.

It can be shown that the projection alignments (1)–(13) define a set X of seven three-dimensional paths $\gamma_{\bullet}$ through the lattice L:

(i) path γ_1 and multiple alignment A_1 with projections (1), (3), and (13), and cost $S(\gamma_1) = S(1) + S(3) + S(13) = 21$:

$$
\begin{array}{llllllll}
(A_1) & C & T & C & A & C & A \\
      & C & - & - & A & C & - \\
      & G & T & - & A & C & -
\end{array}
$$

(ii) path γ_2 and multiple alignment A_2 with projections (2), (9), and (4), and cost $S(\gamma_2) = S(2) + S(9) + S(4) = 21$:

$$
\begin{array}{llllllll}
(A_2) & C & T & C & A & C & A \\
      & - & - & C & A & C & - \\
      & - & G & T & A & C & -
\end{array}
$$

(iii) path γ_3 and multiple alignment A_3 with projections (6), (3), and (4), and cost $S(\gamma_3) = S(6) + S(3) + S(4) = 22$:

$$
\begin{array}{llllllll}
(A_3) & C & T & C & A & C & A \\
      & - & C & - & A & C & - \\
      & G & T & - & A & C & -
\end{array}
$$

(iv) path γ_4 and multiple alignment A_4 with projections (6), (9), and (13), and cost $S(\gamma_4) = 24$:

$$
\begin{array}{llllllll}
(A_4) & C & T & C & A & C & A \\
      & - & C & - & A & C & - \\
      & - & G & T & A & C & -
\end{array}
$$

(v) path $\gamma_5 = \gamma^e$ and multiple alignment A_5 with projections (1), (10), and (13), and cost $S(\gamma_5) = S(\gamma^e) = 22$:

$$
\begin{array}{cccccccc}
(A_5) & C & T & C & A & C & A \\
& C & - & - & A & C & - \\
& G & - & T & A & C & -
\end{array}
$$

(vi) path γ_6 and multiple alignment A_6 with projections (2), (10), and (4), and cost $S(\gamma_6) = S(2) + S(10) + S(4) = 21$:

$$
\begin{array}{cccccccc}
(A_6) & C & T & C & A & C & A \\
& - & - & C & A & C & - \\
& G & - & T & A & C & -
\end{array}
$$

(vii) path γ_7 and multiple alignment A_7 with projections (8), (12), and (13), and cost $S(\gamma_7) = S(8) + S(12) + S(13) = 25$:

$$
\begin{array}{cccccccc}
(A_7) & C & T & C & A & C & A \\
& C & - & - & A & - & C \\
& G & T & - & A & - & C
\end{array}
$$

Three paths γ_1, γ_2, and γ_6 have cost 21, the minimal cost among paths from X; hence, A_1, A_2, and A_6 are the optimal multiple alignments of sequences x_1, x_2, x_3. ☐

6.2.2 Progressive alignments: the Feng–Doolittle algorithm

Theoretical introduction to Problem 6.3

One of the frequently used multiple alignment methods is the *progressive alignment*, the principal idea of which is to use a succession of pairwise alignments to align pairs of sequences, a sequence to a group, and pairs of sequence groups.

Here we remind ourselves of the major steps of the pioneering progressive alignment algorithm by Feng and Doolittle (1987). We assume that we are dealing with N protein sequences $x_1, \ldots, x_N$.

(1) The distance D_{ij} between sequences x_i and x_j, $i, j = 1, \ldots, N$, is defined using the score S_{ij} of the optimal global pairwise alignment of x_i and x_j determined by the Needleman–Wunsch algorithm:

$$
D_{ij} = -\ln \frac{S_{ij} - S_{rand}}{S_{max} - S_{rand}}. \tag{6.5}
$$

Here $S_{max} = (S_{ii} + S_{jj})/2$ is the average of the scores of either sequence aligned to itself and S_{rand} is the expected score for aligning two random sequences of the

172   *Multiple sequence alignment methods*

same length and residue composition,

$$S_{rand} = (1/L) \sum_{a \in i} \sum_{b \in j} S(a,b) N_i(a) N_j(b) - N(g)d. \qquad (6.6)$$

In Equation (6.6) $N_i(a)$ ($N_j(b)$) is the number of times that residue a (b) appears in sequence x_i (x_j); $S(a,b)$ is a substitution score of amino acids a and b; $N(g)$ is the number of gaps in the optimal alignment of sequences x_i and x_j; d is the gap penalty; L is the overall length of the alignment of sequences x_i and x_j. A matrix of $N(N-1)/2$ distances between all pairs of N sequences is calculated.

(2) A *guide tree* is built from the matrix of distances D_{ij} using the clustering algorithm by Fitch and Margoliash (1967).

(3) The multiple alignment is constructed in the order suggested by the guide tree, starting with the pairwise alignment of the two closest sequences. A sequence addition to an already built alignment is guided by the highest scoring pairwise alignment of the "new" sequence to a sequence from the aligned group. In the course of algorithm implementation the guide tree may require the alignment of two already aligned groups. Then the group-to-group alignment is guided by the highest scoring alignment between a pair of sequences – one from the first group and one from the second group. Note that the order of steps in alignment construction and the resulting multiple alignment itself does not depend on the edge lengths of the guide tree; only the labeled history associated with the tree matters (see Chapter 7).

A more recent and currently widely used multiple alignment algorithm, CLUSTALW (Thompson, Higgins, and Gibson, 1994b) uses the progressive alignment scheme with the pairwise distances D_{ij} as suggested by Kimura (1983):

$$D_{ij} = -\ln(1 - d_{ij} - d_{ij}^2/5).$$

Here d_{ij} is the number of mismatches in the optimal alignment of sequences x_i and x_j divided by the alignment length, not counting positions with gaps. The guide tree in CLUSTALW is built from the distance matrix $D = (D_{ij})$ by the neighbor-joining algorithm (Saitou and Nei, 1987). To construct sequence-to-group or group-to-group alignment, CLUSTALW first computes the group profile, the scoring scheme for which includes position-specific gap penalties as well as the weights of sequences in the group.

Problem 6.3 Four fragments of proteins from the I-immunoglobulin superfamily, x_1, x_2, x_3 and x_4, are shown below (in the top down order):

```
ILDMDVVEGSAARFDCKVEGYPDPEVMWFKDDNPVKESRHFQIDYDEEGN
RDPVKTHEGWGVMLPCNPPAHYPGLSYRWLLNEFPNFIPTDGRHFVSQTT
ISDTEADIGSNLRWGCAAAGKPRPMVRWLRNGEPLASQNRVEVLA
RRLIPAARGGEISILCQPRAAPKATILWSKGTEILGNSTRVTVTSD
```

Use the Feng–Doolittle method to construct the multiple alignment of these sequences. Consider the following sets of parameters:

(a) PAM250 (Table 6.1) and linear gap penalty $d = 8$;
(b) BLOSUM62 (Table 6.6) and linear gap penalty $d = 6$.

Both selections (a) and (b) were suggested by Feng and Doolittle (1996).

Solution (a) For each pair of sequences x_i, x_j, $i, j = 1, 2, 3, 4$, the Needleman–Wunsch algorithm produces the optimal alignment A_{ij} with score S_{ij} shown below.
Alignment A_{12} with score $S_{12} = 31$:

```
ILDMDVVEGSAARFDCKVEG-YPDPEVMWFKDDNPVKESRHFQIDYDEEGN
RDPVKTHEGWGVMLPCNPPAHYPGLSYRWLLNEFPNFIPTD-GRHFVSQTT
```

Alignment A_{13} with score $S_{13} = 44$:

```
ILDMDVVEGSAARFDCKVEGYPDPEVMWFKDDNPVKESRHFQIDYDEEGN
ISDTEADIGSNLRWGCAAAGKPRPMVRWLRNGEPL-ASQN-RV--EVLA-
```

Alignment A_{14} with score $S_{14} = 13$:

```
ILDMDVVEGSAARFDCKVEGYPDPEVMWFKDDNPVKESRHFQIDYDEEGN
RRLIPAARGGEISILCQPRAAPKATILWSKGTE-ILGNST-RV--TVTSD
```

Alignment A_{23} with score $S_{23} = 15$:

```
RDPVKTHEGWGVMLPCNPPAHYPGLSYRWLLNEFPNFIPTDGRHFVSQTT
ISDTEADIGSNLRWGCAAAGKPRPMV-RWLRNGEP--LASQNR--VEVLA
```

Alignment A_{24} with score $S_{24} = 16$:

```
RDPVKTHEGWGVMLPCNPPAHYPGLSYRWLLNEFPNFIPTDGRHFVSQTT
RRLIPAARGGEISILCQPRAA-PKATILW-SKG-TEILGNSTRVTVT-SD
```

Alignment A_{34} with score $S_{34} = 45$:

```
ISDTEADIGSNLRWGCAAAGKPRPMVRWLRNGEPLASQNRVEVLA-
RRLIPAARGGEISILCQPRAAPKATILWSKGTEILGNSTRVTVTSD
```

The four sequence self-alignments yield the following scores: $S_{11} = 262$, $S_{22} = 287$, $S_{33} = 222$, $S_{44} = 215$. Next, Equations (6.6) and (6.5) give the values of S_{rand}

Table 6.1. *The 250 PAM log-odds amino acid substitution matrix with elements* $s(i,j) = 3\log_2 M_{ji}/f_j$ *(in 1/3 bit units)*

M_{ji} is the element of the 250 PAM mutation probability matrix and f_j is a frequency of amino acid j (see Section 2.2.1 on the derivation of the PAM substitution matrices).

	A	R	N	D	C	Q	E	G	H	I	L	K	M	F	P	S	T	W	Y	V
A	2	-2	0	0	-2	0	0	1	-1	-1	-2	-1	-1	-3	1	1	1	-6	-3	0
R	-2	6	0	-1	-4	1	-1	-3	2	-2	-3	3	0	-4	0	0	-1	2	-4	-2
N	0	0	2	2	-4	1	1	0	2	-2	-3	1	-2	-3	0	1	0	-4	-2	-2
D	0	-1	2	4	-5	2	3	1	1	-2	-4	0	-3	-6	-1	0	0	-7	-4	-2
C	-2	-4	-4	-5	12	-5	-5	-3	-3	-2	-6	-5	-5	-4	-3	0	-2	-8	0	-2
Q	0	1	1	2	-5	4	2	-1	3	-2	-2	1	-1	-5	0	-1	-1	-5	-4	-2
E	0	-1	1	3	-5	2	4	0	1	-2	-3	0	-2	-5	-1	0	0	-7	-4	-2
G	1	-3	0	1	-3	-1	0	5	-2	-3	-4	-2	-3	-5	-1	1	0	-7	-5	-1
H	-1	2	2	1	-3	3	1	-2	6	-2	-2	0	-2	-2	0	-1	-1	-3	0	-2
I	-1	-2	-2	-2	-2	-2	-2	-3	-2	5	2	-2	2	1	-2	-1	0	-5	-1	4
L	-2	-3	-3	-4	-6	-2	-3	-4	-2	2	6	-3	4	2	-3	-3	-2	-2	-1	2
K	-1	3	1	0	-5	1	0	-2	0	-2	-3	5	0	-5	-1	0	0	-3	-4	-2
M	-1	0	-2	-3	-5	-1	-2	-3	-2	2	4	0	6	0	-2	-2	-1	-4	-2	2
F	-3	-4	-3	-6	-4	-5	-5	-5	-2	1	2	-5	0	9	-5	-3	-3	0	7	-1
P	1	0	0	-1	-3	0	-1	0	0	-2	-3	-1	-2	-5	6	1	0	-6	-5	-1
S	1	0	1	0	0	-1	0	1	-1	-1	-3	0	-2	-3	1	2	1	-2	-3	-1
T	1	-1	0	0	-2	-1	0	0	-1	0	-2	0	-1	-3	0	1	3	-5	-3	0
W	-6	2	-4	-7	-8	-5	-7	-7	-3	-5	-2	-3	-4	0	-6	-2	-5	17	0	-6
Y	-3	-4	-2	-4	0	-4	-4	-5	0	-1	-1	-4	-2	7	-5	-3	-3	0	10	-2
V	0	-2	-2	-2	-2	-2	-2	-1	-2	4	2	-2	2	-1	-1	-1	0	-6	-2	4

Table 6.2. *The BLOSUM62 log-odds amino acid substitution matrix with elements given in 1/2 bit units*

	A	R	N	D	C	Q	E	G	H	I	L	K	M	F	P	S	T	W	Y	V
A	4	-1	-2	-2	0	-1	-1	0	-2	-1	-1	-1	-1	-2	-1	1	0	-3	-2	0
R	-1	5	0	-2	-3	1	0	-2	0	-3	-2	2	-1	-3	-2	-1	-1	-3	-2	-3
N	-2	0	6	1	-3	0	0	0	1	-3	-3	0	-2	-3	-2	1	0	-4	-2	-3
D	-2	-2	1	6	-3	0	2	-1	-1	-3	-4	-1	-3	-3	-1	0	-1	-4	-3	-3
C	0	-3	-3	-3	9	-3	-4	-3	-3	-1	-1	-3	-1	-2	-3	-1	-1	-2	-2	-1
Q	-1	1	0	0	-3	5	2	-2	0	-3	-2	1	0	-3	-1	0	-1	-2	-1	-2
E	-1	0	0	2	-4	2	5	-2	0	-3	-3	1	-2	-3	-1	0	-1	-3	-2	-2
G	0	-2	0	-1	-3	-2	-2	6	-2	-4	-4	-2	-3	-3	-2	0	-2	-2	-3	-3
H	-2	0	1	-1	-3	0	0	-2	8	-3	-3	-1	-2	-1	-2	-1	-2	-2	2	-3
I	-1	-3	-3	-3	-1	-3	-3	-4	-3	4	2	-3	1	0	-3	-2	-1	-3	-1	3
L	-1	-2	-3	-4	-1	-2	-3	-4	-3	2	4	-2	2	0	-3	-2	-1	-2	-1	1
K	-1	2	0	-1	-3	1	1	-2	-1	-3	-2	5	-1	-3	-1	0	-1	-3	-2	-2
M	-1	-1	-2	-3	-1	0	-2	-3	-2	1	2	-1	5	0	-2	-1	-1	-1	-1	1
F	-2	-3	-3	-3	-2	-3	-3	-3	-1	0	0	-3	0	6	-4	-2	-2	1	3	-1
P	-1	-2	-2	-1	-3	-1	-1	-2	-2	-3	-3	-1	-2	-4	7	-1	-1	-4	-3	-2
S	1	-1	1	0	-1	0	0	0	-1	-2	-2	0	-1	-2	-1	4	1	-3	-2	-2
T	0	-1	0	-1	-1	-1	-1	-2	-2	-1	-1	-1	-1	-2	-1	1	5	-2	-2	0
W	-3	-3	-4	-4	-2	-2	-3	-2	-2	-3	-2	-3	-1	1	-4	-3	-2	11	2	-3
Y	-2	-2	-2	-3	-2	-1	-2	-3	2	-1	-1	-2	-1	3	-3	-2	-2	2	7	-1
V	0	-3	-3	-3	-1	-2	-2	-3	-3	3	1	-2	1	-1	-2	-2	0	-3	-1	4

and pairwise distances D_{ij}:

$$
\begin{aligned}
S_{rand}(1,2) &= -66.94, & D_{12} &= 1.25; \\
S_{rand}(1,3) &= -80.28, & D_{13} &= 0.95; \\
S_{rand}(1,4) &= -70.48, & D_{14} &= 1.31; \\
S_{rand}(2,3) &= -82.86, & D_{23} &= 1.24; \\
S_{rand}(2,4) &= -72.52, & D_{24} &= 1.30; \\
S_{rand}(3,4) &= -37.85, & D_{34} &= 1.13.
\end{aligned}
$$

The Fitch–Margoliash algorithm of tree building works as follows. At each step it selects and joins a pair of closest nodes from the set of available current nodes. The effective distance between two branch nodes k and l is defined as an average of distances D_{ij} between all possible pairs i and j, where i is a descendant leaf of node k and j is a descendant leaf of node l. The node joining step is repeated until the last two nodes are joined into the root node.

Here we start with the leaf nodes 1, 2, 3, and 4 for sequences x_1, x_2, x_3, and x_4, respectively. The closest leaves are leaves 1 and 3 separated by distance $D_{13} = 0.95$. Their joining creates node 5, and for a new set of nodes 2, 4, and 5 the effective distances are as follows: $D_{25} = (D_{21} + D_{23})/2 = 1.245$ and $D_{45} = (D_{41} + D_{43})/2 = 1.22$. Distance $D_{24} = 1.30$ does not change. Next, we join nodes 4 and 5 and create a new node 6. The last step joins remaining nodes 2 and 6 and produces the following guide tree:

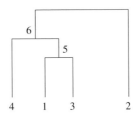

Now we should add the sequences to the growing multiple alignment in the same order as the leaves are joined in the guide tree. Alignment A_{13} of sequences x_1 and x_3 is taken from the list of optimal alignments and serves as the initial seed. To align sequence x_4 to A_{13}, we will use as a guidance alignment A_{34}, rather than A_{14}, since $S_{34} > S_{14}$. Gaps in x_4 should be inserted in correspondence with A_{34}. This operation leads to the alignment of x_1, x_3, x_4 (in top down order):

```
ILDMDVVEGSAARFDCKVEGYPDPEVMWFKDDNPVKESRHFQIDYDEEGN
ISDTEADIGSNLRWGCAAAGKPRPMVRWLRNGEPL-ASQN-RV--EVLA-
RRLIPAARGGEISILCQPRAAPKATILWSKGTEIL-GNST-RV--TVTSD
```

Finally, sequence x_2 is added to the group above as determined by A_{12}, the highest scoring alignment among A_{i2}, 1, 3, 4. Therefore, the multiple alignment of four sequences (in the initial order) is given by

```
ILDMDVVEGSAARFDCKVEG-YPDPEVMWFKDDNPVKESRHFQIDYDEEGN
RDPVKTHEGWGVMLPCNPPAHYPGLSYRWLLNEFPNFIPTD-GRHFVSQTT
ISDTEADIGSNLRWGCAAAG-KPRPMVRWLRNGEPL-ASQN-RV--EVLA-
RRLIPAARGGEISILCQPRA-APKATILWSKGTEIL-GNST-RV--TVTSD
```

(b) For another set of the alignment parameters, the amino acid substitution matrix BLOSUM62 and the linear gap penalty $d = 6$, the same procedure takes the following steps. As initial pairwise alignments we have the following:
Alignment A_{12} with score $S_{12} = 4$:

```
ILDMDVVEGSAARFDCKVEG-YPDPEVMWFKDDNP--V-KESRHFQIDYDEEGN
RDPVKTHEGWGVMLPCNPPAHYPGLSYRWLLNEFPNFIPTDGRHF-V--SQT-T
```

Alignment A_{13} with score $S_{13} = 37$:

```
ILDMDVVEGSAARFDCKVEGYPDPEVMWFKDDNPVKESRHFQIDYDEEGN
ISDTEADIGSNLRWGCAAAGKPRPMVRWLRNGEPL-ASQN-RVEV--LA-
```

Alignment A_{14} with score $S_{14} = -4$:

```
ILDMDVVEGSAARFDCKVEGYPDPEVMWFKDDNPVKESRHFQIDYDEEGN
RRLIPAARGGEISILCQPRAAPKATILWSKGTEILGNSTRVTVTSD----
```

Alignment A_{23} with score $S_{23} = 3$:

```
RDPVKTHEGWGVMLPCNPPAHYPGLSYRWLLNEFPNFIPTDGRHFVSQTT
ISDTEADIGSNLRWGC-AAAGKPRPMVRWLRNGEP--LASQNR--VEVLA
```

Alignment A_{24} with score $S_{24} = 9$:

```
RDPVKTHEGWGVMLPCNPPAHYPGLSYRWLLNEFPNFIPTDGRHFVSQTT
RRLIPAARGGEISILCQPRA-APKATILW--SKGTEILGNSTRVTVT-SD
```

Alignment A_{34} with score $S_{34} = 24$:

```
ISDTEADIGSNLRWGCAAAGKPRPMVRWLRNGEPLASQNRVEVLA-
RRLIPAARGGEISILCQPRAAPKATILWSKGTEILGNSTRVTVTSD
```

The self-alignment scores are $S_{11} = 277$, $S_{22} = 294$, $S_{33} = 238$, $S_{44} = 232$. Then we use Equations (6.6) and (6.5) to calculate S_{rand} and D_{ij} for each pair of sequences:

$$S_{rand}(1,2) = -101.65, \quad D_{12} = 1.30;$$

$$S_{rand}(1,3) = -78.64, \quad D_{13} = 1.07;$$

$$S_{rand}(1,4) = -75.04, \quad D_{14} = 1.53;$$

$$S_{rand}(2,3) = -81.38, \quad D_{23} = 1.42;$$

$$S_{rand}(2,4) = -75.12, \quad D_{24} = 1.39;$$

$$S_{rand}(3,4) = -45.83, \quad D_{34} = 1.39.$$

To construct a guide tree, first we join the closest leaves 1 and 3 ($D_{13} = 1.07$) at node 5 and calculate new pairwise distances: $D_{25} = (D_{21} + D_{23})/2 = 1.36$ and $D_{45} = (D_{41} + D_{43})/2 = 1.46$ ($D_{24} = 1.39$ does not change). At the last step, joining remaining nodes 4 and 6 completes the construction of the guide tree as follows:

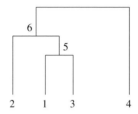

The labeled history of this tree differs from the labeled history of the guide tree built in part (a). Finally, we align the sequences in the order defined by the guide tree: we add sequence x_2 to alignment A_{13} (as directed by A_{12}); then sequence x_4 is aligned to the group of three sequences (as directed by A_{34}). The resulting multiple alignment is as follows:

```
ILDMDVVEGSAARFDCKVEG-YPDPEVMWFKDDNP--V-KESRHFQIDYDEEGN
RDPVKTHEGWGVMLPCNPPAHYPGLSYRWLLNEFPNFIPTDGRHF-V--SQT-T
ISDTEADIGSNLRWGCAAAG-KPRPMVRWLRNGEP--L--ASQN-RVEV--LA-
RRLIPAARGGEISILCQPRA-APKATILWSKGTEI--L--GNST-RVTV--TSD
```

Remark Comparison of (a) and (b) is instructive as it shows not only that a lower gap penalty tolerates extra gaps in the case (b), but also that a change in scoring matrix can alter the labeled history of the guide tree and bring changes into the resulting multiple alignment as a whole. ☐

6.2.3 *Gibbs sampling algorithm for local multiple alignment*

The Gibbs sampling multiple alignment algorithm introduced by Lawrence *et al.* (1993) identifies a pattern (ungapped motif of length w) contained in each of the protein sequences $S_1, S_2, \ldots, S_N$. It is assumed that the occurrences of symbols within the motif are described by a positional independence model with parameters $q_{i\alpha}$ (the probability of occurrence of symbol α in site i of the motif, $i = 1, \ldots, w$). The background independence model with parameters p_α describes the symbol occurrences outside the motif. The initial step of the algorithm is the assignment of the motif start positions $a_1, a_2, \ldots, a_N$ in each of sequences $S_1, S_2, \ldots, S_N$, respectively. The sequences are aligned against positions $a_1, a_2, \ldots, a_N$, therefore, the motifs are aligned as one ungapped block. Then the algorithm proceeds as follows:

(1) The estimation step. One of the sequences, say $S_j = (x_1, \ldots, x_m)$, chosen either at random or in a specified order, is removed from the set. For the remaining $N-1$ sequences, parameters $q_{i\alpha}$ and p_α are re-estimated by the maximum likelihood method from the counts of symbol α in the corresponding sites within or outside the motif, respectively. Pseudocounts b_α could be used for small N.

(2) The sampling step. A new starting position a_j of the motif in sequence S_j is selected with probability $P(a_j)$ equal to the normalized value of the odds ratio R_{a_j}:

$$R_{a_j} = \frac{q_{1x_{a_j}} q_{2x_{a_j+1}} \times \cdots \times q_{wx_{a_j+w-1}}}{p_{x_{a_j}} p_{x_{a_j+1}} \times \cdots \times p_{x_{a_j+w-1}}},$$

$$P(a_j) = \frac{R_{a_j}}{\sum_{a_j'} R_{a_j'}}.$$

Thus, the set of starting positions $a_1, \ldots, a_N$ of the motif is redefined, and the algorithm returns to step (1).

As the likelihood $L = \prod_{j=1}^{N} R_{a_j}$ is expected to increase at each step of the iterative procedure, the algorithm converges to the set of starting positions $a_1^*, a_2^*, \ldots, a_N^*$ of the motif, which maximizes the likelihood L, although this could be a local maximum. Simultaneously, the alignment finds the ungapped optimal local alignment associated with numbers $a_1^*, a_2^*, \ldots, a_N^*$, which mark the starting positions of the motif in sequences $S_1, S_2, \ldots, S_N$.

The algorithm can be modified to deal with multiple motifs within sequences and to determine the optimal motif width w in the process of iterations. To avoid being locked into a non-optimal local maximum, the authors suggest that shifts in all current starting positions $a_1, \ldots, a_N$ should be introduced after a fixed number of iterations. These shifts are made at random with the probability of selecting a new set of starting positions proportional to the likelihood L of alignment associated with the set. The running time of the algorithm is $M \times N \times L \times w$, where L is the average length of aligned sequences and M is the number of iterations (for a single motif the average number of iterations before convergence is expected to be at most one hundred).

A detailed description of the Gibbs sampling can be found in Liu (2001).

Problem 6.4 Show that an ungapped local multiple alignment with maximum likelihood possesses the minimum value of the sum of the positional entropies. Assume that a positional independence model is chosen as the statistical model for a motif associated with the ungapped local multiple alignment.

Solution The likelihood of the sequence alignment block of size w (without gaps) is defined as a product of the likelihoods of sequences x_k, $k = 1, 2, \ldots, N$, participating in the alignment:

$$L = \prod_{k=1}^{N} P(x_k) = \prod_{k=1}^{N} p^1_{x^1_k} p^2_{x^2_k} p^3_{x^3_k} \cdots p^w_{x^w_k}.$$

Here p^i_x, the parameter of the positional independence model, is the probability of occurrence of symbol x from the alphabet of size M in position i. We transform the expression for L from a "row-based" product to a "column-based" product as follows:

$$L = p^1_{x^1_1} p^2_{x^2_1} \cdots p^w_{x^w_1} p^1_{x^1_2} p^2_{x^2_2} \cdots p^w_{x^w_2} p^1_{x^1_N} p^2_{x^2_N} \cdots p^w_{x^w_N}$$

$$= \prod_{i=1}^{N} p^1_{x^1_i} \prod_{i=1}^{N} p^2_{x^2_i} \cdots \prod_{i=1}^{N} p^w_{x^w_i}$$

$$= \prod_{m=1}^{M} (p^1_m)^{n^1_m} \prod_{m=1}^{M} (p^2_m)^{n^2_m} \cdots \prod_{m=1}^{M} (p^w_m)^{n^w_m}.$$

Here n^j_m stands for the number of occurrences of symbol m in position j, and $n^j_m \simeq p^j_m N$ for sufficiently large N. Then

$$\log_2 L \simeq N \sum_{m=1}^{M} p^1_m \log_2 p^1_m + N \sum_{m=1}^{M} p^2_m \log_2 p^2_m + \cdots + N \sum_{m=1}^{M} p^w_m \log_2 p^w_m$$

$$= -N \sum_{i=1}^{w} H_i. \tag{6.7}$$

Equation (6.7) implies that $-\log_2 L/N = H$, where H is the sum of positional entropy values H_i. Therefore, for the ungapped alignment with the maximum likelihood L the entropy value H is minimal.

Remark The problem statement was proved for sequences described by the inhomogeneous Markov chain by Borodovsky and Peresetsky (1994). □

6.3 Further reading

Several new multiple sequence alignment algorithms have been introduced since the mid 1990s. Brocchieri and Karlin (1998) proposed the symmetric-iterative method for multiple alignment of protein sequences. This method combines a statistical motif-finding procedure with a local dynamic programming. In the DIALIGN algorithm developed by Morgenstern *et al.* (1998), a multiple alignment is based on a set of ungapped pairwise local alignments (associated with matrix diagonals). The score of the whole alignment is defined as the sum of weights of the participating diagonals (no gap penalties are employed). The new version of this program, DIALIGN 2 (Morgenstern, 1999), selecting fewer but longer diagonals, works considerably faster than DIALIGN. The T-Coffee algorithm of progressive type was proposed by Notredame, Higgins, and Heringa (2000). This algorithm uses a position-specific scoring scheme to align the sequences instead of the substitution matrix used in traditional progressive alignment methods. Löytynoja and Milinkovitch (2003) introduced an algorithm combining a progressive alignment algorithm with a profile HMM and a probabilistic model of DNA or protein evolution. Reinert, Stoye, and Will (2000b) proposed an iterative weighted sum-of-pairs algorithm that uses the divide-and-conquer strategy (DCA; see Tönges *et al.*, 1996) together with dynamic programming on the reduced search space. Brudno *et al.* (2003) developed Multi-LAGAN, the progressive alignment algorithm for homologous genomic sequences. The pairwise alignments in Multi-LAGAN are constructed by the LAGAN program (Brudno *et al.*, 2003) which selects an optimal set of non-overlapping local alignments with subsequent alignment of intermediate sequence fragments by the Needleman–Wunsch algorithm. Knudsen (2003) proposed an algorithm for optimal parsimony alignment with affine gap cost for any number of sequences related by a tree (to reduce the required memory this algorithm uses the linear space algorithm by Hirschberg, 1975). *Probabilistic consistency*, a modification of the sum-of-pairs scoring system, incorporating HMM-derived posterior probabilities and three-way alignment consistency, was introduced by Do *et al.* (2005). The same group proposed a protein progressive multiple alignment algorithm, ProbCons, based on the probabilistic consistency scoring function.

Another objective function for a multiple alignment, norMD, was introduced by Thompson *et al.* (2001). The norMD, a column-based scoring scheme, uses the mean distance (MD) scores employed in CLUSTALX. The MD scores are normalized with respect to the number of aligned sequences and their lengths.

The propagation model of multiple alignment presented by Liu, Neuwald, and Lawrence (1999) combines a profile HMM approach with a block-based Gibbs sampling.

Thompson, Plewniak, and Poch (1999a) evaluated ten frequently used multiple alignment algorithms. The authors used as a benchmark the BAliBASE set of structurally verified alignments (Thompson, Plewniak, and Poch, 1999b). An alignment produced by each algorithm in the study was evaluated by several objective functions assessing the closeness of the alignment to the corresponding benchmark alignment. This comparative assessment of multiple alignment algorithms allowed the identification of alignment strategies most suitable for a given set of sequences.

The features of the multiple alignment programs, such as robustness, portability, and user-friendliness, undergo continuous improvement (see, for example, Chenna *et al.* (2003) regarding the latest updates in the CLUSTAL series of programs).

7

Building phylogenetic trees

The discussion of the concept of pairwise sequence alignment and algorithms for finding optimal alignments was divided in *BSA* into two chapters (Chapters 2 and 3). Chapter 2 was devoted to conventional non-probabilistic dynamic programming algorithms and Chapter 3 described the full probabilistic approach to the pairwise sequence alignment using a pair HMM.

Similarly, the discussion of phylogenetic tree concepts and algorithms of phylogenetic tree building was divided in *BSA* between Chapters 7 and 8. Chapter 7 introduced the non-probabilistic methods and included basic definitions, an inventory of tree topologies, the elucidation of evolutionary distance properties, and a description of conventional non-probabilistic algorithms such as clustering (UPGMA), neighbor-joining, and parsimony algorithms. This comprehensive material was completed by the description of non-probabilistic algorithms of simultaneous alignment and phylogeny. The translation of the phylogenetic tree building theory into probabilistic terms was left for Chapter 8.

The problems included in Chapter 7 require knowledge of combinatorics and graph theory. The level of difficulty of some of these problems is perhaps one of the highest in *BSA*.

The additional problems offer practice with the UPGMA and the neighbor-joining tree building methods and a chance to derive a useful combinatorial formula for the number of labeled histories.

7.1 Original problems

Problem 7.1 Draw the rooted trees obtained by adding the root in all seven possible positions to the unrooted tree in the picture on the following page.

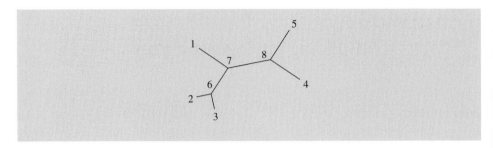

Solution Adding a root in turn to each of the seven edges generates seven trees with a distinct labeled patterns, as shown in Figure 7.1.   ☐

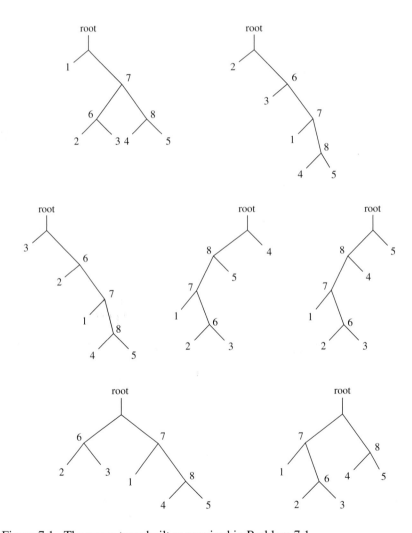

Figure 7.1. The seven trees built as required in Problem 7.1

Problem 7.2 For both rooted and unrooted trees, how many leaves do there have to be to obtain more than one unlabeled branching pattern? Find a recurrence relation for the number of rooted trees.

Solution Let $\mathcal{N}_n$ be the number of rooted trees with n leaves. Obviously, $\mathcal{N}_1 = \mathcal{N}_2 = \mathcal{N}_3 = 1$, and these unlabeled branching pattern types (topologies) are as follows:

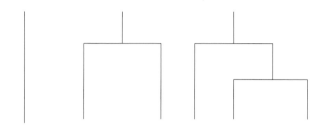

For a rooted tree with four leaves, there are two distinct topologies ($\mathcal{N}_4 = 2$):

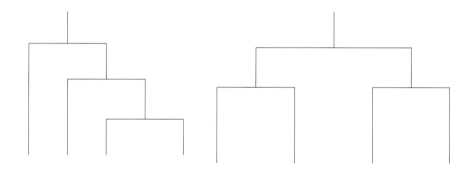

Therefore, to allow for more than one unlabeled branching pattern, a *rooted tree* must have at least four leaves.

Let us consider *unrooted trees*. Trees with three, four, and five leaves all have only one unlabeled branching pattern:

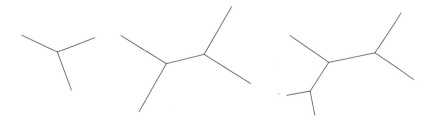

For a tree with six leaves there are two possible unlabeled branching patterns, as follows:

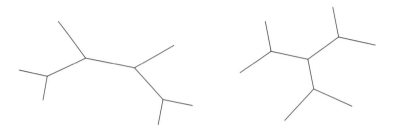

Hence, to allow for more than one topology, an *unrooted tree* must have at least six leaves.

To derive the recurrence formula for $\mathcal{N}_n$, one can count all the possible ways to obtain a rooted tree with n leaves by joining two rooted trees with l and $n - l$ leaves ($l \leq n - l$) at their roots.

If n is an odd number, $n = 2k + 1$, $k > 0$, any two "subtrees" will have a different number of leaves l and $n - l$ ($l < n - l$) and necessarily will have different topologies just because of the different number of leaves. Therefore, joining a tree with l leaves and a tree with $n - l$ leaves at their roots will produce $\mathcal{N}_l \mathcal{N}_{n-l}$ different topologies. Hence, the recurrence formula is given by

$$\mathcal{N}_n = \mathcal{N}_{2k+1} = \sum_{l=1}^{k} \mathcal{N}_l \mathcal{N}_{n-l}.$$

For $n = 2k$, $k > 1$, in a similar formula we have to make an adjustment for the $\mathcal{N}_k^2$ term, corresponding to joining "subtrees" with k leaves each, to avoid multiple counting of mirror symmetrical topologies. The term $\mathcal{N}_k^2$ should be replaced by $\binom{\mathcal{N}_k}{2} + \mathcal{N}_k$, where $\binom{\mathcal{N}_k}{2}$ is the number of possible pairs of distinct topologies among $\mathcal{N}_k$, and $\mathcal{N}_k$ is the number of pairs of trees with the same topology types. Thus, the recurrence relation is given by

$$\mathcal{N}_n = \mathcal{N}_{2k} = \sum_{l=1}^{k-1} \mathcal{N}_l \mathcal{N}_{n-l} + \frac{\mathcal{N}_k(\mathcal{N}_k - 1)}{2} + \mathcal{N}_k$$

$$= \sum_{l=1}^{k-1} \mathcal{N}_l \mathcal{N}_{n-l} + \frac{\mathcal{N}_k(\mathcal{N}_k + 1)}{2}.$$

For small n the recurrence formulas yield: $\mathcal{N}_5 = 3$, $\mathcal{N}_6 = 6$, $\mathcal{N}_7 = 11$, $\mathcal{N}_8 = 23$, $\mathcal{N}_9 = 46$, $\mathcal{N}_{10} = 98$.   □

Problem 7.3 All trees considered so far have been binary, but one can envisage ternary trees that, in their rooted form, have *three* branches descending from a branch node. The unrooted trees therefore have four edges radiating from every branch node. If there are m branch nodes in an unrooted ternary tree, how many leaves are there and how many edges?

Solution We designate the numbers of leaves and edges of an unrooted ternary tree with m branch nodes as L_m and E_m, respectively. A minimal ternary tree with $m = 1$ internal node has four leaves and four edges, thus $L_1 = E_1 = 4$. A ternary tree with $m - 1$ nodes can be transformed into a ternary tree with m nodes by replacing one of the leaves into a new node with three leaves. Therefore, $L_m = L_{m-1} + 2$ and $E_m = E_{m-1} + 3$, and we have

$$L_m = L_{m-1}+2 = (L_{m-2}+2)+2 = \cdots = L_1+2(m-1) = 4+2m-12 = 2m+2,$$

$$E_m = E_{m-1}+3 = (E_{m-2}+3)+3 = \cdots = E_1+3(m-1) = 4+3m-3 = 3m+1.$$

□

Problem 7.4 Consider a composite unrooted tree with m ternary branch nodes and n binary branch nodes. How many leaves are there, and how many edges? Let $N_{m,n}$ denote the number of distinct labeled branching patterns of this tree. Extend the counting argument for binary trees to show that

$$N_{m,n} = (3m + 2n - 1)N_{m,n-1} + (n + 1)N_{m-1,n+1}. \tag{7.1}$$

Solution First, for a binary tree with n branch nodes the number of leaves L_n and the number of edges E_n are defined by $L_n = n+2, E_n = 2n+1$, which immediately follow from the recurrence equations $L_n = L_{n-1}+1, E_n = E_{n-1}+2$, with $L_1 = 3$, $E_1 = 3$.

Now, we designate the numbers of leaves and edges of a composite unrooted tree with m ternary branch nodes and n binary branch nodes as $L_{m,n}$ and $E_{m,n}$, respectively. By adding a root node to a particular edge, we get a rooted tree with m ternary nodes, n binary nodes, and a root node (the root node is not counted as a binary node). By enumeration of the edges via traversing down the tree (starting with the two edges descending from the root), we will pass through three edges descending from each ternary node, with a total of $3m$ edges, and two edges descending from each binary node with a total of $2n$ edges. Finally, by adding the edge topped by the root we obtain

$$E_{m,n} = E_m + E_n - 1 = 3m + 1 + 2n + 1 - 1 = 3m + 2n + 1.$$

The same logic allows us to derive the formula for the number of leaves $L_{m,n}$. Again by traversing down the tree from the root we will descend along $3m + 2n + 2$ edges of the tree (the edge with the root is counted twice). Descending along an edge will bring us either to a branch node or to a leaf node. We will encounter branch nodes $m + n$ times and leaves $(3m + 2n + 2) - (m + n)$ times. Therefore,

$$L_{m,n} = (3m + 2n + 2) - (m + n) = 2m + n + 2.$$

Now let $N_{m,n}$ denote the number of distinct labeled branching patterns of the composite tree in question. This tree can be built either (i) from a composite tree with m ternary nodes and $n - 1$ binary nodes by adding a new edge to one of $E_{m,n-1} = 3m + 2n - 1$ existing edges, thus creating an additional binary node; or (ii) from a composite tree with $m - 1$ ternary nodes and $n + 1$ binary nodes by adding an edge at one of $n + 1$ binary nodes, thus producing a new ternary node. Therefore, we have the recurrence formula for $N_{m,n}$ as follows:

$$N_{m,n} = (3m + 2n - 1)N_{m,n-1} + (n + 1)N_{m-1,n+1}.$$

Here $3m + 2n - 1$ is the number of ways to add a new edge to an existing edge of a tree with m ternary nodes and $n - 1$ binary nodes, i.e. the number of the tree edges $E_{m,n-1} = 3m + 2(n - 1) + 1 = 3m + 2n - 1$; similarly, $n + 1$ is the number of ways to add a new edge at a binary node of a tree with $m - 1$ ternary nodes and $n + 1$ binary nodes, which is obviously $n + 1$.

Remark Note that for binary trees with n branch nodes ($m = 0$) we have $N_{0,n} = (2n - 1)N_{0,n-1}$ with $N_{0,1} = 1$. Then $N_{0,n} = (2n - 1)!!$, or, in terms of the number of leaves, $N_{E_n} = (2E_n - 5)!!$ $\qquad\square$

Problem 7.5 Use the recurrence relationship from Problem 7.4 to calculate $N_{m,0}$, the number of distinct pure ternary trees with m branch nodes, for small values of m. Check that the calculated numbers satisfy

$$N_{m,0} = \prod_{i=1}^{m}\left(1 + \frac{9i(i - 1)}{2}\right). \tag{7.2}$$

Prove formula (7.2).

Solution First, we use the recurrence formula (7.1) to determine the number of distinct labeled branching patterns of ternary trees with m branch nodes for several small values of m. Since $N_{m,-1} = 0$, we have

$$N_{1,0} = N_{0,1} = (2 - 1)!! = 1,$$

$$N_{2,0} = N_{1,1} = 4N_{1,0} + 2N_{0,2} = 4 + 2(3!!) = 10,$$

$$N_{3,0} = N_{2,1} = 7N_{2,0} + 2N_{1,2} = 70 + 2(6N_{1,1} + 3N_{0,3})$$

$$= 70 + 2(60 + 3(5!!)) = 280,$$

$$N_{4,0} = N_{3,1} = 10N_{3,0} + 2N_{2,2} = 2800 + 2(9N_{2,1} + 3N_{1,3})$$

$$= 2800 + 2[9(7N_{2,0} + 2N_{1,2}) + 3(8N_{1,2} + 4N_{0,4})]$$

$$= 2800 + (9 \times 280 + 3 \times 1260) = 15\,400.$$

Obviously, $N_{m,0}$ grows very quickly as m increases. It is easy to check that the calculated values of $N_{m,0}$ satisfy the equation $N_{m,0} = \prod_{i=1}^{m}(1 + 9i(i - 1)/2)$:

$$N_{1,0} = 1 = (1 + 9 \times 1 \times 0/2),$$

$$N_{2,0} = 10 = N_{1,0}(1 + 9 \times 2 \times 1/2),$$

$$N_{3,0} = 280 = N_{2,0}(1 + 9 \times 2 \times 3/2),$$

$$N_{4,0} = 15\,400 = N_{3,0}(1 + 9 \times 3 \times 4/2).$$

Secondly, we show how to prove Equation (7.2). We will verify that for $n \geq 3$ there is a one-to-one correspondence between trees T_n with n labeled (branch and leaf) nodes and n^{n-2} sequences $(a_1, a_2, \ldots, a_{n-2})$ that can be formed from numbers $1, 2, \ldots, n$. This statement appears as Theorem 2.1 in Moon (1970) with a reference to much earlier work by Cayley (1889). We begin the a proof of the one-to-one correspondence by verifying that a tree with n nodes defines a sequence $(a_1, a_2, \ldots, a_{n-2})$. The procedure of generating this sequence for tree T_n with nodes labeled by integers $1, 2, \ldots, n$ is called the Prüfer construction (Prüfer, 1918). We identify a leaf node with the smallest label and remove this node and its incident edge from T_n. The label of the node to which the removed node was attached becomes a_1. We repeat this process for the reduced tree T_{n-1} to determine a_2 and continue until only two nodes, joined by an edge, are left. Different trees T_n, T'_n determine different sequences $(a_1, a_2, \ldots, a_{n-2})$, $(B_1, B_2, \ldots, B_{n-2})$.

It remains to show that a sequence $A_{n-2} = (a_1, a_2, \ldots, a_{n-2})$ of $n - 2$ numbers sampled from numbers $1, 2, \ldots, n$ uniquely defines tree T_n. This can be shown by a step-by-step reconstruction of tree edges that we have earlier removed one after another upon implementation of the Prüfer algorithm. We define a list of integers $L_{n-2} = (b_1, \ldots, b_k)$, $1 \leq b_1 < b_2 < \cdots < b_k \leq n$, consisting of numbers absent in A_{n-2} (there are at least two elements in L_{n-2}). Obviously, due to the rules of the Prüfer construction, b_1 is a leaf connected to a_1. At the first step we remove a_1 from A_{n-2}, b_1 from L_{n-2}, and obtain $A_{n-3} = (a_2, \ldots, a_{n-2})$. If a_1 appears in A_{n-3}, we define a new list as $L_{n-3} = (b_2, \ldots, b_k)$; otherwise $L_{n-3} = (b_2, \ldots, b_k, a_1)$. After rearranging elements of sequence L_{n-3} (if necessary) in increasing order, we obtain $L_{n-3} = (b'_1, \ldots, b'_k)$ (or $L_{n-3} = (b'_1, \ldots, b'_{k-1})$) and infer that b'_1 is a node connected to a_2. At the next step we remove a_2 from A_{n-3}, b'_1 from L_{n-3}, repeat the procedure described above to determine a node b''_1 attached to a_3 in tree T_n, and so on. When this algorithm is advanced to the step $n - 3$, we will have the one-element sequence $A_1 = (a_{n-2})$, and we can prove that a list L_1 will have exactly two elements b^*_1, b^*_2. Indeed,

$$|L_1| = k - (n - 3) + (n - k + 1) = 2,$$

where k is the size of initial list L_{n-2}, $n - 3$ is the number of mandatory removals of the smallest members of the lists, and $n - k + 1$ is the number of insertions of elements $a_\bullet$ to the lists $L_{n-3}, \ldots, L_2$, equal to the number of different elements among $(a_1, a_2, \ldots, a_{n-3})$ not equal to a_{n-2}. At this (next to the final) step, b^*_1 joins a_{n-2}. Then a_{n-2} is reinserted in A_0 and b^*_2 joins a_{n-2} as well. As a result we have a sequence of $n - 1$ edges $(b_1, a_1), (b'_1, a_2), (b''_1, a_3), \ldots, (b^*_1, a_{n-2}), (b^*_2, a_{n-2})$, that uniquely defines tree T_n (all edges are different, because at each step after joining $a_\bullet$ the minimal $b_\bullet$ is removed from the list; each number from 1 to n is the end of at least one edge in the set $(b_1, a_1), \ldots, (b^*_2, a_{n-2})$). Moreover, by using the Prüfer algorithm for T_n, we should get the same sequence $A_{n-2} = (a_1, a_2, \ldots, a_{n-2})$. Hence, Prüfer's construction defines the one-to-one correspondence between these sequences and the trees T_n.

From the proved statement we can derive a quite general corollary about the number of labeled trees in terms of *degrees* of their nodes. The *degree* of a node is defined as the number of edges incident to the node. We can associate with tree T_n a degree sequence $(d_1, d_2, \ldots, d_n)$, where d_i is the degree of node i, $i = 1, \ldots, n$. In the sequence $(a_1, a_2, \ldots, a_{n-2})$ built for tree T_n by Prüfer's construction, each number i, $i = 1, \ldots, n$, will appear exactly $d_i - 1$ times. For tree T_n with a given sequence of node degrees the number of ways to choose positions for labels is equal to the multinomial coefficient

$$\binom{n - 2}{d_1 - 1, d_2 - 1, \ldots, d_n - 1}.$$

Therefore, due to one-to-one correspondence between trees and sequences, the number $\mathcal{N}$ of trees with n labeled (branch and leaf) nodes $1, 2, \ldots, n$, and a degree sequence $(d_1, d_2, \ldots, d_n)$ is given by the same multinomial coefficient:

$$\mathcal{N} = \binom{n - 2}{d_1 - 1, d_2 - 1, \ldots, d_n - 1}. \tag{7.3}$$

Let us consider a ternary tree T with m branch nodes. According to the formula proved in Problem 7.3, this tree has $2m + 2$ leaf nodes. The total number of (branch and leaf) nodes of T is $3m + 2$; let the leaves be labeled by numbers from 1 to $2m + 2$ and the branch nodes by numbers from $2m + 3$ to $3m + 2$. Then the degree sequence corresponding to tree T is $(1, \ldots, 1, 4, \ldots, 4)$ with $2m + 2$ ones and m fours. According to Equation (7.3), the number of trees with such degree sequence is given by

$$\binom{(3m+2)-2}{0,\ldots,0,3,\ldots,3} = \binom{3m}{0,\ldots,0,3,\ldots,3} = \frac{(3m)!}{(3!)^m} = \frac{(3m)!}{6^m}.$$

To determine the number N_m^* of ternary trees with labeled leaves (and unlabeled branch nodes), we note that among all $(3m)!6^{-m}$ trees there are groups of size $m!$ with fixed leaf nodes and all possible permutations of labels for branch nodes that identify the same tree with labeled leaves and unlabeled branch nodes. Hence, we have

$$N_m^* = \frac{(3m)!}{6^m m!}. \tag{7.4}$$

It is not obvious, but formulas (7.2) and (7.4) are equivalent. Indeed,

$$N_{m,0} = \prod_{i=1}^{m} \left(1 + \frac{9i(i-1)}{2}\right)$$

$$= \prod_{i=1}^{m} \frac{2 + 9i^2 - 9i}{2} = \prod_{i=1}^{m} \frac{(3i-2)(3i-1)}{2}$$

$$= \frac{1 \times 2 \times 4 \times 5 \times 7 \times 8 \times \cdots \times (3m-2)(3m-1)}{2^m}$$

$$= \frac{(3m)!}{(2^m)3 \times 6 \times 9 \times \cdots \times (3m)} = \frac{(3m)!}{6^m m!} = N_m^*.$$

Remark 1 Interestingly, formula (7.3) for the number of trees in terms of degrees of tree nodes can be used to derive an explicit formula for the number $N_{m,n}$ of composite trees with m ternary and n binary branch nodes, and labeled leaves (see Problem 7.4).

Any composite tree with m ternary and n binary branch nodes has $L_{m,n} = 2m + n + 2$ leaves (it was proved in Problem 7.3). The total number of (branch and leaf) nodes of the tree is $3m + 2n + 2$; we assign to leaves numbers from 1 to $2m + n + 2$ and to branch nodes numbers from $2m + n + 3$ to $3m + 2n + 2$. There are $\binom{m+n}{m}$ degree sequences $(d_1, d_2, \ldots, d_{3m+2n+2})$ corresponding to composite trees with such labelling of nodes (since there are $\binom{m+n}{m}$ possibilities to choose m positions to place fours among $d_{2m+n+3}, \ldots, d_{3m+2n+2}$; $d_1 = \cdots = d_{2m+n+2} = 1$). According to Equation (7.3), each of these degree sequences defines

$$\frac{(3m+2n)!}{(3!)^m (2!)^n} = \frac{(3m+2n)!}{6^m 2^n}$$

composite trees with labeled branch and leaf nodes. To determine the number $N_{m,n}$ of composite trees with labeled leaves (and unlabeled branch nodes), we note that $(m+n)!$ different trees with fixed labels for leaves and all possible permutations of labels for branch nodes (among $\binom{m+n}{m}(3m+2n)!6^{-m}2^{-n}$ trees with labeled branch and leaf nodes) identify the same tree with labeled leaves and unlabeled branch nodes. Thus we obtain

$$N_{m,n} = \frac{\binom{m+n}{m}(3m+2n)!}{6^m 2^n (m+n)!} = \frac{(3m+2n)!}{6^m 2^n m! n!}. \tag{7.5}$$

Remark 2 Equation (7.4) is a special case of Equation (7.5) for $n = 0$.

Remark 3 The recurrence relationship given by Equation (7.1) can also be derived from Equation (7.5):

$$(3m+2n-1)N_{m,n-1} + (n+1)N_{m-1,n+1}$$

$$= (3m+2n-1)\frac{(3m+2n-2)!}{6^m 2^{n-1} m!(n-1)!} + (n+1)\frac{(3m+2n-1)!}{6^{m-1}2^{n+1}(m-1)!(n+1)!}$$

$$= \frac{(3m+2n-1)!}{6^m 2^{n-1} m!(n-1)!} + \frac{(3m+2n-1)!}{6^{m-1}2^{n+1}(m-1)!n!}$$

$$= \frac{(3m+2n-1)!}{6^{m-1}2^{n-1}(m-1)!(n-1)!}\left(\frac{1}{6m} + \frac{1}{4n}\right)$$

$$= \frac{(3m+2n-1)!}{6^{m-1}2^{n-1}(m-1)!(n-1)!}\frac{3m+2n}{12mn} = \frac{(3m+2n)!}{6^m 2^n m! n!} = N_{m,n}.$$

A further discussion on counting multifurcating trees can be found in Felsenstein (2004). □

Problem 7.6 Show that, if distances between clusters are defined by

$$d_{ij} = \frac{1}{|C_i||C_j|} \sum_{p \in C_i, \, q \in C_j} d_{pq} \tag{7.6}$$

and if $C_k = C_i \cup C_j$, then d_{kl} for any l is given by

$$d_{kl} = \frac{d_{il}|C_i| + d_{jl}|C_j|}{|C_i| + |C_j|}.$$

Solution If $C_k = C_i \cup C_j$ and C_l is any other cluster, then from the definition of the distance between clusters given above, we have

$$\sum_{p \in C_l, \, q \in C_k} d_{pq} = \sum_{p \in C_l, \, q \in C_i} d_{pq} + \sum_{p \in C_l, \, q \in C_j} d_{pq}$$

$$= d_{il}|C_i||C_l| + d_{jl}|C_j||C_l|. \tag{7.7}$$

Since $|C_k| = |C_i| + |C_j|$, Equation (7.7) allows to rewrite the formula for d_{kl} as follows:

$$d_{kl} = \frac{1}{|C_l||C_k|} \sum_{p \in C_l, \, q \in C_k} d_{pq} = \frac{d_{il}|C_i||C_l| + d_{jl}|C_j||C_l|}{|C_l|(|C_i| + |C_j|)}$$

$$= \frac{d_{il}|C_i| + d_{jl}|C_j|}{|C_i| + |C_j|}. \qquad \square$$

Problem 7.7 Show that in a tree constructed by UPGMA, a node always lies above its daughter nodes.

Solution Upon building a rooted tree by UPGMA, at each step we choose two clusters i and j with minimal distance d_{ij}, join them into new cluster k, and place a new node at height $h_{ij} = d_{ij}/2$. Then we remove the minimal value d_{ij} from a set of all pairwise distances and substitute all distances d_{il}, d_{jl} (l is any other cluster) by the distances d_{kl} between the new cluster k and cluster l defined by Equation (7.6). We will show that

$$d_{i'j'} \geq d_{ij} \tag{7.8}$$

for any distance $d_{i'j'}$ in the set of pairwise distances defined for the next step of UPGMA. If $d_{i'j'}$ does not involve a new cluster k, the inequality (7.8) holds because d_{ij} was selected as the minimum distance after comparing all $d_{i'j'}$, while $d_{i'j'}$ stays the same for pairs i', j' such that $i', j' \neq i, j$. Now

$$d_{kl} \geq \min(d_{il}, d_{jl})$$

for any cluster l, since d_{kl} is a weighted average of distances d_{il} and d_{jl} (see Problem 7.6). On the other hand, inequalities $d_{il} \geq d_{ij}$, $d_{jl} \geq d_{ij}$ hold for any l, as d_{ij} was chosen as the minimum distance, thus $d_{kl} \geq d_{ij}$ for any l. This proves inequality (7.8). Finding the minimum among distances $d_{i'j'}$ (say, d_{i*j*}) will define a new pair of clusters i^* and j^* to join into a new node with height $h_{i*j*} = d_{i*j*}/2$. We have $h_{i*j*} \geq h_{ij}$, and see that the sequence of heights of the nodes built in subsequent steps of UPGMA is not decreasing. For any daughter node the parent node will be found at one of the later steps of the algorithm and, hence, no daughter node can lie above its parent node.

Remark Here is a simple example when the daughter node height is equal to the height of the parent node. Assume that three sequences x_1, x_2, x_3 satisfy $d_{12} = d_{23} = d_{13} = d$. At the first step of the clustering procedure, we can choose any pair of leaves to join in new cluster 4. We choose leaves 1 and 2, and place node 4 at height $d/2$. A new distance $d_{43} = d_{13} = d_{23} = d$; thus, next we join cluster 4 and leaf 3, and place node 5 at height $d/2$. Since the daughter node 4 and the parent node 5 have equal heights, they merge and create a ternary node. Thus, sequences x_1, x_2, x_3 become daughters of the ternary node, which would be the best way to build a tree for these sequences. $\square$

Problem 7.8   The distances between sequences are said to be ultrametric if, for any triplet of sequences x^i, x^j, x^k, the distances d_{ij}, d_{jk}, d_{ik} are either all equal, or two are equal and the remaining one is smaller. It can be shown that if the distances d_{ij} are ultrametric, and if a tree is constructed from these distances by UPGMA, then the distances obtained from this tree by taking twice the heights of the node on the path between i and j are identical to the d_{ij}. Check that this is true in the example of UPGMA applied to five sequences if the distances are ultrametric.

Solution To check this property, we consider three possible distinct unlabeled patterns of rooted trees with five leaves. All trees are assumed to be constructed by UPGMA from distances satisfying the ultrametric condition. In each case the tree leaves are labeled for convenience.

(a) For the following branching pattern below (denoted as $(((2 + 1) + 1) + 1)$):

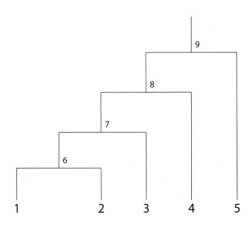

the closest clusters at the first pass of UPGMA are leaves 1 and 2. Distance d_{12} is twice as large as the height of node 6. Next, leaves 1 and 2 are joined into new cluster 6, and all distances d_{6i}, $i = 3, 4, 5$, satisfy $d_{6i} = \frac{1}{2}(d_{1i}+d_{2i}) = d_{1i} = d_{2i}$ due to the ultrametric property. The second pass joins the closest clusters 6 and 3, while the distance $d_{63} = d_{13} = d_{23}$ is twice as large as the height of node 7. Now we join cluster 6 and leaf 3 into the new cluster 7, and for distances d_{7i}, $i = 4, 5$, we have $d_{7i} = \frac{1}{3}(d_{1i}+d_{2i}+d_{3i}) = d_{1i} = d_{2i} = d_{3i}$ due to the ultrametric property. The next pair of closest clusters are cluster 7 and leaf 4, and distance $d_{74} = d_{14} = d_{24} = d_{34}$ is as large as the height of node 8. Cluster 7 and leaf 4 are joined into the new cluster 8, while $d_{85} = \frac{1}{4}(d_{15} + d_{25} + d_{35} + d_{45}) = d_{15} = d_{25} = d_{35} = d_{45}$ due to the ultrametric property. The last node, 9, is placed at height $\frac{1}{2}d_{85} = \frac{1}{2}d_{15} = \frac{1}{2}d_{25} = \frac{1}{2}d_{35} = \frac{1}{2}d_{45}$. Thus, we have verified that distance d_{ij} between leaves i and j is twice as large as the height of the highest node of the tree between i and j.

(b) For the second branching pattern $((2 + 2) + 1)$:

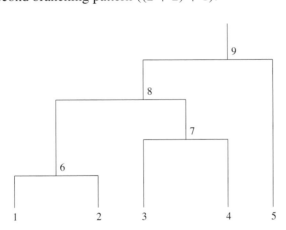

the closest clusters at the first pass of UPGMA are leaves 1 and 2. Distance d_{12} is twice as large as the height of node 6. We join leaves 1 and 2 into the new cluster 6, and all distances d_{6i}, $i = 3, 4, 5$, satisfy $d_{6i} = \frac{1}{2}(d_{1i} + d_{2i}) = d_{1i} = d_{2i}$ due to the ultrametric property. At the second pass the closest clusters are leaves 3 and 4, and distance d_{34} is twice as large as the height of node 7. Next, leaves 3 and 4 are joined into the new cluster 7, and then $d_{76} = \frac{1}{4}(d_{13} + d_{23} + d_{14} + d_{24}) = d_{13} = d_{23} = d_{14} = d_{24}$ with $d_{75} = \frac{1}{2}(d_{35} + d_{45}) = d_{35} = d_{45}$ due to the ultrametric property. The next pair of closest clusters are clusters 6 and 7, and distance $d_{67} = d_{13} = d_{23} = d_{14} = d_{24}$ is twice as large as the height of node 8. We define a new cluster 8 by joining clusters 6 and 7, while $d_{85} = \frac{1}{4}(d_{15}+d_{25}+d_{35}+d_{45}) = d_{15} = d_{25} = d_{35} = d_{45}$. Therefore, for the last node, 9, placed at height $\frac{1}{2}d_{85}$, we have $d_{85} = d_{15} = d_{25} = d_{35} = d_{45}$, which completes the proof for case (b).

(c) For the last branching pattern $((2 + 1) + 2)$:

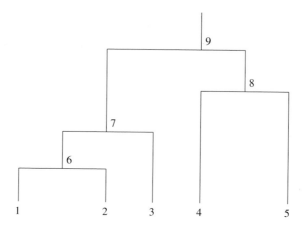

the closest clusters at the first pass of UPGMA are leaves 1 and 2. The distance d_{12} is twice as large as the height of node 6. We join leaves 1 and 2 into the new cluster 6, and for all distances d_{6i}, $i = 3, 4, 5$, equality $d_{6i} = \frac{1}{2}(d_{1i} + d_{2i}) = d_{1i} = d_{2i}$ holds due to the ultrametric property. At the second pass the closest clusters are 6 and 3, and distance $d_{63} = d_{13} = d_{23}$ is twice as large as the height of node 7. We introduce a new cluster, 7, by joining cluster 6 and leaf 3, and for both distances we have $d_{7i} = \frac{1}{3}(d_{1i} + d_{2i} + d_{3i}) = d_{1i} = d_{2i} = d_{3i}$ due to the ultrametric property. The next pair of closest clusters are leaves 4 and 5, and distance d_{45} is twice as large as the height of node 8. Leaves 4 and 5 are joined into a new cluster, 8. Now for distance d_{78} we have $d_{78} = \frac{1}{6}(d_{14} + d_{15} + d_{24} + d_{25} + d_{34} + d_{35}) = d_{14} = d_{15} = d_{24} = d_{25} = d_{34} = d_{35}$. Therefore, for the last node, 9, placed at height $\frac{1}{2}d_{78}$, we have $d_{78} = d_{14} = d_{15} = d_{24} = d_{25} = d_{34} = d_{35}$. The proof is complete.   □

Problem 7.9 For a tree with additive edge lengths d we define the distance D_{ij} between leaves i and j as follows:

$$D_{ij} = d_{ij} - (r_i + r_j), \tag{7.9}$$

where

$$r_i = \frac{1}{|L| - 2} \sum_{k \in L} d_{ik}$$

and $|L|$ denotes the size of the set of leaves L. Show that the smallest distances D_{ij} in the tree below correspond to neighboring leaves.

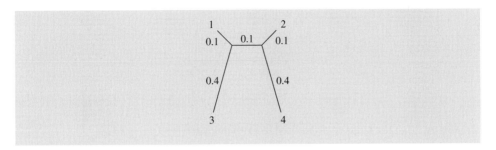

Solution Here $|L| = 4$, and we proceed by using Equation (7.9):

$$r_1 = \frac{1}{2}(0.5 + 0.3 + 0.6) = 0.7,$$

$$r_2 = \frac{1}{2}(0.5 + 0.3 + 0.6) = 0.7,$$

$$r_3 = \frac{1}{2}(0.5 + 0.6 + 0.9) = 1,$$

$$r_4 = \frac{1}{2}(0.5 + 0.6 + 0.9) = 1.$$

Therefore

$$D_{12} = d_{12} - (r_1 + r_2) = 0.3 - (0.7 + 0.7) = -1.1,$$
$$D_{13} = d_{13} - (r_1 + r_3) = 0.5 - (0.7 + 1) = -1.2,$$
$$D_{14} = d_{14} - (r_1 + r_4) = 0.6 - (0.7 + 1) = -1.1,$$
$$D_{23} = d_{23} - (r_2 + r_3) = 0.6 - (0.7 + 1) = -1.1,$$
$$D_{24} = d_{24} - (r_2 + r_4) = 0.5 - (0.7 + 1) = -1.2,$$
$$D_{34} = d_{34} - (r_3 + r_4) = 0.9 - (1 + 1) = -1.1.$$

We can see that two equal minimum values of D, $D_{13} = D_{24} = -1.2$, indeed correspond to the pairs of neighboring leaves: $(1, 3)$ and $(2, 4)$.

Remark Note that D defined by Equation (7.9) does have the usual distance properties. As we saw, D_{ij} can be negative and, surprisingly, $D_{ii} = d_{ii} - 2r_i = -2r_i \neq 0$.

☐

Problem 7.10 Show that, for a tree with four leaves, values D_{ij} for a pair of neighbors is less than D for all other pairs by the "bridge length," i.e. the length of the edge joining the two branch nodes in the tree.

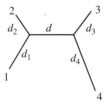

Figure 7.2. The edge notation used in Problem 7.10.

Solution Note that in Problem 7.9 the values D_{13} and D_{24} for pairs of neighbors are indeed less than D_{ij} for other pairs of leaves by the "bridge length," 0.1. Proof of this property for a general four-leaf tree, $|L| = 4$, is straightforward (see Figure 7.2). Given the equalities

$$r_1 = \frac{1}{2}(d_1 + d_2 + d_1 + d + d_3 + d_1 + d + d_4) = d + \frac{3d_1 + d_2 + d_3 + d_4}{2},$$

$$r_2 = \frac{1}{2}(d_2 + d_1 + d_2 + d + d_3 + d_2 + d + d_4) = d + \frac{d_1 + 3d_2 + d_3 + d_4}{2},$$

$$r_3 = \frac{1}{2}(d_3 + d_4 + d_3 + d + d_1 + d_3 + d + d_2) = d + \frac{d_1 + d_2 + 3d_3 + d_4}{2},$$

$$r_4 = \frac{1}{2}(d_4 + d_3 + d_4 + d + d_1 + d_4 + d + d_2) = d + \frac{d_1 + d_2 + d_3 + 3d_4}{2},$$

we find

$$D_{12} = d_{12} - (r_1 + r_2) = d_1 + d_2 - r_1 - r_2 = -2d - d_1 - d_2 - d_3 - d_4,$$

$$D_{13} = d_{13} - (r_1 + r_3) = d_1 + d + d_3 - r_1 - r_3 = -d - d_1 - d_2 - d_3 - d_4,$$

$$D_{14} = d_{14} - (r_1 + r_4) = d_1 + d + d_4 - r_1 - r_4 = -d - d_1 - d_2 - d_3 - d_4,$$

$$D_{23} = d_{23} - (r_2 + r_3) = d_2 + d + d_3 - r_2 - r_3 = -d - d_1 - d_2 - d_3 - d_4,$$

$$D_{24} = d_{24} - (r_2 + r_4) = d_2 + d + d_4 - r_2 - r_4 = -d - d_1 - d_2 - d_3 - d_4,$$

$$D_{34} = d_{34} - (r_3 + r_4) = d_3 + d_4 - r_3 - r_4 = -2d - d_1 - d_2 - d_3 - d_4.$$

Thus, the property holds for this tree: if D' is the value for two neighbors (D_{12} or D_{34}) and D'' is the value for non-neighbors, then $D'' - D' = d$, where d is the length of the edge joining the two branch nodes of the tree. □

Problem 7.11 Show that the traditional parsimony algorithm (Sankoff and Cedergren, 1983) gives the same cost as that for weighted parsimony (Fitch, 1971) using weights $S(a, a) = 0$ for all a, and $S(a, b) = 1$ for all $a \neq b$.

Solution We will use the mathematical induction with respect to the number of tree leaves. For a given sequence site u we will compare the tree costs defined by the weighted parsimony and by the traditional parsimony.

First, we consider tree T_2 with two leaves. If both sequences x_1 and x_2 have residue a at site u, the traditional parsimony algorithm assigns the list of residues $R_{C_3} = a \cap a = \{a\}$ to root node 3, and the traditional parsimony cost of tree T_2, the minimal number of substitutions at site u, is zero ($C_{T_2} = C_3 = 0$) for the ancestral residue a at the root. The weighted parsimony algorithm works for T_2 as follows. At each leaf the cost of residue a is zero, and the cost of any other residue is $+\infty$. Then $S_3(a) = 0$, and $S_3(b) = 2$, $b \neq a$. Thus, the list of minimal cost residues at the root is $R_{S_3} = \{a\}$ and the weighted parsimony cost of T_2 is $S_{T_2} = S_3(a) = 0$. Hence, $S_{T_2} = C_{T_2}$.

If residues a and b at site u are different, then for traditional parsimony the list of residues at root node 3 is $R_{C_3} = a \cup b = \{a, b\}$, and the traditional cost C_{T_2} is 1, the minimal number of substitutions. The weighted parsimony gives $S_3(a) = S_3(b) = 1$, $S_3(x) = 2$, if $x \neq a, b$. Then the weighted parsimony cost of the tree is the same as the cost of any residue from list R_{C_3} at the root: $S_{T_2} = S_3(a) = S_3(b)$. Once again, $S_{T_2} = C_{T_2}$.

Similarly, the equality $S_{T_3} = C_{T_3}$ can be verified for tree T_3 with three leaves. We consider now tree T_N with N leaves. Suppose that for two daughters, i and j, of the root we have $S_i = C_i$, $S_j = C_j$, and for each node the list of minimal cost and list of parsimonious residues kept by both algorithms are identical: $R_{S_i} = R_{C_i} = R_i$, $R_{S_j} = R_{C_j} = R_j$. We will show that at the root the list of ancestral residues with the minimum number of substitutions and the list of minimum cost S_k residues are the same and we prove that $S_T = C_T$.

If $R_i \cap R_j = \emptyset$, then the list R_k of residues for the traditional parsimony algorithm is $R_k = R_i \cup R_j$ and the cost of tree T_N is $C_T = C_k = C_i + C_j + 1$. We suppose that $R_i = \{a_1, \ldots, a_m\}$, $R_j = \{b_1, \ldots, b_n\}$, and therefore that

$$S_k(a_l) = S_i(a_l) + S_j(b_1) + 1 = C_i + C_j + 1,$$
$$S_k(b_q) = S_i(a_1) + 1 + S_j(b_q) = C_i + C_j + 1,$$
$$S_k(x) = S_i(a_1) + 1 + S_j(b_1) + 1 = C_i + C_j + 2,$$

for all $x \notin R_k$, $l = 1, \ldots, m$, $q = 1, \ldots, n$. Thus, the weighted parsimony cost of tree T_N is equal to

$$S_T = S_k(a_1) = \cdots = S_k(a_m) = S_k(b_1) = \cdots = S_k(b_n) = C_i + C_j + 1 = C_T$$

and the list of the minimal cost residues at the root is given by

$$R_{S_k} = \{a_1, \ldots, a_m, b_1, \ldots, b_n\} = R_i \cup R_j = R_{C_k}.$$

Finally, we assume that lists R_i and R_j have a non-empty intersection: $R_i \cap R_j = \{c_1, \ldots, c_t\}$. Then list R_k for traditional parsimony algorithm is $R_k = \{c_1, \ldots, c_t\}$, and the cost of tree T_N is $C_T = C_k = C_i + C_j$. At the same time, the weighted cost of residues at the root is given by

$$S_k(c_p) = S_i(c_p) + S_j(c_p) = C_i + C_j,$$

$$S_k(a_l) = S_i(a_l) + S_j(b_1) + 1 = C_i + C_j + 1,$$

$$S_k(b_q) = S_i(a_1) + 1 + S_j(b_q) = C_i + C_j + 1,$$

$$S_k(x) = S_i(a_1) + 1 + S_j(b_1) + 1 = C_i + C_j + 2,$$

for all $p = 1, \ldots, t$, $a_l \notin R_k$, $b_q \notin R_k$, $x \notin R_i \cup R_j$. Thus, the list of the minimal weighted cost residues at the root is given by

$$R_{S_k} = \{c_1, \ldots, c_t\} = R_k$$

and the weighted parsimony cost of tree T_N is

$$S_T = S_k(c_1) = \cdots = S_k(c_t) = C_i + C_j = C_T.$$

Therefore, we have proved that the traditional parsimony algorithm gives the same cost as that for the weighted parsimony algorithm with the weight function S such that $S(a, a) = 0$ and $S(a, b) = 1$ for $a \neq b$. □

Problem 7.12 Show that the minimal cost with weighted parsimony is independent of the position of the root, provided the substitution cost is a metric, i.e. that it satisfies $S(a, a) = 0$, symmetry $S(a, b) = S(b, a)$, and the triangle inequality

$$S(a, c) \leq S(a, b) + S(b, c),$$

for all a, b, c.

Solution First, we derive a formula for cost S_T of a rooted tree T under the assumption that we have completed the post-order traversal procedure of weighted parsimony. We assume that i and j are daughter nodes of root k as shown in the following:

Assume also that we know costs $S_i(a)$ and $S_j(b)$ of subtrees descending from i and j for any residues a and b. Then

$$S_k(c) = \min_a(S_i(a) + S(c,a)) + \min_b(S_j(b) + S(b,c))$$

and

$$S_T = \min_c S_k(c) = S_k(c^*) = S_i(a^*) + S(c^*, a^*) + S_j(b^*) + S(b^*, c^*). \qquad (7.10)$$

Note that c^* should necessarily be either a^* or b^* to minimize the right-hand side in (7.10). Indeed, if $c^* = a^*$ (or $c^* = b^*$) we have, for any other c',

$$S_T = S_i(a^*) + S(c^*, a^*) + S_j(b^*) + S(b^*, c^*)$$
$$\leq S_i(a^*) + S(c', a^*) + S_j(b^*) + S(b^*, c')$$

due to the triangle inequality. Therefore we deduce that

$$S_T = \min_{a,b}(S_i(a) + S_j(b) + S(a,b)).$$

Now we want to show that the cost of tree will not change if we move the root to another edge. Without loss of generality we consider moving the root to one of the edges descending from node i with two daughter nodes l and m:

With the new root k' located at edge (i,l) the cost of tree T' becomes

$$S_{T'} = \min_{a,d}(S_i'(a) + S_l(d) + S(a,d)).$$

Here d designates a symbol attached to node l and $S_i'(a)$ is the cost of the subtree descending from node i (a different subtree in T' compared with the subtree descending from i in T). Now we will work with the expressions for S_T and $S_{T'}$ using recurrence equations (e is the symbol located at node m) as follows :

$$S_T = \min_{a,b}\left(\min_e(S_m(e) + S(e,a)) + \min_d(S_l(d) + S(d,a)) + S_j(b) + S(a,b)\right),$$
$$(7.11)$$

$$S_{T'} = \min_{a,d}\left(\min_e(S_m(e) + S(e,a)) + \min_b(S_j(b) + S(a,b)) + S_l(d) + S(a,d)\right).$$
$$(7.12)$$

Both right-hand expressions in Equations (7.11) and (7.12) are equivalent to

$$\min_a \left(\min_e (S_m(e) + S(e,a)) + \min_d (S_l(d) + S(d,a)) + \min_b (S_j(b) + S(a,b)) \right).$$

Therefore $S_T = S_{T'}$, which means that the minimal cost of a tree defined by weighted parsimony is independent of the root position if the substitution cost is a metric. □

Problem 7.13 Hein's algorithm (Hein, 1989) can be extended to general weights $S(a,b)$ by attaching a set of minimal cost $S_l(a)$ (as in the weighted parsimony algorithm) to each edge in a sequence graph instead of the set R_l. (a) Show that the equality

$$S(x,z) + S(z,y) = S(x,y)$$

can be satisfied by having z share a residue with x or y provided that $S(a,a) = 0$ for all a. (b) Evaluate the minimal cost (assuming a nucleotide or DNA alphabet) of a tree T

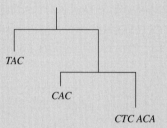

using Hein's algorithm for the sequence graphs.

Solution (a) Suppose that sequences x and y are associated with the daughter nodes of node l, and the minimal cost alignment of x and y is found by using the dynamic programming equations with weights $S(a,b)$. At node l we determine the minimal cost ancestral sequence z aligned to both sequences x and y as follows. At any site, z shares a symbol with either x or y (or both of them). Suppose that at a given site the optimal alignment of x and y has a match, with residue x_i aligned to y_j. Then, if $z_k = x_i$, we obtain

$$S(x_i, z_k) + S(z_k, y_j) = S(x_i, x_i) + S(x_i, y_j) = S(x_i, y_j);$$

if $z_k = y_j$, we have

$$S(x_i, z_k) + S(z_k, y_j) = S(x_i, y_j) + S(y_j, y_j) = S(x_i, y_j).$$

If the alignment of the sequences has a gap at a site (say, y-gap), we choose either a gap or residue x_i for the ancestral sequence z. Then at this site the contributions to both sides of the equality $S(x, z) + S(z, y) = S(x, y)$ are the same and equal to either the gap-open cost d (if the gap follows a match) or the gap-extension cost e (if the gap follows another y-gap). One more rule controls the choice of z-residues. If several consecutive gaps in one sequence occur in the optimal alignment of x and y, then the ancestral sequence z must either skip the entire set of residues aligned to the gap or include them all.

(b) Now we will find the minimal cost S of tree T with leaves $z = TAC$, $y = CAC$, and $x = CTCACA$ by the dynamic programming algorithm for sequence graphs introduced by Hein (1989). We have to calculate current costs, compare them ,and choose the minimum value to fill the cells of dynamic programming matrices. The algorithm parameters are selected as follows: substitution costs $S(A, T) = S(T, A) = S(A, C) = S(C, A) = S(G, C) = S(C, G) = S(G, T) = S(T, G) = 2$, $S(C, T) = S(T, C) = S(A, G) = S(G, A) = 1$; gap-open cost $d = 3$, and gap-extension cost $e = 2$.

To implement this procedure, we associate sequences x, y, and z with graphs G_1, G_2, and G_3, respectively, defined as follows. Each of these graphs is a simple linear graph with the number of edges equal to the length of the corresponding sequence. Additionally, we attach to edge i, $i = 1, \ldots, 6$, of graph G_1 the vector of substitution costs $[S(A, x_i), S(C, x_i), S(G, x_i), S(T, x_i)]$; to edge j, $j = 1, 2, 3$, of graph G_2 the vector of substitution costs $[S(A, y_j), S(C, y_j), S(G, y_j), S(T, y_j)]$; and to edge l, $l = 1, 2, 3$, of graph G_l the vector of substitution costs $[S(A, z_l), S(C, z_l), S(G, z_l), S(T, z_l)]$.

Then we obtain the following simple linear graphs:

$$G_1 \quad \bullet\!\!\xrightarrow{[2,0,2,1]}\!\!\bullet\!\!\xrightarrow{[2,1,2,0]}\!\!\bullet\!\!\xrightarrow{[2,0,2,1]}\!\!\bullet\!\!\xrightarrow{[0,2,1,2]}\!\!\bullet\!\!\xrightarrow{[2,0,2,1]}\!\!\bullet\!\!\xrightarrow{[0,2,1,2]}\!\!\bullet$$

$$G_2 \quad \bullet\!\!\xrightarrow{[2,0,2,1]}\!\!\bullet\!\!\xrightarrow{[0,2,1,2]}\!\!\bullet\!\!\xrightarrow{[2,0,2,1]}\!\!\bullet$$

$$G_3 \quad \bullet\!\!\xrightarrow{[2,1,2,0]}\!\!\bullet\!\!\xrightarrow{[0,2,1,2]}\!\!\bullet\!\!\xrightarrow{[2,0,2,1]}\!\!\bullet$$

The cost $S(G, G^*)$ between sequence graphs G and G^* is defined as the minimal cost of pairwise alignment between sequences s and s^*, $s \in G$, $s^* \in G^*$. For simple graphs (identifying one sequence each) such as G_1 and G_2, the cost $S(G_1, G_2)$ is just the cost of optimal alignment of these two sequences. To determine $S(G_1, G_2)$ and construct the optimal alignments of x and y, we use dynamic programming for graphs G_1 and G_2. Let $S^M(i, j)$, $S^X(i, j)$, $S^Y(i, j)$ denote the minimum cost of the graph alignments up to elements x_i, y_j, with $S^M(i, j)$ corresponding to match (x_i, y_j), $S^X(i, j)$ to x_i aligned to a gap, and $S^Y(i, j)$ to y_j aligned to a gap. The recurrence

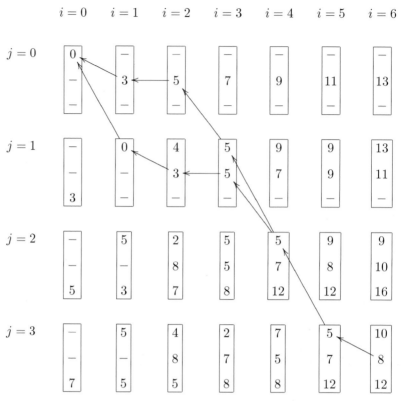

Figure 7.3. The matrix of current costs $S^\bullet(i,j)$ for graphs G_1 and G_2. Each cell (i,j) has three entries $S^M(i,j)$, $S^X(i,j)$, and $S^Y(i,j)$ shown in top down order. Two optimal paths are shown by arrows.

equations are as follows:

$$S^M(i,j) = \min_{K=A,C,G,T} (S(K,x_i) + S(K,y_j)) + \min \begin{cases} S^M(i-1,j-1), \\ S^X(i-1,j-1), \\ S^Y(i-1,j-1); \end{cases} \quad (7.13)$$

$$S^X(i,j) = \min \begin{cases} S^M(i-1,j)+d, \\ S^X(i-1,j)+e; \end{cases} \quad (7.14)$$

$$S^Y(i,j) = \min \begin{cases} S^M(i,j-1)+d, \\ S^Y(i,j-1)+e. \end{cases} \quad (7.15)$$

We apply these equations to fill out the matrix of current costs $S^\bullet(i,j)$ shown in Figure 7.3. We find the minimum cost $S(G_1,G_2) = 8$ as a minimum entry in the right bottom cell of Figure 7.3 ($i = 6, j = 3$) and use the traceback procedure to determine the optimal path through the cost matrix. There are two optimal paths

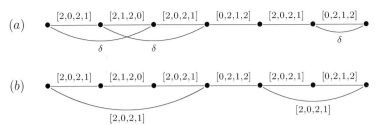

Figure 7.4. The sequence graph G_4 derived from the paths through the matrix in Table 7.1. (a) The graph with the dummy edges (marked by δ). (b) The same graph with the dummy edges replaced by edges labeled by vectors of substitution costs.

that define the following two optimal alignments of sequences x and y:

(1) $\quad C \quad - \quad - \quad A \quad C \quad -$ (2) $\quad - \quad - \quad C \quad A \quad C \quad -$
$\quad\quad C \quad T \quad C \quad A \quad C \quad A$ $\quad\quad\quad C \quad T \quad C \quad A \quad C \quad A$

with costs $S(1) = S(2) = S(G_1, G_2) = 2d + e = 8$. To find the possible ancestral sequences of x and y, we derive a new sequence graph G_4 (shown in Figure 7.4) from alignments (1) and (2). Each path through graph G_4 (Figure 7.4b) uniquely defines a nucleotide sequence by vectors of substitution costs attached to the edges on this path. There are four paths through G_4 identifying four sequences *CTCACA*, *CAC*, *CTCAC*, and *CACA*. Any sequence $s \in G_4$, satisfies the condition $S(s, x) + S(s, y) = S(G_1, G_2) = 8$. These sequences can be considered as ancestor-candidates for x and y in the minimum cost tree T.

Now we repeat the dynamic programming for graphs G_4 and G_3 with modified recurrence equations (7.13)–(7.15): for $i = 1, \ldots, 6, j = 1, 2, 3$, as follows:

$$S^M(i,j) = \min_{i',j'} \left(\min_{K=A,C,G,T} (S(K, x_{i'}) + S(K, y_{j'})) + \min \left\{ \begin{array}{l} S^M(i',j'), \\ S^X(i',j'), \\ S^Y(i',j'); \end{array} \right. \right)$$

$$S^X(i,j) = \min_{i'} \min \left\{ \begin{array}{l} S^M(i',j) + d, \\ S^X(i',j) + e; \end{array} \right.$$

$$S^Y(i,j) = \min_{j'} \min \left\{ \begin{array}{l} S^M(i,j') + d, \\ S^Y(i,j') + e. \end{array} \right.$$

Here the minimum is taken over all predecessor nodes i' of node i in graph G_4 and all predecessor nodes j' of node j in graph G_3. Since G_3 is a single sequence graph, $j' = j-1$. Upon calculation of the current costs $S^\bullet(i,j)$, we fill out cells of the matrix shown in Figure 7.5. The minimum cost $S(G_3, G_4) = 1$ is the minimal entry in the right bottom cell of Figure 7.5 ($i = 6, j = 3$). The traceback procedure recovers

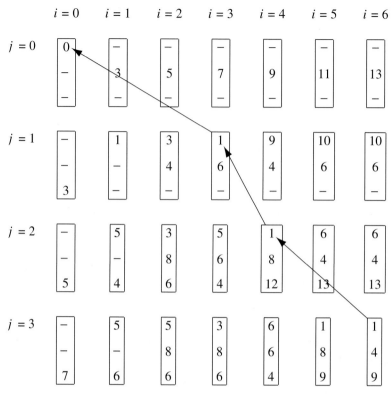

Figure 7.5. The matrix of current costs $S^{\bullet}(i,j)$ for graphs G_4 and G_3. Each cell (i,j) has three entries $S^M(i,j)$, $S^X(i,j)$, and $S^Y(i,j)$ shown in top down order. The optimal backtracking path is shown by arrows.

only one optimal path through the matrix of current costs. This path corresponds to the optimal alignment of sequences $z \in G_3$ and $y \in G_4$:

$$(3) \quad \begin{array}{ccc} T & A & C \\ C & A & C \end{array}$$

with cost $S(3) = S(G_3, G_4) = S(T, C) = 1$. To find a set of possible ancestral sequences for the root node of tree T, we derive from alignment (3) a new sequence graph G_5:

[2,0,2,1]   [0,2,1,2]   [2,0,2,1]

[2,1,2,0]

Two paths through G_5 identify two sequences $y = CAC$ and $z = TAC$ as possible ancestral sequences associated with the root node. Graph G_5 derived from the optimal path for G_3 and G_4 possesses the important property that there exist

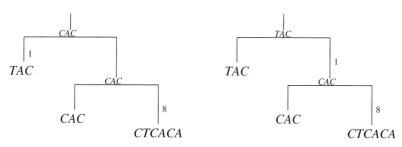

Figure 7.6. Two possible choices of ancestral sequences which produce the minimum tree cost $S_T = 9$.

sequences $s \in G_3$ and $s^* \in G_4$ such that $S(y, s) + S(y, s^*) = S(G_3, G_4) = 1$. The same is true for z.

Now we see that an arrangement of ancestral sequences always includes sequence y at the branch node and y or z at the root node. These two possible choices of ancestors are shown in Figure 7.6. Each choice minimizes the cost of the tree: $\min S_T = S(G_1, G_2) + S(G_3, G_4) = 9$. □

Problem 7.14 A neighbor-joining procedure for reconstructing an unrooted binary tree T proposed by Saitou and Nei (1987) at each iteration finds the minimum of function D_{ij} and joins corresponding leaves i and j into a single parental node. The function D is defined by Equations (7.9) for given distances d_{kl}, $k, l = 1, \ldots, N$, possessing an additive property. Show that this procedure reconstructs true neighbors, i.e. that the minimal value D_{ij} identifies neighboring leaves i and j in the case when the parental node of j (or i) is a parent to two leaves.

Solution The problem has already been solved directly for a four-leaf tree directly (Problem 7.10). Now we consider a tree T with $N \geq 5$ leaves.

Let us assume that the value D_{ij} is the minimal one, but that i and j are not neighbors. We seek a contradiction.

It is stated that the other daughter of the parental node k of leaf j is a leaf, say leaf m. Since i and j are not neighboring leaves, there are at least two nodes on the path between them, say l and k, as shown in the following diagram:

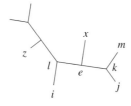

Let L_l be a set of N_l leaves ($N_l > 0$) derived from the third edge radiating from l (neither edge (l, i) nor path (l, k) of edges). We define set L_k containing just leaf m, and set L_x with N_x leaves ($N_x \geq 0$) containing all leaves x derived from path (l, k). Obviously, $N = N_l + N_x + 3$.

We will show that $D_{ij} > D_{jm}$. By the definition (7.9) of value D we have $D_{ij} = d_{ij} - (r_i + r_j)$, $D_{jm} = d_{jm} - (r_j + r_m)$, where

$$r_i = \frac{1}{N-2} \sum_{\text{all leaves } u} d_{iu}.$$

We consider the difference between D_{ij} and D_{jm} and use the additivity property of the distance d:

$$D_{ij} - D_{jm} = d_{il} + d_{lk} + d_{kj} - d_{jk} - d_{km} - r_i - r_j + r_j + r_m$$

$$= d_{il} + d_{lk} + d_{km} - \frac{1}{N-2} \left(\sum_{\text{all leaves } u} d_{iu} - \sum_{\text{all leaves } u} d_{mu} \right).$$

The sum

$$\Sigma_1 = \sum_{\text{all leaves } u} d_{iu}$$

has a term d_{ii} equal to zero; the same is true for d_{mm} from the second sum

$$\Sigma_2 = \sum_{\text{all leaves } u} d_{mu}.$$

Sums Σ_1 and Σ_2 also have a pair of equal terms (d_{im} and d_{mi}). Other terms in Σ_1 are as follows:

$$d_{ij} = d_{il} + d_{lk} + d_{kj},$$
$$d_{iz} = d_{il} + d_{lz},$$
$$d_{ix} = d_{il} + d_{le} + d_{ex},$$

for any $z \in L_l$, $x \in L_x$. Similarly, for the terms in Σ_2,

$$d_{mj} = d_{mk} + d_{kj},$$
$$d_{mz} = d_{mk} + d_{kl} + d_{lz},$$
$$d_{mx} = d_{mk} + d_{ke} + d_{ex}.$$

Thus, $\Sigma_1 - \Sigma_2$ becomes

$$\Sigma_1 - \Sigma_2 = (N-2)d_{il} - (N-2)d_{mk} - (N_l - 1)d_{lk} + \sum_{x \in L_x} d_{le} - \sum_{x \in L_x} d_{ke}$$

(note that distances d_{le} and d_{ke} in last two sums depend on x). Therefore,

$$D_{ij} - D_{jm} = d_{il} + d_{lk} + d_{km} - \frac{1}{N-2}(\Sigma_1 - \Sigma_2)$$

$$= \frac{1}{N-2}\left((N - N_l - 1)d_{lk} - \sum_{x \in L_x} d_{le} + \sum_{x \in L_x} d_{ke}\right)$$

$$= \frac{1}{N-2}\left(N_x d_{lk} + 2d_{lk} - \sum_{x \in L_x} d_{le} + \sum_{x \in L_x} d_{ke}\right)$$

$$= \frac{1}{N-2}\left(2\sum_{x \in L_x} d_{ke} + 2d_{lk}\right) > 0.$$

We have proved that $D_{ij} > D_{jm}$, which contradicts the initial assumption that while the value D_{ij} is minimal, leaves i and j are not neighbors. Hence, leaves i and j are neighboring leaves. $\qquad\square$

Problem 7.15 Complete the proof of the theorem stated in Problem 7.14 for the case when the set of leaves L_k contains at least two leaves m and n, as shown in the following diagram:

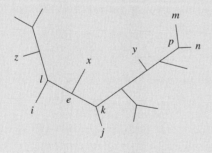

Solution We will show again that $D_{ij} > D_{mn}$. We will assume that $N_l \geq N_k$ (for designations, see the solution of Problem 7.14); otherwise we would consider two neighbors from set L_l instead of leaves m and n from N_k.

For any leaf y in L_k, $y \neq m$, $y \neq n$, it holds by additivity that $d_{iy} + d_{jy} = d_{ij} + 2d_{ky}$. Similarly, $d_{my} + d_{ny} = d_{mn} + 2d_{py}$. Thus,

$$d_{iy} + d_{jy} - d_{my} - d_{ny} = d_{ij} + 2d_{ky} - 2d_{py} - d_{mn}. \qquad (7.16)$$

If $y = n$, $y = m$, we have, respectively, $d_{in} + d_{jn} = d_{ij} + 2d_{kn}$, $d_{im} + d_{jm} = d_{ij} + 2d_{km}$ and

$$d_{in} + d_{jn} - d_{mn} = d_{ij} + 2d_{kn} - d_{mn},$$
$$d_{im} + d_{jm} - d_{mn} = d_{ij} + 2d_{km} - d_{mn}. \qquad (7.17)$$

For any leaf z in set L_l it holds that $d_{iz} + d_{jz} = d_{ij} + 2d_{lz}$, $d_{mz} + d_{nz} = d_{mn} + 2d_{pk} + 2d_{kl} + 2d_{lz}$, and then

$$d_{iz} + d_{jz} - d_{mz} - d_{nz} = d_{ij} - 2d_{pk} - 2d_{kl} - d_{mn}. \qquad (7.18)$$

Finally, for any leaf x from L_x we have $d_{ix} + d_{jx} = d_{ij} + 2d_{ex}$, $d_{mx} + d_{nx} = d_{mn} + 2d_{pk} + 2d_{ke} + 2d_{ex}$, thus

$$d_{ix} + d_{jx} - d_{mx} - d_{nx} = d_{ij} - d_{mn} - 2d_{pk} - 2d_{ke}. \qquad (7.19)$$

From the definition (7.9) of D_{ij},

$$D_{ij} - D_{mn} = d_{ij} - d_{mn} - \frac{1}{N-2} \sum_{\text{all leaves } u} (d_{iu} + d_{ju} - d_{mu} - d_{nu}).$$

By using Equations (7.16)–(7.19) for leaves u and the fact that for $u = i$ and $u = j$

$$d_{ij} - d_{mi} - d_{ni} + d_{ij} - d_{mj} - d_{nj} = -4d_{pk} - 2d_{mn},$$

we derive the following:

$$D_{ij} - D_{mn} = d_{ij} - d_{mn} - \frac{1}{N-2} \left(\sum_{y \in L_k} (d_{ij} + 2d_{ky} - d_{mn} - 2d_{py}) + 2d_{mn} \right.$$

$$-4d_{pk} - 2d_{mn} + \sum_{z \in L_l} (d_{ij} - 2d_{pk} - 2d_{kl} - d_{mn})$$

$$\left. + \sum_{x \in L_x} (d_{ij} - d_{mn} - 2d_{pk} - 2d_{ke}) \right).$$

We have used the fact that the term $2d_{py}$ is absent in Equations (7.17), which implies that terms $-2d_{py}$ for $y = m$, $y = n$, could be included in the sum over all leaves in L_k, provided that we add $2d_{pm} + 2d_{pn} = 2d_{mn}$ to the sum. The coefficients of d_{ij} and d_{mn}, summed over all leaves, are $N - 2$ and $-(N - 2)$, respectively. Thus,

$$D_{ij} - D_{mn}$$

$$= \frac{1}{N-2} \left(\sum_{y \in L_k} (2d_{py} - 2d_{ky}) + 4d_{pk} + \sum_{z \in L_l} (2d_{pk} + 2d_{kl}) + \sum_{x \in L_x} (2d_{pk} + 2d_{ke}) \right).$$

(Note that distances d_{py} and d_{ky} depend on y and that d_{ke} depends on x). Recalling the inequality $d_{py} - d_{ky} > -d_{pk}$, we have

$$D_{ij} - D_{mn} > \frac{2d_{pk}(-N_k + 2 + N_l + N_x)}{N - 2} > 0$$

for any $N_x \geq 0$ due to the assumption $N_l \geq N_k$. Therefore, the inequality $D_{ij} > D_{mn}$ contradicts the assumption that the D_{ij} value is minimal while i and j are not neighbors. Hence, leaves i and j must be neighbors. $\square$

7.2 Additional problems

In this section we deal with the construction of a phylogenetic tree for a given set of protein fragments: we have to build a rooted tree with the molecular clock property by the UPGMA (Problem 7.16) or we have to find an unrooted tree by the neighbor-joining algorithm (Problem 7.17). Also we derive a formula for the number of labeled histories with n leaves (Problem 7.18).

Problem 7.16   An ungapped multiple alignment of the fragments of cytochrome c from four different species – *Rickettsia conorii*, *Rickettsia prowazekii*, *Bradyrhizobium japonicum* and *Agrobacterium tumefaciens* – is shown below (with the sequences in top down order):

```
NIPELMKTANADNGREIAKK
NIQELMKTANANHGREIAKK
PIEKLLQTASVEKGAAAAKK
PIAKLLASADAAKGEAVFKK
```

Use the evolutionary distance, defined by the formula $d_{ij} = -\ln(1 - p_{ij})$, where p_{ij} is the fraction of mismatches in the pairwise alignment of sequences i and j, to build a rooted phylogenetic tree for the given sequences by the UPGMA. Show that distances d_{ij}^* defined by a tree built by the UPGMA do not always coincide with the initial distances d_{ij}.

Solution Let x_i, $i = 1, 2, 3, 4$, denote the ith protein sequence of the given alignment in top down order. From the counts of mismatches in the pairwise alignment of sequences x_i and x_j, we determine the pairwise distance d_{ij}, $i, j = 1, \ldots, 4$

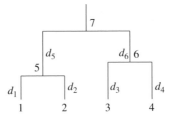

Figure 7.7. The rooted tree constructed by the UPGMA for the sequences given in Problem 7.16.

as follows:

$$d_{12} = -\ln(1 - 3/20) = 0.1625,$$

$$d_{13} = -\ln(1 - 12/20) = 0.9163,$$

$$d_{14} = -\ln(1 - 13/20) = 1.0498,$$

$$d_{23} = -\ln(1 - 12/20) = 0.9163,$$

$$d_{24} = -\ln(1 - 13/20) = 1.0498,$$

$$d_{34} = -\ln(1 - 9/20) = 0.5978.$$

To implement the UPGMA, we assign each sequence x_i to its own cluster C_i, $i = 1, \ldots, 4$, as well as to leaf i of a four-leaf tree. The minimal distance among d_{ij} is $d_{12} = 0.1625$, thus we define a new cluster $C_5 = C_1 \cap C_2$ and place node 5 at height $h_5 = d_{12}/2 = 0.0813$. Then, we remove clusters C_1 and C_2 from the list of clusters and calculate the following new distances:

$$d_{53} = \frac{1}{2}(d_{13} + d_{23}) = 0.9163,$$

$$d_{54} = \frac{1}{2}(d_{14} + d_{24}) = 1.0498.$$

The minimal distance among the distances d_{53}, d_{54}, and d_{34} is $d_{34} = 0.5978$. So, we introduce $C_6 = C_3 \cap C_4$, place a new node 7 at height $h_6 = d_{34}/2 = 0.2989$, and determine the distance between clusters C_5 and C_6 as follows:

$$d_{56} = \frac{1}{4}(d_{13} + d_{14} + d_{23} + d_{24}) = 0.9831.$$

With only two clusters C_5 and C_6 remaining, we define $C_7 = C_5 \cap C_6$ and place root node 7 at height $h_7 = d_{56}/2 = 0.4915$. The tree is completed (see Figure 7.7).

The lengths of the tree edges are now defined as follows:

$$d_1 = d_2 = h_5 = 0.0813,$$
$$d_3 = d_4 = h_6 = 0.2989,$$
$$d_5 = h_7 - d_1 = 0.4102,$$
$$d_6 = h_7 - d_3 = 0.1926.$$

Now, it is easy to see that the pairwise distances d_{ij}^* defined by the newly built tree are not always the same as the evolutionary distances d_{ij}:

$$d_{12}^* = d_1 + d_2 = 0.1625 = d_{12},$$
$$d_{13}^* = d_1 + d_5 + d_6 + d_3 = 2h_7 = 0.9831 \neq 0.9163 = d_{13},$$
$$d_{14}^* = d_1 + d_5 + d_6 + d_4 = 2h_7 = 0.9831 \neq 1.0498 = d_{14},$$
$$d_{23}^* = d_2 + d_5 + d_6 + d_3 = 2h_7 = 0.9831 \neq 0.9163 = d_{23},$$
$$d_{24}^* = d_2 + d_5 + d_6 + d_4 = 2h_7 = 0.9831 \neq 1.0498 = d_{24},$$
$$d_{34}^* = d_3 + d_4 = 0.5978 = d_{34}. \qquad \square$$

Problem 7.17 Given the protein sequences in Problem 7.16 with evolutionary distances d_{ij} (a) build an unrooted tree T by the neighbor-joining method and show that the additivity property holds; (b) show that a molecular clock property fails for any rooted tree derived from tree T by adding a root.

Solution (a) We calculate values of r_i and D_{ij} for $i, j = 1, \ldots, 4$ by Equation (7.9):

$$r_1 = \frac{1}{2} \sum_{j=1}^{4} d_{1j} = \frac{1}{2}(0.1625 + 0.9163 + 1.0498) = 1.0643,$$

$$r_2 = \frac{1}{2} \sum_{j=1}^{4} d_{2j} = \frac{1}{2}(0.1625 + 0.9163 + 1.0498) = 1.0643,$$

$$r_3 = \frac{1}{2} \sum_{j=1}^{4} d_{3j} = \frac{1}{2}(0.9163 + 0.9163 + 0.5978) = 1.2152,$$

$$r_4 = \frac{1}{2} \sum_{j=1}^{4} d_{4j} = \frac{1}{2}(1.0498 + 1.0498 + 0.5978) = 1.3487;$$

$$D_{12} = d_{12} - (r_1 + r_2) = -1.9661,$$

$$D_{13} = d_{13} - (r_1 + r_3) = -1.3632,$$

$$D_{14} = d_{14} - (r_1 + r_4) = -1.3432,$$

$$D_{23} = d_{23} - (r_2 + r_3) = -1.3632,$$

$$D_{24} = d_{24} - (r_2 + r_4) = -1.3632,$$

$$D_{34} = d_{34} - (r_3 + r_4) = -1.9661.$$

To start building a tree, we take the pair of leaves 1 and 2 with minimal D_{12} and join them into a new node 5. The new edges have lengths $d_{51} = \frac{1}{2}(d_{12} + r_1 - r_2) = 0.08125$ and $d_{52} = d_{12} - d_{51} = 0.08125$. Then we the find distances from node 5 to leaves 3 and 4:

$$d_{53} = \frac{1}{2}(d_{13} + d_{23} - d_{12}) = 0.83505,$$

$$d_{54} = \frac{1}{2}(d_{14} + d_{24} - d_{12}) = 0.96855.$$

We remove leaves 1 and 2, add node 5 to the set of leaves, and determine r_5 and D_{ij} for $i, j = 3, 4, 5$:

$$r_5 = \frac{1}{2}(d_{53} + d_{54}) = 0.9018,$$

$$D_{34} = d_{34} - (r_3 + r_4) = -1.9661,$$

$$D_{35} = d_{35} - (r_3 + r_5) = -1.28195,$$

$$D_{45} = d_{45} - (r_4 + r_5) = -1.28195.$$

Next, we choose leaves 3 and 4 for which D_{34} is minimal and join them into a new node 6 by edges with lengths $d_{36} = \frac{1}{2}(d_{34} + r_3 - r_4) = 0.23215$ and $d_{46} = d_{34} - d_{36} = 0.36565$. Finally, we join the two remaining nodes 5 and 6 by a new edge of length $d_{56} = \frac{1}{2}(d_{53} + d_{54} - d_{34}) = 0.6029$ to obtain the tree T (Figure 7.8).

The additivity property holds for tree T, as it can be directly verified that:

$$d_{12} = d_{15} + d_{52},$$

$$d_{13} = d_{15} + d_{56} + d_{63},$$

$$d_{14} = d_{15} + d_{56} + d_{64},$$

$$d_{23} = d_{25} + d_{56} + d_{63},$$

$$d_{24} = d_{25} + d_{56} + d_{64}.$$

(b) Obviously, if a new root 7 is added to one of the edges $(1, 5)$, $(2, 5)$, or $(5, 6)$, then $d_{73} < d_{74}$. If root 7 is added to edge $(3, 6)$, then $d_{73} < d_{74}$. If root 7 is added

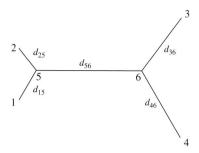

Figure 7.8. The unrooted tree constructed using the neighbor-joining method for the same sequences as in Figure 7.7.

to edge $(4, 6)$, then $d_{74} < d_{71} = d_{56} + d_{51}$. Thus, any choice of position of the root creates a rooted tree that does not possess the molecular clock property. □

Problem 7.18 Edwards (1970) introduced the following notion of a *labeled history*. A set of trees with the molecular clock property defines the same labeled history if these trees have the same topology and the same order of branch nodes in time. Show that the number L_n of labeled histories for trees with n leaves is given by the following formula:

$$L_n = \frac{n!(n-1)!}{2^{n-1}}.$$

Solution The labeled history is defined if the order of branch nodes in time is chosen. We consider the set of branch nodes as a set of common ancestors that we have to trace back in time. For n leaves, the number of ways to choose the first two to join (back in time) is equal to $\binom{n}{2} = (n(n-1))/2$. The next two nodes to join are chosen among the remaining $n - 2$ leaf nodes and the branch node (the ancestor) obtained at the first step. The number of ways to do so is $\binom{n-1}{2} = ((n-1)(n-2))/2$. We continue the process of joining the parentless nodes and counting the number of choices at each step, until the last two parentless nodes remain, and we join them in the only one possible way. Thus, the total number of possible orders of joining all the nodes of the tree is given by

$$L_n = \frac{n(n-1)}{2} \times \frac{(n-1)(n-2)}{2} \times \cdots \times \frac{2 \times 1}{2} = \frac{n!(n-1)!}{2^{n-1}}. \qquad \square$$

7.3 Further reading

The neighbor-joining (NJ) algorithm by Saitou and Nei (1987) has been frequently used for the reconstruction of phylogenetic trees. Several modifications of the NJ have been proposed: Gascuel (1997) suggested a generalization of the formula for evolutionary distance reducing the variance of the estimates of the new distances at each iteration. Bruno, Socci, and Halpern (2000) introduced a version of the NJ algorithm with distance weights compensating for errors in the estimates of the longer distances and increasing the likelihood of a reconstructed tree.

The quartet puzzling, a maximum likelihood based algorithm for the reconstruction of the topology of a phylogenetic tree, was presented by Strimmer and von Haeseler (1996). This algorithm is a three-step procedure, first reconstructing all possible quartet maximum likelihood trees, then repeatedly incorporating the quartet trees in an overall tree, and finally computing the majority rule consensus of all intermediate trees, the quartet puzzling tree. Parallelization of the quartet puzzling algorithm was performed by Schmidt *et al.* (2002). One more algorithm, IQPNNI, based on the quartet concept was introduced by Vinh and von Haeseler (2004). Guindon and Gascuel (2003) proposed a simple algorithm, implemented in the computer program PHYML. At each step PHYML modifies the current tree for the given data to increase the likelihood of the tree; only a few iterations are sufficient to reach the convergence.

The algorithms mentioned above (Strimmer and von Haeseler, 1996; Guindon and Gascuel, 2003; Vinh and von Haeseler, 2004) converge quickly; however, the space of trees examined by these algorithms is still just a small subset of the space of all possible trees. For example, the PHYML algorithm considers at most $3(n-3)$ unrooted topologies for n leaf sequences upon adjusting branch lengths to increase the likelihood, while there exist $(2n-5)!!$ distinct topologies (Problem 7.4). Even for small n ($n > 4$) the algorithm traverses less than half of the topologies, while for large n it stops upon examining a vanishingly small fraction of all possible topologies: for $n = 20$ we have $3(n-3) = 51$ and $(2n-5)!! \approx 2.22 \times 10^{20}$ (Felsenstein, 2004). Therefore, the success of such algorithms in building a tree close to the optimal one (with the maximum likelihood) critically depends on a good choice of the initial tree.

Grishin (1995, 1999) suggested new formulas to estimate the evolutionary distances between homologous protein sequences (traditionally defined as the average number of substitutions per site). Under the assumption of the independence of substitutions at different sites, the author provided a rigorous treatment of multiple and back substitutions as well as substitution rate variability among amino acids and among sequence sites. Using these "corrected" evolutionary distances, Feng, Cho, and Doolittle (1997) calculated the divergence times of the principal groups of eukaryotes and reconstructed the principal eukaryotic phylogeny. Huynen and Bork (1998) applied Grishin's formulas to determine divergence times from the complete genome sequences of nine archaeal and bacterial species. The comparison of complete genomes has led to new ways of measuring evolutionary distance and phylogenetic tree construction (Fitz-Gibbon and House, 1999; Snel, Bork, and Huyen, 1999; Tekaia, Lazcano, and Dujon, 1999; Brown *et al.*, 2001; Li *et al.*, 2001; Gubser *et al.*, 2004; Qi, Wang, and Hao, 2004). A comprehensive review of the whole genome-based methods of the phylogenetic tree building can be found in Snel, Huynen, and Dutilh (2005).

Issues of estimation of the divergence times and tree reconstruction were addressed by the following groups: Kumar and Hedges (1998) (for some vertebrate species); Arnason, Gullberg, and Janke (1998) (for primate species); Nei, Xu, and Glazko (2001) (for some mammalian species); Glazko and Nei (2003) (for major lineages of primate species); Korber *et al.* (2003) (for viruses from HIV-1 M group); Wolf, Rogozin, and Koonin (2004) (for plants, animals, and fungi species).

A comprehensive review by Arbogast *et al.* (2002) described recent advances in estimating the divergence time from molecular data on both phylogenetic and population genetic time scales. The authors emphasized the importance of model testing for divergence time estimation.

Compatibility of the molecular clock hypothesis with estimates of divergence times for the metazoan species derived from paleontology data has been shown by Ayala, Rzhetsky, and Ayala (1998). An extended discussion of the molecular clock property can be found in Bromham and Penny (2003).

The tree building methods are typically used not only for reconstructing phylogenies, but also for visualizing relationships among biological sequences. A new approach to such a visualization was proposed by Grishin and Grishin (2002). This method provides an isometric transformation of a set of protein sequences with evolutionary distances between them (defined as the average number of substitutions per site) into the points of a multidimensional Euclidean space. Then the points are grouped together according to the newly designed clustering procedure utilizing the mixture of multidimensional Gaussian probability densities.

8

Probabilistic approaches to phylogeny

Establishing phylogenetic relationships between species is one of the central problems of biological science. While in Chapter 7 the reader was introduced to non-probabilistic methods of building phylogenetic trees for DNA and protein sequences, Chapter 8 continues the subject from the standpoint of consistent probabilistic methodology. The evolution of biological sequences has been largely viewed as a random process, and several probabilistic models with varying levels of complexity have been proposed. Therefore, the reconstruction of phylogenetic relationships can be formulated in probabilistic terms as well.

Several introductory *BSA* problems in Chapter 8 are concerned with the properties of the simplest probabilistic models of evolution, such as the Jukes–Cantor and the Kimura models.

Given a set of sequences (associated with the leaves of a tree) and a model of the process of substitutions in a DNA or protein sequence, it is important to know how to compute the likelihood of a tree with a given topology. The Felsenstein algorithm addresses this issue using the post-order traversal. Felsenstein also developed an EM-type algorithm for finding the optimal (maximum likelihood) lengths of the tree edges. However, as the number of leaves increases, the number of tree topologies grows too quickly to be processed in a reasonable time.

Therefore, finding the optimal tree among all possible trees for a rather large number of sequences (leaves) is one of the major challenges. The mainstream approach to managing such a problem is sampling from the posterior distribution on the space of trees.

The tree HMM concept described in *BSA* could be used for phylogenetic tree construction utilizing most general models of the sequence evolution. However, full implementation of this impressive theoretical approach as an efficient computational tool presents a yet unsurmounted challenge due to the complexity of the algorithm.

Most of the *BSA* problems in this chapter are concerned with properties of the simple probabilistic models of evolution of biological sequences. Realistic models, which allow for variable rates at different sequence sites as well as insertions and deletions, are becoming too complicated in practice without excessive use of a computer. However, one of the models of this kind, the TKF model (Thorne, Kishino, and Felsenstein, 1991) is used in one of the additional problems. Finally, two more theoretical problems focus on the probabilistic interpretation of the algorithms described in Chapter 7, the parsimony and distance method, as well as the algorithm of simultaneous tree reconstruction and alignment (Sankoff and Cedergren, 1983; Hein, 1989).

8.1 Original problems

Problem 8.1 Show that the Jukes–Cantor and Kimura nucleotide substitution matrices introduced by Jukes and Cantor (1969) and Kimura (1980), respectively, are multiplicative: $S(t)S(s) = S(t + s)$ for all values of s and t.

Solution (a) The Jukes–Cantor substitution matrix for time t has the following form:

$$
\begin{pmatrix}
\frac{1}{4}(1 + 3e^{-4\alpha t}) & \frac{1}{4}(1 - e^{-4\alpha t}) & \frac{1}{4}(1 - e^{-4\alpha t}) & \frac{1}{4}(1 - e^{-4\alpha t}) \\[2ex]
\frac{1}{4}(1 - e^{-4\alpha t}) & \frac{1}{4}(1 + 3e^{-4\alpha t}) & \frac{1}{4}(1 - e^{-4\alpha t}) & \frac{1}{4}(1 - e^{-4\alpha t}) \\[2ex]
\frac{1}{4}(1 - e^{-4\alpha t}) & \frac{1}{4}(1 - e^{-4\alpha t}) & \frac{1}{4}(1 + 3e^{-4\alpha t}) & \frac{1}{4}(1 - e^{-4\alpha t}) \\[2ex]
\frac{1}{4}(1 - e^{-4\alpha t}) & \frac{1}{4}(1 - e^{-4\alpha t}) & \frac{1}{4}(1 - e^{-4\alpha t}) & \frac{1}{4}(1 + 3e^{-4\alpha t})
\end{pmatrix}
$$

Here $S_{ij}(t) = P(A_j|A_i, t)$ is the probability of a nucleotide of type A_i being substituted by a nucleotide of type A_j over time t. To prove multiplicativity, we have to show that for any $i, j = 1, \ldots, 4$

$$(S(t)S(s))_{ij} = S_{ij}(t + s).$$

This equality holds for the diagonal elements as follows:

$$(S(t)S(s))_{ii} = \sum_{k=1}^{4} S_{ik}(t)S_{ki}(s)$$

$$= \frac{1}{16}(1 + 3e^{-4\alpha t})(1 + 3e^{-4\alpha s}) + \frac{3}{16}(1 - e^{-4\alpha t})(1 - e^{-4\alpha s})$$

$$= \frac{1}{16}(1 + 3e^{-4\alpha t} + 3e^{-4\alpha s} + 9e^{-4\alpha(t+s)} + 3 - 3e^{-4\alpha t} - 3e^{-4\alpha s}$$

$$+ 3e^{-4\alpha(t+s)}) = \frac{1}{4}(1 + 3e^{-4\alpha(t+s)}) = S_{ii}(t + s).$$

Similarly, for the non-diagonal elements,

$$(S(t)S(s))_{ij} = \sum_{k=1}^{4} S_{ik}(t)S_{kj}(s)$$

$$= \frac{1}{16}(1 + 3e^{-4\alpha t})(1 - e^{-4\alpha s}) + \frac{1}{16}(1 + 3e^{-4\alpha s})(1 - e^{-4\alpha t})$$

$$+ \frac{2}{16}(1 - e^{-4\alpha t})(1 - e^{-4\alpha s})$$

$$= \frac{1}{16}(1 - e^{-4\alpha t} + 3e^{-4\alpha s} - 3e^{-4\alpha(s+t)}$$

$$+ 1 - e^{-4\alpha s} + 3e^{-4\alpha t} - 3e^{-4\alpha(s+t)} + 2 - 2e^{-4\alpha t} - 2e^{-4\alpha s}$$

$$+ 2e^{-4\alpha(s+t)}) = \frac{1}{4}(1 - e^{-4\alpha(t+s)}) = S_{ij}(t + s).$$

Hence, the family of Jukes–Cantor matrices is multiplicative.

(b) The Kimura nucleotide substitution matrix

$$\begin{pmatrix} r_t & q_t & u_t & q_t \\ q_t & r_t & q_t & u_t \\ u_t & q_t & r_t & q_t \\ q_t & u_t & q_t & r_t \end{pmatrix}$$

has elements

$$q_t = \frac{1}{4}(1 - e^{-4\beta t}),$$

$$u_t = \frac{1}{4}(1 + e^{-4\beta t} - 2e^{-2(\alpha+\beta)t}),$$

$$r_t = 1 - 2q_t - u_t.$$

Again we need to show that for any $i, j = 1, \ldots, 4$

$$(S(t)S(s))_{ij} = S_{ij}(t + s).$$

The equality holds for the diagonal elements as follows:

$$(S(t)S(s))_{ii} = \sum_{k=1}^{4} S_{ik}(t)S_{ki}(s) = r_t r_s + 2q_t q_s + u_t u_s$$

$$= \frac{2}{16}(1 - e^{-4\beta t})(1 - e^{-4\beta s})$$

$$+ \frac{1}{16}(1 + e^{-4\beta t} - 2e^{-2(\alpha+\beta)t})(1 + e^{-4\beta s} - 2e^{-2(\alpha+\beta)s})$$

$$+ \frac{1}{16}(1 + e^{-4\beta t} + 2e^{-2(\alpha+\beta)t})(1 + e^{-4\beta s} + 2e^{-2(\alpha+\beta)s})$$

$$= \frac{1}{4} + \frac{1}{4}e^{-4\beta(t+s)} + \frac{1}{2}e^{-2(\alpha+\beta)(t+s)} = S_{ii}(t+s).$$

For the non-diagonal entries S_{ij}, such that $i + j$ is an odd number, we have

$$S_{12}(t+s) = S_{14}(t+s) = S_{21}(t+s) = S_{23}(t+s) = S_{32}(t+s)$$

$$= S_{34}(t+s) = S_{41}(t+s) = S_{43}(t+s) = q_{t+s}$$

$$= \frac{1}{4}(1 - e^{-4\beta(t+s)}).$$

On the other hand,

$$(S(t)S(s))_{12} = (S(t)S(s))_{14} = (S(t)S(s))_{21} = (S(t)S(s))_{23}$$

$$= (S(t)S(s))_{32} = (S(t)S(s))_{34} = (S(t)S(s))_{41} = (S(t)S(s))_{42}$$

$$= r_t q_s + q_t r_s + u_t q_s + q_t u_s$$

$$= \frac{1}{16}(1 + e^{-4\beta t} + 2e^{-2(\alpha+\beta)t})(1 - e^{-4\beta s})$$

$$+ \frac{1}{16}(1 + e^{-4\beta s} + 2e^{-2(\alpha+\beta)s})(1 - e^{-4\beta t})$$

$$+ \frac{1}{16}(1 + e^{-4\beta t} - 2e^{-2(\alpha+\beta)t})(1 - e^{-4\beta s})$$

$$+ \frac{1}{16}(1 + e^{-4\beta s} - 2e^{-2(\alpha+\beta)s})(1 - e^{-4\beta t})$$

$$= \frac{1}{4}(1 - e^{-4\beta(t+s)}).$$

Finally, for the non-diagonal entries S_{ij}, such that $i + j$ is an even number, we have

$$S_{13}(t+s) = S_{24}(t+s) = S_{31}(t+s) = S_{42}(t+s) = u_{t+s}$$

$$= \frac{1}{4}(1 + e^{-4\beta(t+s)} - 2e^{-2(\alpha+\beta)(t+s)}),$$

and corresponding elements of the product of matrices $S(t)$ and $S(s)$ are as follows:

$$(S(t)S(s))_{13} = (S(t)S(s))_{24} = (S(t)S(s))_{31} = (S(t)S(s))_{42}$$

$$= r_t u_s + 2q_t q_s + u_t r_s$$

$$= \frac{1}{16}(1 + e^{-4\beta t} + 2e^{-2(\alpha+\beta)t})(1 + e^{-4\beta s} - 2e^{-2(\alpha+\beta)s})$$

$$+ \frac{2}{16}(1 - e^{-4\beta t})(1 - e^{-4\beta s})$$

$$+ \frac{1}{16}(1 + e^{-4\beta t} - 2e^{-2(\alpha+\beta)t})(1 + e^{-4\beta s} + 2e^{-2(\alpha+\beta)s})$$

$$= \frac{1}{4}(1 + e^{-4\beta(t+s)} - 2e^{-2(\alpha+\beta)(t+s)}).$$

We have directly demonstrated that the matrix product $S(t)S(s)$ has the same elements as the matrix $S(t + s)$; therefore, the multiplicative property holds for the family of the Kimura substitution matrices. □

Problem 8.2 Let $P(a(t_2)|b(t_1))$ denote the probability of a residue b, present at time t_1, having been substituted by a residue a by time t_2. The stationarity means that we can write this as $P(a|b, t_2 - t_1)$. The Markov property means that $P(a(t_2)|b(t_1), c(t_0)) = P(a(t_2)|b(t_1))$ if $t_0 < t_1$, i.e. that the probability of the substitution of b by a is not influenced by the event that the residue was c at the earlier time t_0. Show that

$$\sum_{b(t)} P(a(s+t)|b(t), c(0))P(b(t)|c(0)) = P(a(s+t)|c(0)), \qquad (8.1)$$

and deduce that multiplicativity holds.

Solution The Markov property allows us to transform each term on the left-hand side of Equation (8.1):

$$P(a(s+t)|b(t), c(0))P(b(t)|c(0)) = P(a(s+t)|b(t))P(b(t)|c(0)).$$

The stationarity condition allows us to write

$$P(a(s+t)|b(t))P(b(t)|c(0)) = P(a|b, s+t-t)P(b|c, t-0) = P(a|b, s)P(b|c, t).$$

By returning to the sum over all possible types of residues $b(t)$ at moment t, we obtain the following expression:

$$\sum_{b(t)} P(a|b, s)P(b|c, t) = P(a|c, s+t),$$

which is the probability that residue c will be substituted by residue a over time $s + t$. Once again, use of the stationarity condition leads to

$$P(a|c, s + t) = P(a|c, s + t - 0) = P(a(s + t)|c(0)),$$

which completes the proof of equality (8.1).

The multiplicative property $S(s)S(t) = S(s + t)$ immediately follows from

$$\sum_b P(a|b, s)P(b|c, t) = P(a|c, s + t),$$

which shows that an element of the substitution matrix $S(s+t)$ can be obtained from the elements of matrices $S(s)$ and $S(t)$ by the rule of matrix multiplication. □

Problem 8.3 Let r_t denote a diagonal element of the Jukes–Cantor substitution matrix, $r_t = \frac{1}{4}(1 + 3e^{-4\alpha t})$, and let s_t denote a non-diagonal element of this matrix, $s_t = \frac{1}{4}(1 - e^{-4\alpha t})$. Show that $r_{t_1} r_{t_2} + 3 s_{t_1} s_{t_2}$ and $2 s_{t_1} s_{t_2} + s_{t_1} r_{t_2} + s_{t_2} r_{t_1}$ are terms arising from the product of the Jukes–Cantor matrices for times t_1 and t_2, and deduce that they can be written as $r_{t_1 + t_2}$ and $s_{t_1 + t_2}$, respectively.

Solution Let a matrix

$$A = \begin{pmatrix} a & b & b & b \\ b & a & b & b \\ b & b & a & b \\ b & b & b & a \end{pmatrix}$$

designate the product of the Jukes–Cantor substitution matrices $S(t_1)$ and $S(t_2)$ for times t_1 and t_2, respectively, i.e. $A = S(t_1)S(t_2)$. The rule of matrix multiplication implies $a = r_{t_1} + 3 s_{t_1} s_{t_2}$ and $b = r_{t_1} s_{t_2} + r_{t_2} s_{t_1} + 2 s_{t_1} s_{t_2}$. On the other hand, $S(t_1)S(t_2) = S(t_1 + t_2)$ since matrices $S(t)$ possess the multiplicative property (see Problem 8.1). Therefore,

$$r_{t_1 + t_2} = a = r_{t_1} + 3 s_{t_1} s_{t_2},$$

$$s_{t_1 + t_2} = b = r_{t_1} s_{t_2} + r_{t_2} s_{t_1} + 2 s_{t_1} s_{t_2}. \qquad □$$

Problem 8.4 Let $P(x^1, x^2, x^3 | T, t_1, t_2, t_3)$ be a likelihood of the trifurcating tree T with edge lengths t_1, t_2, t_3 and leaves x^1, x^2, x^3, where x^1, x^2, x^3 are three nucleotide sequences composed only of C's and G's:

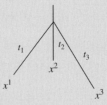

Show that if an ungapped alignment of sequences x^1, x^2, x^3 contains n_1 sites with residues of the same type, n_2 sites with residues of type CCG or GGC, n_3 sites with residues of type CGC or GCG, and n_4 sites with residues of type GCC or CGG, then

$$P(x^1, x^2, x^3 | T, t_1, t_2, t_3) = 4^{-3(n_1+n_2+n_3+n_4)} a(t_1, t_2, t_3)^{n_1} b(t_1, t_2, t_3)^{n_2}$$
$$\times b(t_1, t_3, t_2)^{n_3} b(t_3, t_2, t_1)^{n_4},$$

where $a(t_1, t_2, t_3)$ and $b(t_1, t_2, t_3)$ are given by

$$a(t_1, t_2, t_3) = 1 + 3e^{-4\alpha(t_1+t_2)} + 3e^{-4\alpha(t_1+t_3)} + 3e^{-4\alpha(t_2+t_3)} + 6e^{-4\alpha(t_1+t_2+t_3)}$$

and

$$b(t_1, t_2, t_3) = 1 + 3e^{-4\alpha(t_1+t_2)} - e^{-4\alpha(t_1+t_3)} - 3e^{-4\alpha(t_2+t_3)} - 2e^{-4\alpha(t_1+t_2+t_3)}.$$

Solution Suppose that the three given nucleotide sequences have length N; then, under the assumption of the site independence, we can write

$$P(x^1, x^2, x^3 | T, t_1, t_2, t_3) = \prod_{u=1}^{N} P(x_u^1, x_u^2, x_u^3 | T, t_1, t_2, t_3).$$

Now we consider the possible combinations of nucleotides at a given site u. The probability of C occuring at site u in all three leaf sequences expressed in terms of the substitution probabilities, the elements of the Jukes–Cantor matrix, and the probability q_a of a residue a to appear at the tree root is given by

$$P(x_u^1, x_u^2, x_u^3 | T, t_1, t_2, t_3)$$
$$= P(CCC | T, t_1, t_2, t_3)$$
$$= q_C P(C|C, t_1) P(C|C, t_2) P(C|C, t_3) + q_G P(C|G, t_1) P(C|G, t_2) P(C|G, t_3)$$
$$+ q_A P(C|A, t_1) P(C|A, t_2) P(C|A, t_3) + q_T P(C|T, t_1) P(C|T, t_2) P(C|T, t_3)$$
$$= q_C r_{t_1} r_{t_2} r_{t_3} + q_G s_{t_1} s_{t_2} s_{t_3} + q_A s_{t_1} s_{t_2} s_{t_3} + q_C s_{t_1} s_{t_2} s_{t_3}.$$

Under the assumption that the nucleotide frequencies at the root are equal to the equilibrium frequencies of the Jukes–Cantor matrix, $q_A = q_C = q_G = q_T = 1/4$, we have

$$P(CCC|T, t_1, t_2, t_3) = \frac{1}{4}(r_{t_1} r_{t_2} r_{t_3} + 3s_{t_1} s_{t_2} s_{t_3})$$

$$= 4^{-4}((1 + 3e^{-4\alpha t_1})(1 + 3e^{-4\alpha t_2})$$

$$\times (1 + 3e^{-4\alpha t_3}) + 3(1 - e^{-4\alpha t_1})$$

$$\times (1 - e^{-4\alpha t_2})(1 - e^{-4\alpha t_3}))$$

$$= 4^{-3}(1 + 3e^{-4\alpha(t_1 + t_2)} + 3e^{-4\alpha(t_1 + t_3)}$$

$$+ 3e^{-4\alpha(t_2 + t_3)} + 6e^{-4\alpha(t_1 + t_2 + t_3)}).$$

Due to the substitution process symmetry, the expression for the probability $P(GGG|T, t_1, t_2, t_3)$ of G occurring at site u will be just the same. The likelihoods for the other combinations of the nucleotides at site u will be given by the following expressions. For GCC (and CGG):

$$P(GCC|T, t_1, t_2, t_3) = q_C s_{t_1} r_{t_2} r_{t_3} + q_G r_{t_1} s_{t_2} s_{t_3} + q_A s_{t_1} s_{t_2} s_{t_3} + q_T s_{t_1} s_{t_2} s_{t_3}$$

$$= \frac{1}{4}(r_{t_1} s_{t_2} s_{t_3} + s_{t_1} r_{t_2} r_{t_3} + 2s_{t_1} s_{t_2} s_{t_3})$$

$$= 4^{-3}(1 - e^{-4\alpha(t_1 + t_2)} - e^{-4\alpha(t_1 + t_3)}$$

$$+ 3e^{-4\alpha(t_2 + t_3)} - 2e^{-4\alpha(t_1 + t_2 + t_3)});$$

for GGC (and CCG):

$$P(GGC|T, t_1, t_2, t_3) = q_C s_{t_1} s_{t_2} r_{t_3} + q_G r_{t_1} r_{t_2} s_{t_3} + q_A s_{t_1} s_{t_2} s_{t_3} + q_T s_{t_1} s_{t_2} s_{t_3}$$

$$= \frac{1}{4}(s_{t_1} s_{t_2} r_{t_3} + r_{t_1} r_{t_2} s_{t_3} + 2s_{t_1} s_{t_2} s_{t_3})$$

$$= 4^{-3}(1 - e^{-4\alpha(t_1 + t_3)} - e^{-4\alpha(t_2 + t_3)}$$

$$+ 3e^{-4\alpha(t_1 + t_2)} - 2e^{-4\alpha(t_1 + t_2 + t_3)});$$

for GCG (and CGC):

$$P(GCG|T, t_1, t_2, t_3) = q_C s_{t_1} r_{t_2} s_{t_3} + q_G r_{t_1} s_{t_2} r_{t_3} + q_A s_{t_1} s_{t_2} s_{t_3} + q_T s_{t_1} s_{t_2} s_{t_3}$$

$$= \frac{1}{4}(s_{t_1} r_{t_2} s_{t_3} + r_{t_1} s_{t_2} r_{t_3} + 2s_{t_1} s_{t_2} s_{t_3})$$

$$= 4^{-3}(1 - e^{-4\alpha(t_1 + t_2)} - e^{-4\alpha(t_2 + t_3)}$$

$$+ 3e^{-4\alpha(t_1 + t_3)} - 2e^{-4\alpha(t_1 + t_2 + t_3)}).$$

Therefore, if the alignment of sequences x^1, x^2, x^3 contains n_1 sites with residues of the same type (CCC or GGG), n_2 sites of CCG or GGC type, n_3 sites of CGC or

GCG types, and n_4 sites of GCC or CGG types, then the likelihood of the trifurcating tree T becomes

$$P(x^1, x^2, x^3 | T, t_1, t_2, t_3) = 4^{-3(n_1+n_2+n_3+n_4)} a(t_1, t_2, t_3)^{n_1} b(t_1, t_2, t_3)^{n_2}$$
$$\times b(t_1, t_3, t_2)^{n_3} b(t_3, t_2, t_1)^{n_4}.$$

Here

$$a(t_1, t_2, t_3) = 1 + 3e^{-4\alpha(t_1+t_2)} + 3e^{-4\alpha(t_1+t_3)} + 3e^{-4\alpha(t_2+t_3)} + 6e^{-4\alpha(t_1+t_2+t_3)}$$

and

$$b(t_1, t_2, t_3) = 1 + 3e^{-4\alpha(t_1+t_2)} - e^{-4\alpha(t_1+t_3)} - 3e^{-4\alpha(t_2+t_3)} - 2e^{-4\alpha(t_1+t_2+t_3)}.$$

□

Problem 8.5 A family of substitution matrices $S(t)$ for the three-residue alphabet $\{A, B, C\}$ is defined by

$$S(t) = \begin{pmatrix} r_t & s_t & u_t \\ u_t & r_t & s_t \\ s_t & u_t & r_t \end{pmatrix}$$

with elements

$$r_t = \frac{1}{3}\left(1 + 2e^{-3\alpha t/2} \cos\left(\frac{\sqrt{3}\alpha t}{2}\right)\right),$$

$$s_t = \frac{1}{3}\left(1 - e^{-3\alpha t/2} \cos\left(\frac{\sqrt{3}\alpha t}{2}\right) + \sqrt{3}e^{-3\alpha t/2} \sin\left(\frac{\sqrt{3}\alpha t}{2}\right)\right),$$

$$u_t = \frac{1}{3}\left(1 - e^{-3\alpha t/2} \cos\left(\frac{\sqrt{3}\alpha t}{2}\right) - \sqrt{3}e^{-3\alpha t/2} \sin\left(\frac{\sqrt{3}\alpha t}{2}\right)\right).$$

Show that this family is multiplicative, has positive entries, and that the substitution rates are given by the following matrix:

$$\begin{pmatrix} -\alpha & \alpha & 0 \\ 0 & -\alpha & \alpha \\ \alpha & 0 & -\alpha \end{pmatrix}$$

Find the limiting distribution and show that the reversibility property, i.e.

$$P(b|a, t)q_a = P(a|b, t)q_b,$$

fails for almost all $t > 0$.

Solution (1) To prove multiplicativity, we need to show that for any $i, j = 1, 2, 3$,

$$(S(t)S(s))_{ij} = S_{ij}(t + s).$$

By direct multiplication and comparison for the diagonal elements $S_{ii}(\bullet), i = 1, 2, 3$, we have on the right-hand side:

$$S_{ii}(t + s) = r(t + s) = \frac{1}{3}\left(1 + 2e^{-3\alpha(t+s)/2}\cos\left(\frac{\sqrt{3}\alpha(t+s)}{2}\right)\right)$$

$$= \frac{1}{3} + \frac{2}{3}e^{-3\alpha(t+s)/2}\cos\left(\frac{\sqrt{3}\alpha t}{2}\right)\cos\left(\frac{\sqrt{3}\alpha s}{2}\right)$$

$$+ \frac{2}{3}e^{-3\alpha(t+s)/2}\sin\left(\frac{\sqrt{3}\alpha t}{2}\right)\sin\left(\frac{\sqrt{3}\alpha s}{2}\right).$$

On the left-hand side:

$$(S(t)S(s))_{ii} = r_t r_s + s_t u_s + u_t s_s$$

$$= \frac{1}{9}\left(1 + 2e^{-3\alpha t/2}\cos\left(\frac{\sqrt{3}\alpha t}{2}\right)\right)\left(1 + 2e^{-3\alpha s/2}\cos\left(\frac{\sqrt{3}\alpha s}{2}\right)\right)$$

$$+ \frac{1}{9}\left(1 - e^{-3\alpha t/2}\cos\left(\frac{\sqrt{3}\alpha t}{2}\right) + \sqrt{3}e^{-3\alpha t/2}\sin\left(\frac{\sqrt{3}\alpha t}{2}\right)\right)$$

$$\times \left(1 - e^{-3\alpha s/2}\cos\left(\frac{\sqrt{3}\alpha s}{2}\right) - \sqrt{3}e^{-3\alpha s/2}\sin\left(\frac{\sqrt{3}\alpha s}{2}\right)\right)$$

$$+ \frac{1}{9}\left(1 - e^{-3\alpha t/2}\cos\left(\frac{\sqrt{3}\alpha t}{2}\right) - \sqrt{3}e^{-3\alpha t/2}\sin\left(\frac{\sqrt{3}\alpha t}{2}\right)\right)$$

$$\times \left(1 - e^{-3\alpha s/2}\cos\left(\frac{\sqrt{3}\alpha s}{2}\right) + \sqrt{3}e^{-3\alpha st/2}\sin\left(\frac{\sqrt{3}\alpha s}{2}\right)\right)$$

$$= \frac{1}{3} + \frac{2}{3}e^{-3\alpha(t+s)/2}\cos\left(\frac{\sqrt{3}\alpha t}{2}\right)\cos\left(\frac{\sqrt{3}\alpha s}{2}\right)$$

$$+ \frac{2}{3}e^{-3\alpha(t+s)/2}\sin\left(\frac{\sqrt{3}\alpha t}{2}\right)\sin\left(\frac{\sqrt{3}\alpha s}{2}\right).$$

A similar demonstration for the two types of the non-diagonal elements, s_t and u_t, is left to the reader.

(2) To show that elements of $S(t)$ are positive, we introduce functions f_i, $i = 1, 2, 3$, of variable x, $x = \sqrt{3}\alpha t/2 \geq 0$:

$$f_1(x) = 3r_t = 1 + 2e^{-\sqrt{3}x} \cos x,$$

$$f_2(x) = 3s_t = 1 - e^{-\sqrt{3}x} \cos x + \sqrt{3}e^{-\sqrt{3}x} \sin x,$$

$$f_3(x) = 3u_t = 1 - e^{-\sqrt{3}x} \cos x - \sqrt{3}e^{-\sqrt{3}x} \sin x.$$

To prove that f_i is positive, it is sufficient to show that f_i is positive at the point(s) of the global minimum of f_i. For the derivative $f_1'(x)$, we have

$$f_1'(x) = -2\sqrt{3}e^{-\sqrt{3}x} \cos x - 2e^{-\sqrt{3}x} \sin x$$

$$= -4e^{-\sqrt{3}x} \left(\frac{\sqrt{3}}{2} \cos x + \frac{1}{2} \sin x \right) = -4e^{-\sqrt{3}x} \sin \left(\frac{\pi}{3} + x \right).$$

We find that $f_1'(x) = 0$ for $x = -(\pi/3) + \pi n$, n is a positive integer, while $f_1'(x) < 0$ for $x \in (-(\pi/3) + 2\pi n, (2\pi/3) + 2\pi n)$. Therefore, points $x = (2\pi/3) + 2\pi n$ (n is a positive integer) are points of minimum of $f_1(x)$, since f_1' changes sign from negative to positive at these points. Since

$$f_1 \left(\frac{2\pi}{3} + 2\pi n \right) = 1 - e^{-\sqrt{3}(\frac{2\pi}{3} + 2\pi n)} > 0,$$

the diagonal element $r_t = f_1(x)/3$ of matrix $S(t)$ is positive for any $t > 0$.

Similarly, for f_2 we have

$$f_2(x) = 1 - e^{-\sqrt{3}x} \cos x + \sqrt{3}e^{-\sqrt{3}x} \sin x$$

$$= 1 - 2e^{-\sqrt{3}x} \left(\frac{1}{2} \cos x - \frac{\sqrt{3}}{2} \sin x \right)$$

$$= 1 - 2e^{-\sqrt{3}x} \sin \left(\frac{\pi}{6} - x \right),$$

$$f_2'(x) = 2\sqrt{3}e^{-\sqrt{3}x} \sin \left(\frac{\pi}{6} - x \right) + 2e^{-\sqrt{3}x} \cos \left(\frac{\pi}{6} - x \right)$$

$$= 4e^{-\sqrt{3}x} \left(\frac{\sqrt{3}}{2} \sin \left(\frac{\pi}{6} - x \right) + \frac{1}{2} \cos \left(\frac{\pi}{6} - x \right) \right)$$

$$= 4e^{-\sqrt{3}x} \sin \left(\frac{\pi}{3} - x \right).$$

The function f_2 attains the minimum values at points $x = (4\pi/3) + 2\pi n$, n is a positive integer, with

$$f_2 \left(\frac{4\pi}{3} + 2\pi n \right) = 1 - e^{-\sqrt{3}((4\pi/3) + 2\pi n)} > 0.$$

Hence, $s_t > 0$ for any $t > 0$ ($s_0 = 0$).

Finally,

$$f_3(x) = 1 - e^{-\sqrt{3}x}\cos x - \sqrt{3}e^{-\sqrt{3}x}\sin x$$

$$= 1 - 2e^{-\sqrt{3}x}\left(\frac{1}{2}\cos x + \frac{\sqrt{3}}{2}\sin x\right)$$

$$= 1 - 2e^{-\sqrt{3}x}\sin\left(\frac{\pi}{6} + x\right),$$

$$f_3'(x) = 2\sqrt{3}e^{-\sqrt{3}x}\sin\left(\frac{\pi}{6} + x\right) - 2e^{-\sqrt{3}x}\cos\left(\frac{\pi}{6} + x\right)$$

$$= 4e^{-\sqrt{3}x}\left(\frac{\sqrt{3}}{2}\sin\left(\frac{\pi}{6} + x\right) - \frac{1}{2}\cos\left(\frac{\pi}{6} + x\right)\right)$$

$$= 4e^{-\sqrt{3}x}\sin x.$$

Function f_3 reaches minimal values at points $x = 2\pi n$, n is a positive integer, with

$$f_3(2\pi n) = 1 - e^{-2-\sqrt{3}\pi n} > 0.$$

Hence, $u_t > 0$ for any $t > 0$ ($u_0 = 0$). Thus, we have shown that all elements of the substitution matrix $S(t)$ are positive for any $t > 0$ ($s_0 = u_0 = 0$).

(3) If R is a matrix of substitution rates for the family $S(t)$, $t \geq 0$, then $S'(t) = S(t)R$. At $t = 0$ with $S(0) = I$ (I is the identity matrix of order 3), this equation becomes $S'(0) = IR = R$. To compute the derivatives of the elements of matrix S, we once again use the derivatives of $f_i(x)$, $i = 1, 2, 3$, $x = (\sqrt{3}/2)\alpha t$. We have

$$r_t = \frac{1}{3}f_1(x), \qquad\qquad s_t = \frac{1}{3}f_2(x), \qquad\qquad u_t = \frac{1}{3}f_3(x),$$

$$r_0' = -\frac{2}{\sqrt{3}}\alpha\sin\frac{\pi}{3} = -\alpha; \quad s_0' = \frac{2}{\sqrt{3}}\alpha\sin\frac{\pi}{3} = \alpha; \quad u_0' = \frac{2}{\sqrt{3}}\alpha\sin 0 = 0.$$

Therefore,

$$R = S'(0) = \begin{pmatrix} r_0' & s_0' & u_0' \\ u_0' & r_0' & s_0' \\ s_0' & u_0' & r_0' \end{pmatrix} = \begin{pmatrix} -\alpha & \alpha & 0 \\ 0 & -\alpha & \alpha \\ \alpha & 0 & -\alpha \end{pmatrix}.$$

(4) When t tends to infinity, the rows of the substitution matrix $S(t)$ converge to the stationary distribution of residues. From the analytical expressions for elements r_t, s_t, and u_t, we obtain

$$\lim_{t\to\infty} S(t) = \begin{pmatrix} \frac{1}{3} & \frac{1}{3} & \frac{1}{3} \\ \frac{1}{3} & \frac{1}{3} & \frac{1}{3} \\ \frac{1}{3} & \frac{1}{3} & \frac{1}{3} \end{pmatrix} = \begin{pmatrix} q_A & q_B & q_C \\ q_A & q_B & q_C \\ q_A & q_B & q_C \end{pmatrix},$$

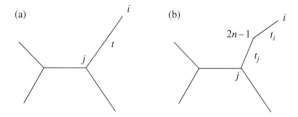

Figure 8.1. (a) Initial unrooted tree T. (b) Addition of the root node $(2n-1)$ to the edge (i,j).

where A, B, and C designate the residue types. Thus, all equilibrium frequencies are equal, i.e. $q_A = q_B = q_C = \frac{1}{3}$.

Let a and b be two different types of residues. The reversibility property holds if and only if

$$P(b|a,t)q_a = P(a|b,t)q_b,$$

for all $t > 0$, which leads to $P(b|a,t) = P(a|b,t)$ if $q_a = q_b$. This requires that $s_{ij} = s_{ji}$ for any $i \neq j$, $i,j = 1,2,3$, i.e. that $S(t)$ must be a symmetric matrix. However, the symmetry does not take place in our case. For instance, $P(B|A,t) = u_t$ and $P(A|B,t) = s_t$ and the values of these two matrix elements are not the same for $t \neq 2\pi n/\sqrt{3}\alpha$, n is a positive integer. Hence, the reversibility property fails.   □

Problem 8.6   The reversibility property of the probabilistic model of the mutation process means that

$$P(b|a,t)q_a = P(a|b,t)q_b$$

for any residues a, b, and any time t. It can be shown that this property and the multiplicativity of the substitution matrix make a tree likelihood independent of the root position. What happens when the root is moved to one of the leaf nodes?

Solution   Assume that n sequences of length N are associated with the leaves of an unrooted tree T. We also assume that sequence x^i is associated with the leaf node i connected to the branch node j by edge (i,j) with length t (Figure 8.1a).

If the root node $(2n-1)$ is added to the edge (i,j) (Figure 8.1b), thus creating two edges with lengths t_i and t_j, then at site u the likelihood $P(L_{2n-1}|a)$ for tree T given residue a at the root could be computed by Felsenstein's algorithm (Felsenstein, 1981) as follows:

$$P(L_{2n-1}|a) = \sum_{b,c} P(b|a,t_i)P(L_i|b)P(c|a,t_j)P(L_j|c).$$

Then the likelihood of tree T with edge lengths $t_\bullet$ at site u becomes

$$L_T^u = P(x_u^\bullet|T, t_\bullet) = \sum_a P(L_{2n-1}|a)q_a = \sum_{a,c} P(x_u^i|a, t_i)P(c|a, t_j)P(L_j|c)q_a.$$

Here we have taken into account that, since i is the leaf node, $P(L_i|b) = 1$ if $b = x_u^i$, and $P(L_i|b) = 0$ otherwise. The reversibility and multiplicativity properties allow us to eliminate the dependence of the likelihood L_T^u on the root residue a as follows:

$$L_T^u = \sum_{a,c} P(x_u^i|a, t_i)P(a|c, t_j)q_c P(L_j|c)$$

$$= \sum_c \left(\sum_a P(x_u^i|a, t_i)P(a|c, t_j) \right) P(L_j|c)q_c$$

$$= \sum_c P(x_u^i|c, t_i + t_j)P(L_j|c)q_c = \sum_c P(x_u^i|c, t)P(L_j|c)q_c.$$

Assuming independence of sequence sites, we have, for the likelihood of tree T,

$$L_T = \prod_{u=1}^N L_T^u = \prod_{u=1}^N \sum_c P(x_u^i|c, t)P(L_j|c)q_c. \tag{8.2}$$

If the root node coincides with the leaf node i, then a new tree T^* has $n-1$ leaves, and the sequence at the root node i is x^i:

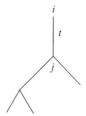

For site u, the probability of appearance of residue x_u^i at the root is $q_{x_u^i}$, and the likelihood of tree T^* with edge lengths $t_\bullet^*$ becomes:

$$L_{T^*}^u = P(L_{2n-1}|x_u^i)q_{x_u^i} = \sum_c P(c|x_u^i, t)q_{x_u^i}P(L_j|c).$$

The likelihood of tree T^* over all sites is given by

$$L_{T^*} = \prod_{u=1}^N L_{T^*}^u = \prod_{u=1}^N q_{x_u^i} \sum_c P(c|x_u^i, t)q_{x_u^i}P(L_j|c).$$

Now, Equation (8.2) and the reversibility property imply

$$L_T = \prod_{u=1}^N \sum_c P(c|x_u^i, t)P(L_j|c)q_{x_u^i} = \prod_{u=1}^N q_{x_u^i} \sum_c P(c|x_u^i, t)P(L_j|c) = L_{T^*}.$$

Therefore, the likelihood of a tree does not change if its root moves to one of the leaf nodes. □

Problem 8.7 Only two types of nucleotides, C and G, are present in sequences x^1 and x^2 of equal length. Given the ungapped alignment of x^1 and x^2, calculate the likelihood of a tree relating these sequences, assuming the Jukes–Cantor model. Show that the maximum likelihood edge lengths, t_1 and t_2, satisfy the following equation:

$$t_1 + t_2 = \frac{1}{4\alpha} \ln \frac{3(n_1 + n_2)}{3n_1 - n_2}.$$

Here n_1 is the number of alignment sites with identical (matching) residues, and n_2 is the number of sites with mismatches.

Solution We consider tree T with leaves x^1 and x^2 and edge lengths t_1 and t_2:

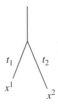

At a site u there are four possible choices of nucleotide pairs on the leaves of the tree: CC, CG, GC, and GG. For pair CC we have

$$P(x^1_u, x^2_u | T, t_1, t_2) = P(CC|T, t_1, t_2)$$

$$= q_C P(C|C, t_1) P(C|C, t_2) + q_G P(C|G, t_1) P(C|G, t_2)$$

$$+ q_A P(C|A, t_1) P(C|A, t_2) + q_T P(C|T, t_1) P(C|T, t_2)$$

$$= q_C r_{t_1} r_{t_2} + q_G s_{t_1} s_{t_2} + q_A s_{t_1} s_{t_2} + q_C s_{t_1} s_{t_2}.$$

Here r_t and s_t are elements of the Jukes–Cantor substitution matrix and q_a is the probability of a nucleotide a at the tree root. Assuming that $q_a = 1/4$ for all nucleotide types, the likelihood at site u with pair CC on the leaves becomes

$$P(CC|T, t_1, t_2) = \frac{1}{4}(r_{t_1} r_{t_2} + 3 s_{t_1} s_{t_2})$$

$$= 4^{-3}((1 + 3e^{-4\alpha t_1})(1 + 3e^{-4\alpha t_2}) + 3(1 - e^{-4\alpha t_1})(1 - e^{-4\alpha t_2}))$$

$$= \frac{1}{16}(1 + 3e^{-4\alpha(t_1 + t_2)}).$$

The likelihood for pair GG will have the same expression due to the substitution process symmetry. If site u has the CG pair on the leaves, then the likelihood is given by

$$P(CG|T,t_1,t_2) = q_C r_{t_1} s_{t_2} + q_G s_{t_1} r_{t_2} + q_A s_{t_1} s_{t_2} + q_T s_{t_1} s_{t_2}$$

$$= \frac{1}{4}(r_{t_1} s_{t_2} + s_{t_1} r_{t_2} + 2 s_{t_1} s_{t_2})$$

$$= \frac{1}{16}(1 - e^{-4\alpha(t_1+t_2)}).$$

Exactly the same expression will appear for $P(GC|T,t_1,t_2)$. Finally, if n_1 sites have matches (CC or GG) and n_2 sites have mismatches (CG or GC), then the likelihood of tree T is given by

$$P(x^1, x^2|T, t_1, t_2) = \prod_{u=1}^{n_1+n_2} P(x_u^1, x_u^2|T, t_1, t_2)$$

$$= \frac{1}{16^{n_1+n_2}}(1 + 3e^{-4\alpha(t_1+t_2)})^{n_1}(1 - e^{-4\alpha(t_1+t_2)})^{n_2}.$$

This likelihood is a function of two variables t_1 and t_2: $P(x^1, x^2|T, t_1, t_2) = P(t_1, t_2)$. Formally, finding the maximum of $P(t_1, t_2)$ would require taking partial derivatives $\partial P/\partial t_1$ and $\partial P/\partial t_2$, setting them to zero, etc. However, it is easy to see that $P(t_1, t_2)$ is, in fact, a function of one variable, $t = t_1 + t_2$.

To simplify further calculations, we switch from $P(t)$ to $\ln P(t)$. The natural logarithm is a monotonically increasing function, thus P and $\ln P$ attain the maximum value at the same point t^*. Thus, we find

$$\ln P(t) = -(n_1 + n_2)\ln 16 + n_1 \ln(1 + 3e^{-4\alpha t}) + n_2 \ln(1 - e^{-4\alpha t}),$$

$$(\ln P(t))' = -\frac{12\alpha n_1 e^{-4\alpha t}}{1 + 3e^{-4\alpha t}} + \frac{4\alpha n_2 e^{-4\alpha t}}{1 - e^{-4\alpha t}}.$$

Solving the equation $(\ln P(t^*))' = 0$ yields

$$t^* = \frac{1}{4\alpha} \ln \frac{3(n_1 + n_2)}{3n_1 - n_2}.$$

The maximum value of $P(t)$ reached at t^* is given by

$$P(t^*) = P(x^1, x^2|T, t_1, t_2) = \frac{1}{16^{n_1+n_2-1}} \frac{n_1 n_2}{3(n_1 + n_2)}.$$

Here the edge lengths t_1, t_2 satisfy the equation $t_1 + t_2 = t^*$. Interestingly, although the point of maximum, $t^* = t_1 + t_2$, depends on the parameter α of the Jukes–Cantor model, the maximum likelihood value does not.

As an example, we consider two nucleotide sequences of length 11:

$$C \quad C \quad G \quad G \quad C \quad C \quad G \quad C \quad G \quad C \quad G$$
$$C \quad G \quad G \quad G \quad C \quad C \quad G \quad G \quad C \quad C \quad G$$

where $n_1 = 8$ and $n_2 = 3$. For the Jukes–Cantor model with $\alpha = 10^{-9}$, we calculate the maximum likelihood value as follows:

$$P(x^1, x^2 | T, t_1, t_2) = \frac{8}{11 \times 16^{10}} = 6.61 \times 10^{-13},$$

which is reached when the evolutionary time $t^* = t_1 + t_2$ between sequences is $t^* = 113$ million years. Under the molecular clock property this means that the divergence time for sequences x^1 and x^2 is $t_1 = t_2 = 56.5$ MYA. ☐

Problem 8.8 Consider a simplified phylogenetic space consisting of two trees T and $\tilde{T}$ with probabilities $P(T)$ and $P(\tilde{T})$. If the proposal procedure always proposes the other tree, i.e. the one that is not the current tree, show that the Metropolis algorithm produces a sequence where the frequencies of T and $\tilde{T}$ converge to their probabilities.

Solution The Metropolis algorithm is a sampling procedure that generates a sequence of trees, each new tree depending on a previous one. Let $P_1 = P(T, t_\bullet | x^\bullet) = P(T)$ (or $P(\tilde{T})$) be the posterior probability of a current tree, then $P_2 = P(\tilde{T}, \tilde{t}_\bullet | x^\bullet) = P(\tilde{T})$ $(P(T))$ is the probability of a proposed new tree. If $P_2 \geq P_1$, the new tree is unconditionally accepted as the next item in the tree sequence; if $P_2 < P_1$, the new tree is accepted with probability P_2/P_1.

First, let us consider the case when $P(T) = P(\tilde{T}) = \frac{1}{2}$. The proposal procedure proposes $\tilde{T}$ as a new tree when T is a current tree. Then, since $P(T) = P(\tilde{T})$, $\tilde{T}$ is accepted as the next tree by the Metropolis rule. At the next step, tree T is proposed and accepted as the next tree for the same reason, and so on. Therefore, the result is a strictly alternating sequence: either $T, \tilde{T}, T, \tilde{T}, T, \tilde{T}, \ldots$ or $\tilde{T}, T, \tilde{T}, T, \tilde{T}, T, \ldots$. In both cases, the frequencies of T and $\tilde{T}$ converge to their probabilities:

$$\lim_{n \to \infty} \frac{n_T}{n} = \frac{1}{2} = P(T), \quad \lim_{n \to \infty} \frac{n_{\tilde{T}}}{n} = \frac{1}{2} = P(\tilde{T}).$$

In the general case, when $P(T) \neq P(\tilde{T})$ (suppose that $P(T) > P(\tilde{T})$), the Metropolis procedure works as follows. If the current tree is T, the proposal procedure offers $\tilde{T}$, and the next tree will be $\tilde{T}$ with probability $q = P(\tilde{T})/P(T)$, and T with probability $p = 1 - q$. However, if the current tree is $\tilde{T}$, the next tree will be T with probability 1. The sequence of trees obtained by the Metropolis algorithm can be described by the Markov chain with two states 1 (T) and 2 $(\tilde{T})$, and the transition probabilities $p_{11} = p$, $p_{12} = q$, $p_{21} = 1$, $p_{22} = 0$. Let n_1 (n_2) be the number of trees T ($\tilde{T}$) in a sequence of length n ($n = n_1 + n_2$). According to the *weak law of large numbers* for an ergodic Markov chain with finite number of states (Freedman, 1983), we have

$$\frac{n_1}{n} \to^P \pi_1, \quad \frac{n_2}{n} \to^P \pi_2,$$

where (π_1, π_2) is a stationary distribution of the states of the Markov chain. The ergodic theorem for the Markov chains states that the stationary probabilities satisfy the following linear system:

$$\pi_1 + \pi_2 = 1,$$

$$\pi_1 = \pi_1 p + \pi_2,$$

$$\pi_2 = \pi_1 q.$$

The solution of this system, $\pi_1 = P(T)$, $\pi_2 = P(\tilde{T})$, indicates that the frequencies of occurrences of trees T and $\tilde{T}$ in the tree sequence generated by the Metropolis algorithm converge in probability to the probabilities of these trees:

$$\frac{n_1}{n} \longrightarrow^P P(T), \quad \frac{n_2}{n} \longrightarrow^P P(\tilde{T}). \qquad \square$$

8.1.1 Bayesian approach to finding the optimal tree and the Mau–Newton–Larget algorithm

Theoretical introduction to Problems 8.9 and 8.10

Finding the maximum likelihood tree requires the solution of an optimization problem for an algorithmically determined function of many variables. Such problems are notoriously computationally expensive unless an efficient strategy is proposed for searching the maximum.

Another alternative can be the Bayesian approach to finding the optimal tree. Mau, Newton, and Larget (1999) adapted the Metropolis algorithm, one of the Markov chain Monte Carlo (MCMC) methods, to sample from the posterior distribution π on the space $\mathcal{T}$ of phylogenetic trees with the molecular clock property. (The algorithm was later generalized by Larget and Simon (1999) to include rooted trees without the molecular clock property.) The approximate posterior distribution of the trees could be determined given sequence data, a stochastic model for the data, and a prior distribution on $\mathcal{T}$. To define the proposal rule Q of the Metropolis algorithm, a tree T with n leaves, $T \in \mathcal{T}$, is identified by its canonical representation $(\sigma, \mathbf{t})$. Here $\sigma = (\sigma(1), \ldots, \sigma(n))$ is the left–right order of leaf nodes of T (the permutation of numbers $1, 2, \ldots, n$), and $\mathbf{t} = (t_1, \ldots, t_{n-1})$ is the vector of associated divergence times: $t_i = D_{\sigma(i),\sigma(i+1)}/2$, $i = 1, \ldots, n-1$, with D_{ij} defined as the distance between leaves i and j in tree T. For example, for tree T_1 shown in Figure 8.1(a) the canonical representation is $(\sigma, \mathbf{t}) = ((1, 2, 3, 4, 5), (h_6, h_9, h_8, h_7))$.

The proposal rule Q is implemented in two steps and works with a current tree T as follows. For the tree T with n leaves there exist 2^{n-1} trees (T included) obtained by all possible permutations of $n-1$ pairs of edges emanating from each internal node of T. The first step of the proposal rule Q, Q_1, selects with equal probability $1/2^{n-1}$ one of these 2^{n-1} trees with permutated edges. Thus, after the step Q_1 a new order of leaves σ^* is generated as the one associated with the newly selected tree. At the second step, Q_2, the components of $\mathbf{t}$ are modified independently from each other. Specifically, value t_i^*, a component of the new vector $\mathbf{t}^*$, is sampled from the uniform distribution on the interval $(t_i - \delta, t_i + \delta)$, where δ is a tuning constant. To reconstruct a proposed tree from the canonical representation $(\sigma^*, \mathbf{t}^*)$, one must place the leaf nodes along a horizontal axis in the order defined by σ^*. Then from each midpoint between adjacent leaves $\sigma^*(i)$ and $\sigma^*(i+1)$ draw a vertical line and place a node at the height equal to the corresponding divergence time t_i^*. The highest constructed node will be the root of the tree. Next, working from top to bottom, branches are drawn from each internal node to the highest parentless nodes on both the left and right sides. For example, the canonical representation of tree T_1 in Figure 8.2(b) can be transformed by the proposal rule Q to the canonical representation $(\sigma, \mathbf{t}^*) = ((1, 2, 3, 4, 5), (h_6, h_9, h_7, h_8))$ (with Q_1 choosing the same orientation for all branches and Q_2 decreasing the divergence time $t_3 = D_{34}/2 = h_8$ by $h_8 - h_7$, and increasing the divergence time $t_4 = D_{45}/2 = h_7$ by $h_8 - h_7$). Proposed tree T_2 reconstructed from the new canonical representation is shown in Figure 8.2(b).

Note that Q_1 does not change the topology of the tree, while Q_2 may or may not change it: a greater value of δ provides more chances to switch to a different topology. The proposal rule $Q = Q_2 \circ Q_1$ is symmetric in the sense that for any two trees T_1 and T_2 from $\mathcal{T}$ the probability to proposing T_2 when a current tree is T_1 is equal to the probability of proposing T_1 when a current tree is T_2.

Finally, the standard acceptance rule of the Metropolis algorithm replaces a current tree T_1 by a newly proposed tree T_2 with probability

$$\min\left(1, \frac{\pi(T_2)}{\pi(T_1)}\right) = \min\left(1, \frac{L_{T_2}P(T_2)}{L_{T_1}P(T_1)}\right).$$

Here L_{T_i} and $P(T_i)$ are the likelihood and the prior probability of tree T_i, respectively. The resulting sequence of trees is the Markov chain that has the important property (Liu (2001), Sect. 5): for almost all realizations of this Markov chain the (relative) frequency of a particular tree T observed along a tree sequence converges to the

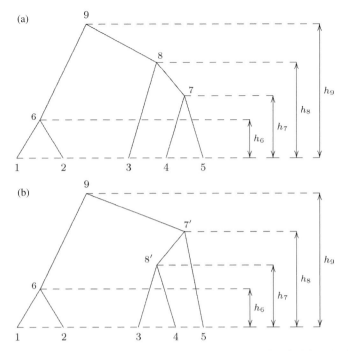

Figure 8.2. Tree T_1 (a) changes to tree T_2 (b) as a result of switching the heights of nodes 7 and 8 (at step Q_2 of the proposal rule Q).

tree posterior probability $\pi(T)$ (see Problem 8.8). (The same statement holds for a particular tree topology or a particular property of the tree.) Therefore, the tree with the highest posterior probability will have the highest frequency in a sufficiently long realization of the Markov chain and, hence, can be detected by the MCMC method.

Problem 8.9 The transformation of a tree from space T resulting from step Q_2 of the proposal procedure suggested by Mau *et al.* (1999) is called the profile change. Consider the profile change transforming tree T_1 to tree T_2 (Figure 8.2). Suppose that nodes 8 and 7 of tree T_1 in Figure 8.2 are at heights h_8 and h_7, and their heights are switched to h_7 and h_8. Show that the resulting change in the likelihood tends to zero as $h_8 - h_7$ tends to zero.

Solution Let h_i designate the height of internal node i of the current tree T_1, $i = 1, 2, 3, 4$. In tree T_2 node $7'$ is placed at height h_8, while node $8'$ is placed at height h_7. Let sequences x^i, $i = 1, \ldots, 5$, be leaves of trees T_1 and T_2. We will use Felsenstein's algorithm to estimate the change in the tree likelihood value resulting from the switch of the node heights. First, we determine the likelihood of tree T_1

at site u. Working it out in bottom-up order, we obtain

$$P(L_6|b) = P(x_u^1|b, h_6)P(x_u^2|b, h_6),$$

$$P(L_7|d) = P(x_u^4|d, h_7)P(x_u^5|d, h_7),$$

$$P(L_8|c) = \sum_d P(x_u^3|c, h_8)P(d|c, h_8 - h_7)P(L_7|d)$$

$$= \sum_d P(x_u^3|c, h_8)P(d|c, h_8 - h_7)P(x_u^4|d, h_7)P(x_u^5|d, h_7),$$

$$P(L_9|a) = \sum_{b,c} P(b|a, h_9 - h_6)P(L_6|b)P(c|a, h_9 - h_8)P(L_8|c)$$

$$= \sum_{b,c,d} P(b|a, h_9 - h_6)P(x_u^1|b, h_6)P(x_u^2|b, h_6)$$

$$\times P(c|a, h_9 - h_8)P(x_u^3|c, h_8)P(d|c, h_8 - h_7)P(x_u^4|d, h_7)P(x_u^5|d, h_7).$$

Therefore, the likelihood at site u is given by

$$P(x_u^\bullet|T_1, t_\bullet) = P(x_u^1, x_u^2, x_u^3, x_u^4, x_u^5|T_1)$$

$$= \sum_a q_a P(L_9|a)$$

$$= \sum_{a,b,c,d} q_a P(b|a, h_9 - h_6)P(x_u^1|b, h_6)P(x_u^2|b, h_6)$$

$$\times P(c|a, h_9 - h_8)P(x_u^3|c, h_8)P(d|c, h_8 - h_7)P(x_u^4|d, h_7)P(x_u^5|d, h_7).$$

We assume that the alignment of sequences x^i, $i = 1, \ldots, 5$, has N independent sites. Then we have for the likelihood of tree T_1

$$L_1 = P(x^1, x^2, x^3, x^4, x^5|T_1) = \prod_{u=1}^N P(x_u^\bullet|T_1)$$

$$= \prod_{u=1}^N \sum_{a,b,c,d} q_a P(b|a, h_9 - h_6)P(x_u^1|b, h_6)P(x_u^2|b, h_6)$$

$$\times P(c|a, h_9 - h_8)P(x_u^3|c, h_8)P(d|c, h_8 - h_7)P(x_u^4|d, h_7)P(x_u^5|d, h_7).$$

Similarly, the likelihood of tree T_2 with the same set of leaf sequences is given by

$$L_2 = P(x^1, x^2, x^3, x^4, x^5 | T_2) = \prod_{u=1}^{N} P(x_u^\bullet | T_2)$$

$$= \prod_{u=1}^{N} \sum_{a,b,c,d} q_a P(b|a, h_9 - h_6) P(x_u^1 | b, h_6) P(x_u^2 | b, h_6)$$

$$\times P(c|a, h_9) P(x_u^5 | c, h_8) P(d|c, h_8 - h_7) P(x_u^3 | d, h_7) P(x_u^4 | d, h_7).$$

The change ΔL in the likelihood can be written as follows:

$$\Delta L = L_1 - L_2$$

$$= \sum_{a_i, b_i, c_i, d_i} \prod_{i=1}^{N} q_{a_i} P(b_i|a_i, h_9 - h_6) P(x_i^1 | b_i, h_6) P(x_i^2 | b_i, h_6)$$

$$\times P(c_i|a_i, h_9 - h_8) P(d_i|c_i, h_8 - h_7) P(x_i^4 | d, h_7) \Delta, \qquad (8.3)$$

where

$$\Delta = P(x_i^3 | c_i, h_8) P(x_i^5 | d_i, h_7) - P(x_i^3 | d_i, h_7) P(x_i^5 | c_i, h_8).$$

Now we find the limit of ΔL as $h_8 - h_7 \to 0$, taking into account that the transition probabilities $P(x|y, t)$, the elements of substitution matrices, are continuous functions of variable t.

The sum in Equation (8.3) has two types of terms. For all terms with $c_i \neq d_i$,

$$\lim_{h_8 - h_7 \to 0} P(d_i|c_i, h_8 - h_7) = P(d_i|c_i, 0) = 0,$$

since the substitution matrix for $t = 0$ is the identity matrix. With all other factors bounded, we have

$$\lim_{h_8 - h_7 \to 0} \sum_{a_i, b_i, c_i, d_i : c_i \neq d_i} \prod_{i=1}^{N} q_{a_i} P(b_i|a_i, h_9 - h_6) P(x_i^1 | b_i, h_6) P(x_i^2 | b_i, h_6)$$

$$\times P(c_i|a_i, h_9 - h_8) P(d_i|c_i, h_8 - h_7) P(x_i^4 | d_i, h_7) \Delta = 0. \qquad (8.4)$$

For all terms with $c_i = d_i$,

$$\lim_{h_8 \to h_7} \Delta = \lim_{h_8 \to h_7} (P(x_i^3 | c_i, h_8) P(x_i^5 | c_i, h_7) - P(x_i^3 | c_i, h_7) P(x_i^5 | c_i, h_8)) = 0.$$

All other factors are bounded; therefore,

$$\lim_{h_8 - h_7 \to 0} \sum_{a_i, b_i, c_i} \prod_{i=1}^{N} q_{a_i} P(b_i|a_i, h_9 - h_6) P(x_i^1 | b_i, h_6) P(x_i^2 | b_i, h_6)$$

$$\times P(c_i|a_i, h_9 - h_8) P(c_i|c_i, h_8 - h_7) P(x_i^4 | c_i, h_7) \Delta = 0. \qquad (8.5)$$

Equalities (8.4) and (8.5) lead to

$$\lim_{h_8 - h_7 \to 0} \Delta L = 0.$$

Hence, the likelihood function changes continuously upon small changes in the node heights, even if these changes of heights lead to "quantum" leaps in the tree topology. □

Problem 8.10 Show that non-adjacent leaves cannot become evolutionary neighbors as a result of profile change.

Solution First, note that at step Q_1 of the proposal rule Q any two leaves may become adjacent, but Q_1 does not change the topology of the tree, thus all the neighbors remain the same. At step Q_2, however, shifting heights of internal nodes up and down may generate a new tree topology and, thus, new neighbors. The example of how adjacent leaves have become neighbors after step Q_2 is shown in Figure 8.2 (leaves 3 and 4).

Let us show that non-adjacent leaves cannot become neighbors after changing node heights. It follows from the rules of reconstruction of a tree from the canonical representation that for adjacent leaves i and j the node on the vertical line drawn from a midpoint between them is the last common ancestor of i and j. Then for any two leaves i and j (adjacent and non-adjacent alike) the last common ancestor will be the highest node situated above interval (i, j). Assuming that there are leaves $l_1, \ldots, l_k$ between i and j, we denote the divergence time between i and l_1 as t_1, the divergence time between l_1 and l_2 as t_2, and so on, and the divergence time between l_k and j as t_{k+1}. Then the divergence time between i and j will be $t = \max(t_1, \ldots, t_{k+1})$. The distance between i and j in the tree with the molecular clock property $d_{ij} = 2 \max(t_1, \ldots, t_{k+1})$. Similarly, $d_{il_1} = 2t_1$, $d_{il_2} = 2 \max(t_1, t_2), \ldots, d_{il_k} = 2 \max(t_1, \ldots, t_k)$; thus

$$d_{il_1} \le d_{il_2} \le \cdots \le d_{il_k} \le d_{ij}.$$

We assume that i and j are neighbors and we seek a contradiction. A necessary condition that the distance d_{ij} between two neighboring leaves should satisfy is $d_{ij} = \min_{m \ne i} d_{im}$. The minimum taken over the set of all leaves except i is not greater than the minimum taken over its subset $\{l_1, l_2, \ldots, l_k, j\}$. Therefore,

$$d_{ij} = \min_{m \ne i} d_{im} \le d_{il_1} < d_{ij},$$

if $t_1 \neq \max\{t_1, \ldots, t_{k+1}\}$. If $t_1 = \max\{t_1, \ldots, t_{k+1}\}$, the same logic as above applied to the distances $d_{jl_k}, d_{jl_{k-1}}, \ldots, d_{ji}$ yields

$$d_{ij} = \min_{m \neq j} d_{jm} \leq d_{jl_k} < d_{ji},$$

since $t_1 \neq t_{k+1}$ and $t_{k+1} < \max\{t_1, \ldots, t_{k+1}\}$. Thus we arrive at the contradiction $d_{ij} < d_{ij}$. Therefore, at step Q_2 of the tree proposal procedure a pair of non-adjacent leaves cannot become evolutionary neighbors.   □

Problem 8.11 The flat prior on edge lengths assigns a prior to any topology that is obtained by integrating over all possible edge lengths for that topology. This integral will be defined if, following Mau *et al.* (1999), we impose a bound on the total edge length from the root to any leaf; call this bound B. Consider the case where there is a molecular clock, and show that the tree with four leaves and topology $((12)(34))$ has integrated prior probability $B^3/3$; show that this integral is $B^3/6$ for the topology $((1(23))4)$. This shows that different topologies can have different priors. Show, however, that if one defines a *labeled history* to be a specific ordering for the times of branch nodes relative to the present time (assuming a molecular clock), then all labeled histories for four leaves have the same prior probability. Extend this to n leaves.

Solution For a four leaf tree T_1 with topology $((12)(34))$ as follows:

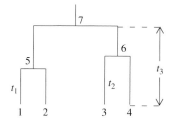

variables t_1, t_2, t_3 are the molecular clock times associated with the branch nodes of tree T_1. A prior probability of T_1 is defined by a triple integral as follows:

$$P(T_1) = \int_0^B \left(\int_0^{t_3} \left(\int_0^{t_3} dt_1 \right) dt_2 \right) dt_3.$$

Here the upper bound for time t_3 is set to B. Integration leads to a simple formula:

$$P(T_1) = \int_0^B \left(\int_0^{t_3} t_3 \, dt_2 \right) dt_3 = \int_0^B t_3^2 \, dt_3 = \frac{B^3}{3}.$$

For a four-leaf tree T_2 with topology $((1(23))4)$ as follows:

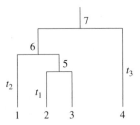

variables t_1, t_2, t_3 are the molecular clock times associated with the branch nodes of tree T_2. Then the prior probability of tree T_2 is given by

$$P(T_2) = \int_0^B \left(\int_0^{t_3} \left(\int_0^{t_2} dt_1 \right) dt_2 \right) dt_3$$

$$= \int_0^B \left(\int_0^{t_3} t_2\, dt_2 \right) dt_3 = \int_0^B \left(\frac{t_3^2}{2} dt_3 \right) = \frac{B^3}{6}.$$

The difference in priors $P(T_1)$ and $P(T_2)$ is an obvious result of the difference in the upper bounds of the integrals dictated by tree topologies. While topology $((1(23))4)$ establishes the dependence between the heights of branch nodes 5 and 6, topology $((12)(34))$ allows either nodes 5 or 6 be higher than the other. However, if an order of node heights or a *labeled history* is defined, then, assuming the molecular clock property, both integrals will have the same set of intervals of integration, and the flat priors for T_1 and T_2 will be equal:

$$P(T_1) = P(T_2) = \int_0^B \left(\int_0^{t_3} \left(\int_0^{t_2} dt_1 \right) dt_2 \right) dt_3 = \frac{B^3}{6}.$$

The same is true for any four-leaf tree T^4: given the labeled history the flat prior for T^4 is $B^3/6$.

The last statement can be extended to an n-leaf tree T^n with molecular clock. If an order of the branch node heights is known, a flat prior for T^n is given by

$$P(T^n) = \int_0^B \left(\int_0^{t_{n-1}} \left(\int_0^{t_{n-2}} \cdots \left(\int_0^{t_2} dt_1 \right) \cdots dt_{n-3} \right) dt_{n-2} \right) dt_{n-1}$$

$$= \int_0^B \frac{t_{n-1}^{n-2}}{(n-2)!} dt_{n-1} = \frac{B^{n-1}}{(n-1)!}. \qquad \square$$

Remark We have shown that all labeled histories with n leaves have the same flat prior $B^{n-1}/(n-1)!$, where B is the upper bound on the total edge length from the root to any leaf of the tree with the molecular clock property. There are

$n!(n-1)!/2^{n-1}$ labeled histories with n leaves (see Problem 7.18). Therefore, to define a proper flat prior distribution on the set of labeled histories, we have to use the flat density function with normalizing constant $C_n = (2/B)(n!)^{-1/(n-1)}$. In this case, we will have for the sum of the prior probabilities over all labeled histories:

$$\frac{n!(n-1)!}{2^{n-1}}P(T^n)$$

$$= \frac{n!(n-1)!}{2^{n-1}} \int_0^B C_n \left(\int_0^{t_{n-1}} C_n \cdots \left(\int_0^{t_2} C_n dt_1 \right) \cdots dt_{n-2} \right) dt_{n-1}$$

$$= \frac{n!(n-1)!}{2^{n-1}} \frac{2^{n-1}}{n!(n-1)!} = 1.$$

However, if the flat prior distribution is defined to be used at the acceptance step of the Metropolis algorithm, the normalization is not necessary. Indeed, if a current labeled history of the generated sequence is T_1, a new labeled history T_2 is accepted as the next element of the sequence with probability equal to $\min(1, L_{T_2}P(T_2)/L_{T_1}P(T_1))$, where L_i and $P(T_i)$ are the likelihood and the prior of tree T_i, respectively. Then the equal normalizing coefficients for the prior probabilities $P(T_1)$ and $P(T_2)$ will be cancelled out. Actually, since flat priors of the labeled histories are equal to each other, the ratio in the acceptance rule becomes the ratio of the likelihoods of labeled histories T_2 and T_1.

Problem 8.12 Under a Yule process, the probability density for no split occurring during the interval 0 to t is given by the limit of $(1 - \lambda\delta t)^{t/\delta t}$ as $\delta t \to 0$, and is therefore $\exp(-\lambda t)$. Deduce that the Yule prior for a tree with n leaves is proportional to $\exp(-\lambda \sum t_i)$, where the t_i are all edge lengths. Following the same reasoning as in the previous problem, show that the priors for all labeled histories on four leaves are equal under the Yule prior. Extend this to the case of n leaves.

Solution For the sake of illustration, we interpret the Yule process $E_t, t \geq 0$, as the size of a population at time t. We assume that each member of the population can divide and produce two descendants at a random moment of time independently of other members. The division rate λ is the same for all members of the population. The probability that no division occurs in a single lineage during time interval $[0, t)$ is $\exp(-\lambda t)$.

Let **T** be a molecular clock tree with n leaves (see Figure 8.3), such that the sum of the edge lengths from the root node $2n - 1$ to each of the leaves $1, 2, \ldots, n$ is T. We can consider such a tree as the realization of the Yule process $E_t, 0 \leq t \leq T$, with initial condition $E_0 = 2$. In tree **T** each branch node $k, k = n + 1, \ldots, 2n - 1$, represents a division event and an appearance of two descendants;

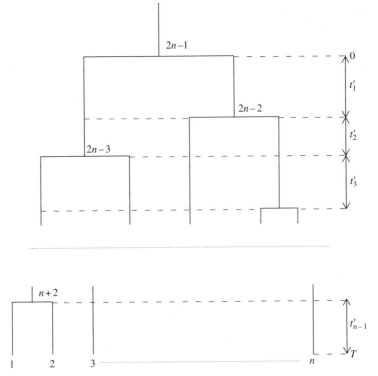

Figure 8.3. Tree **T** with n leaves and the molecular clock property represents a realization of the Yule process through the time interval $[0, T]$: each branch node corresponds to a division (reproduction) event, and n leaves represent a population at time T.

$n - 1$ horizontal dashed lines (except the bottom one) correspond to the times of these division (reproduction) events. The number of intersections of the tree edges with a horizontal line indicating time t, $0 \leq t \leq T$ (not coinciding with any of the dashed lines), is equal to the population size E_t at time t. In particular, $E_T = n$.

We will find the prior probability of tree **T** with edge lengths $t_1, \ldots, t_{2n-1}$. First, we note that n horizontal dashed lines create a partition of the time interval $[0, T]$ into $n-1$ time segments t_i', $i = 1, \ldots, n-1$. Inside the time segment i the population size is equal to $i + 1$. As the history of each member is independent from all the others, the probability P_i of a slice of tree **T** within the ith time segment is the product of $i + 1$ equal probabilities that no descendant is produced during the time interval t_i' by a member of the population:

$$P_i = \prod_1^{i+1} \exp\{-\lambda t_i'\} = \exp\{-\lambda(i + 1)t_i'\}.$$

The probability of division at a given moment t for a single individual is equal to zero (in contrast with a time interval). This implies that the probability of division at a given moment for any finite number of individuals is zero too. Hence, the prior probability of **T** is the product of P_i over all time intervals $t'_1, \ldots, t'_{n-1}$:

$$P(\mathbf{T}) = \prod_{i=1}^{n-1} P_i = \exp\left\{ -\lambda \sum_{i=1}^{n-1} (i+1)t'_i \right\}.$$

We can express $P(\mathbf{T})$ in terms of the lengths of edges $t_1, \ldots, t_{2n-1}$ of tree **T**, since, due to the choice of the partition, there is a simple relationship between $t'_1, \ldots, t'_{n-1}$ and $t_1, \ldots, t_{2n-1}$:

$$\sum_{i=1}^{n-1} (i+1)t'_i = \sum_{j=1}^{2n-1} t_j.$$

Thus, the prior probability of tree **T** with given edge lengths becomes

$$P(\mathbf{T}, t_1, \ldots, t_{2n-1}) = \exp\left(-\lambda \sum_{j=1}^{2n-1} t_j \right). \tag{8.6}$$

Now we will determine the Yule prior for a tree with n leaves and a given labeled history assuming that the upper bound on the time from the root to any leaf is B (similar to Problem 8.11).

We start with the case $n = 4$. Let T^4 be a four-leaf tree, with labeled history $((1(23))4)$ and time variables t_1, t_2, t_3 associated with the branch nodes as follows:

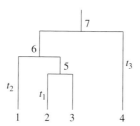

Equation (8.6) gives for the prior probability of tree T^4 with edge lengths $t_1, t_1, t_2, t_2 - t_1, t_3, t_3 - t_2$:

$$P(T^4, t_1, t_1, t_2, t_2 - t_1, t_3, t_3 - t_2)$$
$$= \exp\{-\lambda(t_1 + t_1 + t_2 + (t_2 - t_1) + t_3 + (t_3 - t_2))\}$$
$$= \exp\{-\lambda(t_1 + t_2 + 2t_3)\}.$$

Integration of the probability density function over the values of variables t_1, t_2, t_3 permitted by the labeled history produces the Yule prior for the labeled history as in tree T^4:

$$P(T^4) = \int_0^B 2\lambda e^{-2\lambda t_3} \left(\int_0^{t_3} \lambda e^{-\lambda t_2} \left(\int_0^{t_2} \lambda e^{-\lambda t_1} \, dt_1 \right) dt_2 \right) dt_3$$

$$= \int_0^B 2\lambda e^{-2\lambda t_3} \left(\int_0^{t_3} \lambda e^{-\lambda t_2} (1 - e^{-\lambda t_2}) dt_2 \right) dt_3$$

$$= \int_0^B 2\lambda e^{-2\lambda t_3} \frac{(1 - e^{-\lambda t_3})^2}{2} \, dt_3$$

$$= (1 - e^{-\lambda B})^3 \left(\frac{1}{12} + \frac{1}{4} e^{-\lambda B} \right).$$

These arguments remain valid for a four-leaf tree associated with any other labeled history.

Similarly, we assume that in a tree T^n with n leaves the time variables $t_1, t_2,\ldots, t_{n-1}$ associated with the branch nodes of the tree have been enumerated in the order determined by the labeled history of T^n ($t_i < t_{i+1}$). We apply Equation (8.6) to determine the Yule prior of T^n with edge lengths $t_1, t_1, t_2, t_2 - t_1, \ldots, t_{n-1}, t_{n-1} - t_{n-2}$ as follows:

$$P(T^n, t_1, t_1, t_2, t_2 - t_1, \ldots, t_{n-1}, t_{n-1} - t_{n-2})$$
$$= \exp\{-\lambda(t_1 + t_2 + \cdots + t_{n-2} + 2t_{n-1})\}. \tag{8.7}$$

Integration over all permitted domains of variables $t_1, t_2, \ldots, t_{n-1}$ gives the Yule prior for the labeled history associated with T^n:

$$P(T^n) = \int_0^B 2\lambda e^{-2\lambda t_{n-1}} \left(\int_0^{t_{n-1}} \lambda e^{-\lambda t_{n-2}} \cdots \left(\int_0^{t_2} \lambda e^{-\lambda t_1} dt_1 \right) \cdots dt_{n-2} \right) dt_{n-1}$$

$$= \int_0^B 2\lambda e^{-2\lambda t_{n-1}} \frac{(1 - e^{-\lambda t_{n-1}})^{n-2}}{(n-2)!} \, dt_{n-1}$$

$$= \frac{2}{(n-2)!n} (1 - e^{-\lambda B})^{n-1} \left(\frac{1}{n-1} + e^{-\lambda B} \right).$$

Once again, all the arguments and hence the formula for the Yule prior $P(T^n)$ hold true for a tree with n leaves with any labeled history on the condition that time variables $t_1, t_2, \ldots, t_{n-1}$ are chosen as described above.

Similarly to the case of the flat prior (Problem 8.11), we have shown that the labeled histories of all trees with n leaves and total height at most B have equal Yule priors. It is easy to see that the statement remains true even without restriction on the "age" of a tree, since the integrals of the exponentially decreasing functions are finite over both bounded and unbounded domains of integration. Here is the difference with the flat prior case, where the restriction on the total height of the tree provides a necessary and sufficient condition for the finiteness of integrals and therefore properly defines flat prior probabilities for the labeled histories. For the normalization of the priors and to discover why the normalization is non-necessary for priors used in the Metropolis algorithm, see the Remark to Problem 8.11. □

Remark For the Yule process the transition probability $P_{kn}(t) = P(E_t = n|E_0 = k)$, $n > k$, is given by the following formula (Feller, 1971):

$$P_{kn}(t) = \binom{n-1}{n-k} e^{-k\lambda t}(1 - e^{-\lambda t})^{n-k}.$$

It is interesting to establish a relationship between the transition probability $P_{kn}(t)$ and the Yule prior distribution of trees discussed above.

Note that the Yule process is a special case of the well studied *birth-and-death process* with the death rate set to zero. As we have already seen, any tree with the molecular clock and the probability density $\lambda \exp(-\lambda t)$ along its edges represents a realization of the Yule process with birth (reproduction, splitting) rate λ and beginning state $E_0 = 2$. The transition probability $P_{2n}(t)$ can be found from Equation (8.7) by setting a height t_{n-1} of the root node to t and integrating with respect to all branch node heights $t_1, t_2, \ldots, t_{n-2}$ (corresponding to the further descendants of the root node) over the same interval $[0, t)$. Thus, if time variable t_{n-1} is chosen to be the root node height, we obtain the probability

$$e^{-2\lambda t} \prod_{i=1}^{n-2} \left(\int_0^t \lambda e^{-\lambda t_i} \, dt_i \right) = e^{-2\lambda t}(1 - e^{-\lambda t})^{n-2}.$$

Since there are $(n - 1)$ ways to choose the root node height among time variables $t_1, t_2, \ldots, t_{n-1}$, the transition probability becomes

$$P_{2n}(t) = (n - 1)e^{-2\lambda t}(1 - e^{-\lambda t})^{n-2}.$$

Therefore, we have derived the formula of the transition probability for the Yule process for $k = 2$. In the theory of stochastic processes, such formulas are derived by solving the Kolmogorov system of differential equations (Feller, 1971).

Problem 8.13 Assuming a molecular clock, calculate the expected lengths of all the branches of rooted trees with two, three, or four leaves under the Yule prior with splitting rate λ and the coalescent prior with population size θ.

Solution We begin with a tree T^2 with two leaves. Due to the molecular clock property, two edges of T^2 have equal length; it is defined as a random variable L, taking values $t, 0 \le t < +\infty$. Under the Yule prior (Problem 8.12) the probability of T^2 is $P_Y(T^2, t, t) = e^{-2\lambda t}$ and the probability density function is $p_Y(t) = 2\lambda e^{-2\lambda t}$. Then for the expected value of the edge length we have

$$\mathbf{E}_Y L = \int_0^{+\infty} t p(t)\, dt = \int_0^{+\infty} 2 t \lambda e^{-2\lambda t}\, dt$$

$$= \int_0^{+\infty} t\, d(-e^{-2\lambda t}) = -t e^{-2\lambda t} \Big|_{t=0}^{+\infty} + \int_0^{+\infty} e^{-2\lambda t}\, dt$$

$$= -\frac{e^{-2\lambda t}}{2\lambda} \Big|_{t=0}^{+\infty} = \frac{1}{2\lambda}.$$

Under the coalescent prior with population size θ, the probability of T^2 is $P_c(T^2, t, t|\theta) = (2/\theta) \exp(-2t/\theta)$ and the probability density function is $p_c(t) = (2/\theta) \exp(-2t/\theta)$. Therefore,

$$\mathbf{E}_c L = \int_0^{+\infty} t p(t)\, dt = \int_0^{+\infty} \frac{2t}{\theta} \exp\left(-\frac{2t}{\theta}\right) dt = \frac{\theta}{2}.$$

Now we consider a tree T^3 with three leaves as follows:

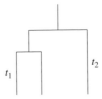

We assume that the lengths of the edges of the tree are defined by the random variables L_1 and L_2 taking values t_1 and t_2, respectively, $0 \le t_2 < +\infty, 0 \le t_1 < t_2$. Under the Yule prior, the probability of T^3 is given by

$$P_Y(T^3, t_1, t_1, t_2, t_2 - t_1) = e^{-2\lambda t_2} e^{-\lambda t_1},$$

and the probability density function is given by

$$p_Y(t_1, t_2) = \lambda e^{-\lambda t_1} 2\lambda e^{-2\lambda t_2}.$$

Then

$$\mathbf{E}_Y L_1 = \int_{t_2=0}^{+\infty} \int_{t_1=0}^{t_2} t_1 p(t_1, t_2) \, dt_1 \, dt_2 = \int_0^{+\infty} 2\lambda e^{-2\lambda t_2} \left(\int_0^{t_2} t_1 \lambda e^{-\lambda t_1} \, dt_1 \right) dt_2$$

$$= \int_0^{+\infty} 2\lambda e^{-2\lambda t_2} \left(-t_2 e^{-\lambda t_2} - \frac{e^{-\lambda t_2}}{\lambda} + \frac{1}{\lambda} \right) dt_2 = -\frac{2}{9\lambda} - \frac{2}{3\lambda} + \frac{1}{\lambda} = \frac{1}{9\lambda};$$

$$\mathbf{E}_Y L_2 = \int_{t_2=0}^{+\infty} \int_{t_1=0}^{t_2} t_2 p(t_1, t_2) \, dt_1 \, dt_2 = \int_0^{+\infty} 2 t_2 \lambda e^{-2\lambda t_2} \left(\int_0^{t_2} \lambda e^{-2\lambda t_1} \, dt_1 \right) dt_2$$

$$= \int_0^{+\infty} 2 t_2 \lambda e^{-2\lambda t_2} \left(1 - e^{-\lambda t_2} \right) dt_2 = \frac{1}{2\lambda} - \frac{2}{9\lambda} = \frac{5}{18\lambda}.$$

Under the coalescent prior, tree T^3 has the probability

$$P_c(T^3, t_1, t_1, t_2, t_2 - t_1 | \theta) = \left(\frac{2}{\theta} \right)^2 \exp\left(-\frac{2t_2}{\theta} \right) \exp\left(-\frac{4t_1}{\theta} \right),$$

and the probability density function is given by

$$p_c(t_1, t_2) = \frac{4}{\theta} \exp\left(-\frac{4t_1}{\theta} \right) \frac{2}{\theta} \exp\left(-\frac{2t_2}{\theta} \right).$$

Then for the expected values of L_1 and L_2 we have

$$\mathbf{E}_c L_1 = \int_{t_2=0}^{+\infty} \int_{t_1=0}^{t_2} t_1 p(t_1, t_2) \, dt_1 \, dt_2$$

$$= \int_0^{+\infty} \frac{2}{\theta} \exp\left(-\frac{2t_2}{\theta} \right) \left(\int_0^{t_2} \frac{4t_1}{\theta} \exp\left(-\frac{4t_1}{\theta} \right) dt_1 \right) dt_2$$

$$= \int_0^{+\infty} \frac{2}{\theta} \exp\left(-\frac{2t_2}{\theta} \right) \left(-t_2 \exp\left(-\frac{4t_2}{\theta} \right) - \frac{\theta}{4} \exp\left(-\frac{4t_2}{\theta} \right) + \frac{\theta}{4} \right) dt_2$$

$$= -\frac{\theta}{18} - \frac{\theta}{12} + \frac{\theta}{4} = \frac{\theta}{9};$$

$$\mathbf{E}_c L_2 = \int_{t_2=0}^{+\infty} \int_{t_1=0}^{t_2} t_2 p(t_1, t_2) \, dt_1 \, dt_2$$

$$= \int_0^{+\infty} \frac{2t_2}{\theta} \exp\left(-\frac{2t_2}{\theta} \right) \left(\int_0^{t_2} \frac{4}{\theta} \exp\left(-\frac{4t_1}{\theta} \right) dt_1 \right) dt_2$$

$$= \int_0^{+\infty} \frac{2t_2}{\theta} \exp\left(-\frac{2t_2}{\theta} \right) \left(1 - \exp\left(-\frac{4t_2}{\theta} \right) \right) dt_2$$

$$= \frac{\theta}{2} - \frac{\theta}{18} = \frac{4\theta}{9}.$$

Finally, we consider four-leaf trees with two possible topologies. Let T_1^4 be a tree with topology $((12)(34))$ and edge lengths $t_1, t_1, t_2, t_2, t_3 - t_1, t_3 - t_2$:

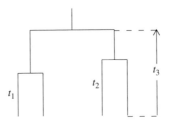

We assume that the random variables L_1, L_2, L_3 define the branch node heights of the tree and take values t_1, t_2, t_3, respectively, $0 \le t_3 < +\infty$, $0 \le t_1 < t_3$, $0 \le t_2 < t_3$. Under the Yule prior, the probability of T_1^4 is given by

$$P_Y(T_1^4, t_1, t_1, t_2, t_2, t_3 - t_1, t_3 - t_2) = e^{-2\lambda t_3} e^{-\lambda t_2} e^{-\lambda t_1},$$

and the probability density function is given by

$$p_Y(t_1, t_2, t_3) = \lambda e^{-\lambda t_1} \lambda e^{-\lambda t_2} 2\lambda e^{-2\lambda t_3}.$$

Then

$$
\begin{aligned}
\mathbf{E}_Y L_1 &= \int_{t_3=0}^{+\infty} \int_{t_2=0}^{t_3} \int_{t_1=0}^{t_3} t_1 p_Y(t_1, t_2, t_3)\, dt_1\, dt_2\, dt_3 \\
&= \int_0^{+\infty} 2\lambda e^{-2\lambda t_3} \left(\int_0^{t_3} t_1 \lambda e^{-\lambda t_1}\, dt_1 \right) \left(\int_0^{t_3} \lambda e^{-\lambda t_2} dt_2 \right) dt_3 \\
&= \int_0^{+\infty} 2\lambda e^{-2\lambda t_3} \left(-t_3 e^{-\lambda t_3} - \frac{e^{-\lambda t_3}}{\lambda} + \frac{1}{\lambda} \right) \left(1 - e^{-\lambda t_3} \right) dt_3 \\
&= -\frac{2}{9\lambda} + \frac{1}{8\lambda} - \frac{4}{3\lambda} + \frac{1}{2\lambda} + \frac{1}{\lambda} = \frac{5}{72\lambda};
\end{aligned}
$$

$$
\begin{aligned}
\mathbf{E}_Y L_2 &= \int_{t_3=0}^{+\infty} \int_{t_2=0}^{t_3} \int_{t_1=0}^{t_3} t_2 p_Y(t_1, t_2, t_3)\, dt_1\, dt_2\, dt_3 \\
&= \int_0^{+\infty} 2\lambda e^{-2\lambda t_3} \left(\int_0^{t_3} \lambda e^{-\lambda t_1}\, dt_1 \right) \left(\int_0^{t_3} t_2 \lambda e^{-\lambda t_2}\, dt_2 \right) dt_3 \\
&= \int_0^{+\infty} 2\lambda e^{-2\lambda t_3} \left(-t_3 e^{-\lambda t_3} - \frac{e^{-\lambda t_3}}{\lambda} + \frac{1}{\lambda} \right) \left(1 - e^{-\lambda t_3} \right) dt_3 \\
&= \frac{5}{72\lambda};
\end{aligned}
$$

$$\mathbf{E}_Y L_3 = \int_{t_3=0}^{+\infty} \int_{t_2=0}^{t_3} \int_{t_1=0}^{t_3} t_3 p_Y(t_1, t_2, t_3) \, dt_1 \, dt_2 \, dt_3$$

$$= \int_0^{+\infty} 2t_3 \lambda e^{-2\lambda t_3} \left(\int_0^{t_3} \lambda e^{-\lambda t_1} \, dt_1 \right) \left(\int_0^{t_3} \lambda e^{-\lambda t_2} dt_2 \right) dt_3$$

$$= \int_0^{+\infty} 2t_3 \lambda e^{-2\lambda t_3} \left(1 - e^{-\lambda t_3} \right)^2 dt_3$$

$$= \frac{1}{2\lambda} - \frac{4}{9\lambda} + \frac{1}{8\lambda} = \frac{11}{72\lambda}.$$

Under the coalescent prior, tree T_1^4 has the probability

$$P_c(T_1^4, t_1, t_1, t_2, t_2, t_3 - t_1, t_3 - t_2 | \theta)$$

$$= \left(\frac{2}{\theta} \right)^3 \exp\left(-\frac{2t_3}{\theta} \right) \exp\left(-\frac{4t_2}{\theta} \right) \exp\left(-\frac{6t_1}{\theta} \right),$$

and the probability density function is given by

$$p_c(t_1, t_2, t_3) = \frac{6}{\theta} \exp\left(-\frac{6t_1}{\theta} \right) \frac{4}{\theta} \exp\left(-\frac{4t_2}{\theta} \right) \frac{2}{\theta} \exp\left(-\frac{2t_3}{\theta} \right).$$

To find the expected values of L_1, L_2, and L_3, we have to complete the following integration:

$$\mathbf{E}_c L_1 = \int_{t_3=0}^{+\infty} \int_{t_2=0}^{t_3} \int_{t_1=0}^{t_3} t_1 p(t_1, t_2, t_3) \, dt_1 \, dt_2 \, dt_3$$

$$= \int_0^{+\infty} \frac{2}{\theta} \exp\left(-\frac{2t_3}{\theta} \right) \left(\int_0^{t_3} \frac{4}{\theta} \exp\left(-\frac{4t_2}{\theta} \right) dt_2 \int_0^{t_3} \frac{6t_1}{\theta} \exp\left(-\frac{6t_1}{\theta} \right) dt_1 \right) dt_3$$

$$= \int_0^{+\infty} \frac{2}{\theta} \exp\left(-\frac{2t_3}{\theta} \right) \left(1 - \exp\left(-\frac{4t_3}{\theta} \right) \right)$$

$$\times \left(-t_3 \exp\left(-\frac{6t_3}{\theta} \right) - \frac{\theta}{4} \exp\left(-\frac{6t_3}{\theta} \right) + \frac{\theta}{6} \right) dt_3$$

$$= -\frac{\theta}{32} - \frac{\theta}{16} + \frac{\theta}{6} + \frac{\theta}{72} + \frac{\theta}{24} - \frac{\theta}{18} = \frac{7\theta}{96};$$

$$\mathbf{E}_c L_2 = \int_{t_3=0}^{+\infty} \int_{t_2=0}^{t_3} \int_{t_1=0}^{t_3} t_2 p(t_1, t_2, t_3)\, dt_1\, dt_2\, dt_3$$

$$= \int_0^{+\infty} \frac{2}{\theta} \exp\left(-\frac{2t_3}{\theta}\right) \left(1 - \exp\left(-\frac{6t_3}{\theta}\right)\right) \left(\int_0^{t_3} \frac{4t_2}{\theta} \exp\left(-\frac{4t_2}{\theta}\right) dt_2\right) dt_3$$

$$= \int_0^{+\infty} \left(-\frac{2t_3}{\theta} \exp\left(-\frac{6t_3}{\theta}\right) - \frac{1}{2} \exp\left(-\frac{6t_3}{\theta}\right) + \frac{1}{2} \exp\left(-\frac{2t_3}{\theta}\right)\right.$$

$$\left. + \frac{2t_3}{\theta} \exp\left(-\frac{12t_3}{\theta}\right) + \frac{1}{2} \exp\left(-\frac{12t_3}{\theta}\right) - \frac{1}{2} \exp\left(-\frac{8t_3}{\theta}\right)\right) dt_3$$

$$= -\frac{\theta}{18} - \frac{\theta}{12} + \frac{\theta}{4} + \frac{\theta}{72} + \frac{\theta}{24} - \frac{\theta}{16} = \frac{\theta}{72};$$

$$\mathbf{E}_c L_3 = \int_{t_3=0}^{+\infty} \int_{t_2=0}^{t_3} \int_{t_1=0}^{t_3} t_3 p(t_1, t_2, t_3)\, dt_1\, dt_2\, dt_3$$

$$= \int_0^{+\infty} \frac{2t_3}{\theta} \exp\left(-\frac{2t_3}{\theta}\right) \left(1 - \exp\left(-\frac{4t_3}{\theta}\right)\right) \left(1 - \exp\left(-\frac{6t_3}{\theta}\right)\right) dt_3$$

$$= \int_0^{+\infty} \left(\frac{2t_3}{\theta} \exp\left(-\frac{2t_3}{\theta}\right) - \frac{2t_3}{\theta} \exp\left(-\frac{8t_3}{\theta}\right)\right.$$

$$\left. - \frac{2t_3}{\theta} \exp\left(-\frac{6t_3}{\theta}\right) + \frac{2t_3}{\theta} \exp\left(-\frac{12t_3}{\theta}\right)\right) dt_3$$

$$= \frac{\theta}{2} - \frac{\theta}{32} - \frac{\theta}{18} + \frac{\theta}{72} = \frac{59\theta}{288}.$$

Note that the expected heights L_1 and L_2 of the branch nodes of T_1^4 are equal under the Yule prior, but have different values ($\mathbf{E}_c L_1 > \mathbf{E}_c L_2$) under the coalescence prior. The symmetry of the probability density function p_Y versus the non-symmetry of p_c with respect to variables t_1 and t_2 explains this difference.

The final case is the tree T_2^4 with topology $(((12)3)4)$ and edge lengths $t_1, t_1, t_2, t_2 - t_1, t_3, t_3 - t_2$:

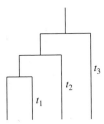

We assume that the random variables L_1, L_2, L_3 define unequal branch node heights and take values t_1, t_2, t_3, respectively, $0 \le t_3 < +\infty, 0 \le t_1 < t_2, 0 \le t_2 < t_3$.

Under the Yule prior, the probability of T_2^4 is given by

$$P_Y(T_2^4, t_1, t_1, t_2, t_2 - t_1, t_3, t_3 - t_2) = e^{-2\lambda t_3} e^{-\lambda t_2} e^{-\lambda t_1},$$

and the density function p_Y is given by

$$p_Y(t_1, t_2, t_3) = \lambda e^{-\lambda t_1} \lambda e^{-\lambda t_2} 2\lambda e^{-2\lambda t_3}.$$

Then

$$
\begin{aligned}
\mathbf{E}_Y L_1 &= \int_{t_3=0}^{+\infty} \int_{t_2=0}^{t_3} \int_{t_1=0}^{t_2} t_1 p_Y(t_1, t_2, t_3) \, dt_1 \, dt_2 \, dt_3 \\
&= \int_0^{+\infty} 2\lambda e^{-2\lambda t_3} \left(\int_0^{t_3} \left(-\lambda t_2 e^{-2\lambda t_2} - e^{-2\lambda t_2} + e^{-\lambda t_2} \right) dt_2 \right) dt_3 \\
&= \int_0^{+\infty} \left(\lambda t_3 e^{-4\lambda t_3} \frac{3}{2} e^{-4\lambda t_3} + \frac{1}{2} e^{-2\lambda t_3} - 2e^{-3\lambda t_3} \right) dt_3 \\
&= \frac{1}{16\lambda} + \frac{3}{8\lambda} + \frac{1}{4\lambda} - \frac{2}{3\lambda} = \frac{1}{48\lambda};
\end{aligned}
$$

$$
\begin{aligned}
\mathbf{E}_Y L_2 &= \int_{t_3=0}^{+\infty} \int_{t_2=0}^{t_3} \int_{t_1=0}^{t_2} t_2 p_Y(t_1, t_2, t_3) \, dt_1 \, dt_2 \, dt_3 \\
&= \int_0^{+\infty} 2\lambda e^{-2\lambda t_3} \left(\int_0^{t_3} \left(\lambda t_2 e^{-\lambda t_2} - \lambda t_2 e^{-2\lambda t_2} \right) dt_2 \right) dt_3 \\
&= \int_0^{+\infty} \left(2t_3 \lambda e^{-3\lambda t_3} - 2e^{-3\lambda t_3} + \frac{3}{2} e^{-2\lambda t_3} + \lambda t_3 e^{-4\lambda t_3} + \frac{1}{2} e^{-4\lambda t_3} \right) dt_3 \\
&= \frac{2}{9\lambda} - \frac{2}{3\lambda} + \frac{3}{4\lambda} + \frac{1}{16\lambda} + \frac{1}{8\lambda} = \frac{11}{144\lambda};
\end{aligned}
$$

$$
\begin{aligned}
\mathbf{E}_Y L_3 &= \int_{t_3=0}^{+\infty} \int_{t_2=0}^{t_3} \int_{t_1=0}^{t_2} t_3 p_Y(t_1, t_2, t_3) \, dt_1 \, dt_2 \, dt_3 \\
&= \int_0^{+\infty} t_3 \lambda e^{-2\lambda t_3} \left(1 - 2\lambda e^{-\lambda t_3} + e^{-2\lambda t_3} \right) dt_3 \\
&= \frac{1}{4\lambda} - \frac{2}{9\lambda} + \frac{1}{16\lambda} = \frac{13}{144\lambda}.
\end{aligned}
$$

Under the coalescent prior, the probability of tree T_2^4 is the same as the probability of tree T_1^4:

$$P_c(T_2^4, t_1, t_1, t_2, t_2 - t_1, t_3, t_3 - t_2 | \theta)$$

$$= \left(\frac{2}{\theta} \right)^3 \exp\left(-\frac{2t_3}{\theta} \right) \exp\left(-\frac{4t_2}{\theta} \right) \exp\left(-\frac{6t_1}{\theta} \right).$$

Thus, the probability density function p_c is the same as for T_1^4:

$$p_c(t_1, t_2, t_3) = \frac{6}{\theta} \exp\left(-\frac{6t_1}{\theta}\right) \frac{4}{\theta} \exp\left(-\frac{4t_2}{\theta}\right) \frac{2}{\theta} \exp\left(-\frac{2t_3}{\theta}\right).$$

Integration yields the following expected values of L_1, L_2, and L_3:

$$E_c L_1 = \int_{t_3=0}^{+\infty} \int_{t_2=0}^{t_3} \int_{t_1=0}^{t_2} t_1 p(t_1, t_2, t_3) \, dt_1 \, dt_2 \, dt_3$$

$$= \int_0^{+\infty} \frac{2}{\theta} \exp\left(-\frac{2t_3}{\theta}\right) \left(\int_0^{t_3} \frac{4}{\theta} \exp\left(-\frac{4t_2}{\theta}\right) \right.$$
$$\times \left. \left(\int_0^{t_2} \frac{6t_1}{\theta} \exp\left(-\frac{6t_1}{\theta}\right) dt_1 \right) dt_2 \right) dt_3$$

$$= \int_0^{+\infty} \frac{2}{\theta} \exp\left(-\frac{2t_3}{\theta}\right) \left(\int_0^{t_3} \frac{4}{\theta} \exp\left(-\frac{4t_2}{\theta}\right) \right.$$
$$\times \left. \left(-t_2 \exp\left(-\frac{6t_2}{\theta}\right) - \frac{\theta}{4} \exp\left(-\frac{6t_2}{\theta}\right) + \frac{\theta}{6} \right) dt_2 \right) dt_3$$

$$= \int_0^{+\infty} \left(\frac{4t_3}{5\theta} \exp\left(-\frac{12t_3}{\theta}\right) - \frac{4}{75} \exp\left(-\frac{12t_3}{\theta}\right) \right.$$
$$\left. - \frac{2}{25} \exp\left(-\frac{2t_3}{\theta}\right) + \frac{1}{3} \exp\left(-\frac{6t_3}{\theta}\right) \right) dt_3$$

$$= \frac{\theta}{60} - \frac{\theta}{225} - \frac{\theta}{25} + \frac{\theta}{18} = \frac{\theta}{25};$$

$$E_c L_2 = \int_{t_3=0}^{+\infty} \int_{t_2=0}^{t_3} \int_{t_1=0}^{t_2} t_2 p(t_1, t_2, t_3) \, dt_1 \, dt_2 \, dt_3$$

$$= \int_0^{+\infty} \frac{2}{\theta} \exp\left(-\frac{2t_3}{\theta}\right) \left(\int_0^{t_3} \frac{4t_2}{\theta} \exp\left(-\frac{4t_2}{\theta}\right) \left(1 - \exp\left(-\frac{6t_1}{\theta}\right)\right) dt_2 \right) dt_3$$

$$= \int_0^{+\infty} \frac{2}{\theta} \exp\left(-\frac{2t_3}{\theta}\right)$$
$$\times \left(\int_0^{t_3} \left(\frac{4t_2}{\theta} \exp\left(-\frac{4t_2}{\theta}\right) - \frac{4t_2}{\theta} \exp\left(-\frac{10t_2}{\theta}\right) \right) dt_2 \right) dt_3$$

$$= \int_0^{+\infty} \left(-\frac{2t_3}{\theta} \exp\left(-\frac{6t_3}{\theta}\right) - \frac{1}{2} \exp\left(-\frac{6t_3}{\theta}\right) \right.$$
$$\left. + \frac{21}{50} \exp\left(-\frac{2t_3}{\theta}\right) + \frac{4t_3}{5\theta} \exp\left(-\frac{12t_3}{\theta}\right) + \frac{2}{25} \exp\left(-\frac{12t_3}{\theta}\right) \right) dt_3$$

$$= -\frac{\theta}{18} - \frac{\theta}{12} + \frac{21\theta}{100} + \frac{\theta}{45} + \frac{\theta}{150} = \frac{\theta}{10};$$

$$\mathbf{E}_c L_3 = \int_{t_3=0}^{+\infty} \int_{t_2=0}^{t_3} \int_{t_1=0}^{t_2} t_3 p(t_1, t_2, t_3) \, dt_1 \, dt_2 \, dt_3$$

$$= \int_0^{+\infty} \frac{2t_3}{\theta} \exp\left(-\frac{2t_3}{\theta}\right) \left(\int_0^{t_3} \frac{4}{\theta} \exp\left(-\frac{4t_2}{\theta}\right) \left(1 - \exp\left(-\frac{6t_1}{\theta}\right) \right) dt_2 \right) dt_3$$

$$= \int_0^{+\infty} \left(\frac{6t_3}{5\theta} \exp\left(-\frac{2t_3}{\theta}\right) - \frac{2t_3}{\theta} \exp\left(-\frac{6t_3}{\theta}\right) + \frac{4t_3}{5\theta} \exp\left(-\frac{12t_3}{\theta}\right) \right) dt_3$$

$$= \frac{3\theta}{10} - \frac{\theta}{18} + \frac{\theta}{15} = \frac{14\theta}{45}. \qquad \square$$

Problem 8.14 For sufficiently large N, the posterior distribution $P(p|mH, nT)$ for the probability of a head p given by a Dirichlet distribution has the following normal approximation:

$$P(p|mH, nT) \simeq \frac{N+1}{\sqrt{2\pi Np(1-p)}} \exp\left(-\frac{(m-Np)^2}{2Np(1-p)}\right).$$

Similarly, the maximum likelihood estimate $P(p^{ML} = p)$ derived from the bootstrap data asymptotically follows the normal approximation, i.e. for sufficiently large N

$$P(p^{ML} = p) \simeq \frac{N+1}{\sqrt{2\pi mn/N}} \exp\left(-\frac{(m-Np)^2}{2mn/N}\right).$$

Show that both probabilities are either very small or else take nearly equal values.

Solution We consider a coin tossing with probability p of getting a head. Suppose that N tosses produced m heads (H) and n tails (T). For large N we have that, almost surely,

$$\frac{m}{N} \simeq p, \quad \frac{n}{N} \simeq 1 - p$$

due to the *strong law of large numbers*. Therefore, for large N

$$\frac{mn}{N} \simeq p(1-p)N$$

and

$$\frac{N+1}{\sqrt{2\pi mn/N}} \exp\left(-\frac{(m-Np)^2}{2mn/N}\right) \simeq \frac{N+1}{\sqrt{2\pi Np(1-p)}} \exp\left(-\frac{(m-Np)^2}{2Np(1-p)}\right).$$

Thus, the distributions $P(p^{ML} = p)$ and $P(p|mH, nT)$ approximate each other for large N. $\qquad \square$

Problem 8.15 Show that finding the most parsimonious tree using weighted parsimony with the costs $S(a, a) = -\log \alpha$, $S(a, b) = -\log \beta$, for $a \neq b$ and the condition $\beta < \alpha$, is equivalent to traditional parsimony with a substitution cost of 1.

Remark Parameters α and β are interpreted as the elements of a substitution matrix, with diagonal elements equal to α and non-diagonal elements equal to β.

Solution First we note that the inequality $\beta < \alpha$ implies that $S(a, a) < S(a, b)$ for $a \neq b$. Suppose we have to identify the most parsimonious tree for n sequences $x^1, \ldots, x^n$ each of length N. For a given tree T we calculate the minimum cost $S_u(T)$ at each site u, $u = 1, \ldots, N$, and then the minimum cost of the tree T is given by

$$S(T) = \sum_{u=1}^{N} S_u(T).$$

There are $(2n - 3)!!$ rooted trees with n leaves, and the most parsimonious tree among them is given by

$$T^* = \operatorname*{argmin}_{T} S(T).$$

We will show that one and the same tree T^* minimizes both the cost of weighted parsimony $S_w(T)$ and the cost of traditional parsimony $S_{tr}(T)$. If k_u is the number of substitutions at site u for tree T, then the weighted parsimony cost at site u is equal to $-(\log \beta)k_u - (\log \alpha)(n - 1 - k_u)$, since a rooted tree with n leaves has $n - 1$ branch nodes, and the total number of matches and substitutions along the edges of tree T is $n - 1$.

For tree T the traditional parsimony cost S_{tr} is equal to the sum of the minimal numbers of substitutions k_u at each site u as follows:

$$S_{tr}(T) = \sum_{u=1}^{N} k_u,$$

while the weighted parsimony cost is given by

$$S_w(T) = \sum_{u=1}^{N} (-\log \beta k_u - \log \alpha (n - 1 - k_u))$$

$$= (\log \alpha - \log \beta) \sum_{u=1}^{N} k_u - \log \alpha (n - 1),$$

with the same numbers of substitutions $k_1, \ldots, k_N$. The set $k_1, \ldots, k_N$ minimizing S_{tr} and S_w is the same since $\log \alpha - \log \beta > 0$.

Therefore, the same tree T^* minimizes the weighted parsimony and the traditional parsimony costs:

$$T^* = \underset{T}{\text{argmin}} \, S_w(T) = \underset{T}{\text{argmin}} \, S_{tr}(T). \qquad \Box$$

Problem 8.16 Obtain the Jukes–Cantor distance from the minimum relative entropy principle.

Remark The relative entropy $H(P|Q)$ for two discrete distributions P and Q taking non-negative values on one and the same set $\{y_j\}$ is defined by the following formula:

$$H(P|Q) = \sum_i P(y_i) \log \frac{P(y_i)}{Q(y_i)},$$

The fact that for any P and Q the relative entropy $H(P|Q)$ is always non-negative and takes zero value if and only if $P = Q$ follows from the inequality $(x > 0)$:

$$\log x \le x - 1,$$

with equality only for $x = 1$. If $x = Q(y_i)/P(y_i)$, then

$$-H(P|Q) = \sum_i P(y_i) \log \frac{Q(y_i)}{P(y_i)} \le \sum_i P(y_i) \left(\frac{Q(y_i)}{P(y_i)} - 1 \right)$$

$$= \sum_i Q(y_i) - \sum_i P(y_i) = 1 - 1 = 0,$$

with equality only for $P = Q$. Therefore,

$$\sum_i P(y_i) \log Q(y_i) \le \sum_i P(y_i) \log P(y_i), \qquad (8.8)$$

with equality if and only if $P = Q$.

Solution We assume that the ungapped alignment of DNA sequences x^1 and x^2 of length N has n_1 matches and n_2 mismatches. The maximum likelihood distance (MLD) between x^1 and x^2 is defined as

$$d_{12}^{ML} = \underset{t}{\text{argmax}} \left(\prod_{u=1}^N P(x_u^2 | x_u^1, t) \right).$$

Under the Jukes–Cantor model the MLD becomes

$$d_{12}^{ML} = \underset{t}{\operatorname{argmax}}(r_t)^{n_1}(s_t)^{n_2},$$

where $r_t = P(a|a,t) = \frac{1}{4}(1+3e^{-4\alpha t})$ and $s_t = P(a|b,t) = \frac{1}{4}(1-e^{-4\alpha t})$ are respectively, the diagonal and the non-diagonal elements of the Jukes–Cantor substitution matrix.

The maximization of function $P(t) = (r_t)^{n_1}(s_t)^{n_2}$ could be achieved by standard methods of calculus (Problem 8.7), and the point of maximum of $P(t)$ is given by

$$t^* = d_{12}^{ML} = \frac{1}{4\alpha}\ln\frac{3(n_1+n_2)}{3n_1-n_2} = -\frac{1}{4\alpha}\ln\left(1-\frac{4}{3}f\right). \qquad (8.9)$$

Here $f = n_2/N$ is the fraction of sites with mismatches (variable sites).

The Jukes–Cantor distance **d** is defined as the expected number of substitutions per site over the time d^{ML}. Since in the Jukes–Cantor model the expected number of substitutions per site per unit time is 3α, we have

$$\mathbf{d}(x^1,x^2) = 3\alpha d_{12}^{ML} = -\frac{3}{4}\ln\left(1-\frac{4}{3}f\right).$$

Now we will show that the non-negative property (8.8) of the relative entropy can be used to obtain the same result. Taking the natural logarithm of $P(t) = (r_t)^{n_1}(s_t)^{n_2}$ yields the following expression:

$$\ln P(t) = n_1 \ln r_t + n_2 \ln s_t$$
$$= N\left(\frac{n_1}{N}\ln r_t + \frac{n_2}{3N}\ln s_t + \frac{n_2}{3N}\ln s_t + \frac{n_2}{3N}\ln s_t\right)$$
$$= \sum_i P(y_i)\ln Q(y_i),$$

where Q and P are discrete distributions defined on the same four-point set with probabilities r_t, s_t, s_t, s_t, and $n_1/N, n_2/3N, n_2/3N, n_2/3N$, respectively. According to Equation (8.8), $\sum_i P(y_i)\ln Q(y_i)$ attains its maximum when $P = Q$. By solving with respect to t the system of two equations

$$r_t = \frac{1}{4}\left(1+3e^{-4\alpha t}\right) = \frac{n_1}{N},$$
$$s_t = \frac{1}{4}\left(1-e^{-4\alpha t}\right) = \frac{n_2}{3N},$$

we find the point of maximum

$$t^{**} = d_{12}^{ML} = -\frac{1}{4\alpha} \ln\left(1 - \frac{4n_2}{3N}\right) = -\frac{1}{4\alpha} \ln\left(1 - \frac{4}{3}f\right).$$

We see that $t^{**} = t^* = d_{12}^{ML}$; therefore, the Jukes–Cantor distance **d** between the sequences x^1 and x^2,

$$\mathbf{d}(x^1, x^2) = 3\alpha d_{12}^{ML} = -\frac{3}{4} \ln\left(1 - \frac{4}{3}f\right),$$

has been derived from the properties of the relative entropy. Whereas finding $t^* = d_{12}^{ML}$ was accomplished by finding the point of maximum of $P(t)$ via the standard necessary conditions of the stationary point, finding t^{**} was achieved directly from condition $P = Q$ by solving algebraic equations. ☐

Remark The meaning of the requirement $P = Q$ is as follows. Probabilities $r_t = P(a|a, t)$ and $s_t = P(b|a, t)$, $a \neq b$, determine expected fractions of "constant" and "variable" sites, respectively, as time t elapses. The maximum likelihood for observed data will be reached when $r_t = n_1/N$ and $s_t = n_2/(3N)$, since n_1/N is the observed fraction of constant sites and $n_2/(3N)$ is the observed average fraction of variable sites for one of the three types of mismatches. ☐

8.2 Additional problems and theory

This section contains theoretical and computational problems dealing with the probabilistic models of molecular evolution.

Several problems are devoted to the Jukes–Cantor model. We show that the maximum likelihood tree for a set of three given nucleotide sequences differs from the tree composed of the pairwise maximum likelihood distances (times) (Problem 8.17). Under the Jukes–Cantor model, we compare likelihoods of the trees corresponding to competing hypotheses on the divergence time of the two species and compute the most probable divergence time (Problem 8.18). In the theoretical introduction to Problem 8.18 we prove that the stochastic process counting the number of substitutions controlled by the stationary Markov process is the Poisson process, if the diagonal elements of the matrix of substitution rates are all the same (both the Jukes–Cantor and the Kimura models satisfy this condition).

The continuous-time Markov process of amino acid substitution similar to the Jukes–Cantor and the Kimura models of DNA evolution (Problem 8.19) has been derived (generalized) from the stationary Markov chain with the matrix of transition probabilities 1 PAM (Section 2.2.1). However, there is an important difference as well: the family of substitution matrices t PAM, $t \geq 0$, and the matrix of rates of substitution (also derived in Problem 8.19) do not depend on parameters, unlike the Jukes–Cantor and the Kimura matrices.

A probabilistic model of evolution of DNA or protein sequences allowing substitutions, insertions, and deletions (the *links* or TKF model by Thorne *et al.* 1991) combines the continuous-time Markov process describing substitutions and the birth–death process with immigration describing insertions and deletions (see the theoretical introduction to Problem 8.20). We use this model to compute the probability of sequence x having transformed to sequence y over time t and to find the likelihood of a tree with given nodes and branch lengths (Problem 8.20).

Finally, a brief description of the models taking into account the variability of the substitution rates among sequence sites is presented in Section 8.2.3.

Problem 8.17 It is assumed that nucleotide sequences x^1, x^2, and x^3 (shown below in top down order)

$$
\begin{array}{ccccccccccc}
C & C & G & G & C & C & G & C & G & C & G \\
C & G & G & G & C & C & G & G & C & C & G \\
G & C & C & G & C & C & G & G & G & C & C
\end{array}
$$

have evolved from a common ancestor. Assuming the Jukes–Cantor model, show that the maximum likelihood tree for these sequences is not composed of their pairwise maximum likelihood distances.

Solution The maximum likelihood distance d_{ij}^{ML} between nucleotide sequences x_i and x_j is given by Equation (8.9) assuming the Jukes–Cantor model (two approaches to the derivation of Equation (8.9) are discussed in Problems 8.7 and 8.16). Therefore, for pairwise maximum likelihood distances of sequences x^1, x^2, and x^3, we have

$$
d_{12}^{ML} = -\frac{1}{4\alpha} \ln\left(1 - \frac{4}{3}f_{12}\right) = -\frac{1}{4\alpha} \ln\left(1 - \frac{4}{3} \times \frac{3}{11}\right) = -\frac{1}{4\alpha} \ln\frac{7}{11},
$$

$$
d_{13}^{ML} = -\frac{1}{4\alpha} \ln\left(1 - \frac{4}{3}f_{13}\right) = -\frac{1}{4\alpha} \ln\left(1 - \frac{4}{3} \times \frac{4}{11}\right) = -\frac{1}{4\alpha} \ln\frac{17}{33},
$$

$$
d_{23}^{ML} = -\frac{1}{4\alpha} \ln\left(1 - \frac{4}{3}f_{23}\right) = -\frac{1}{4\alpha} \ln\left(1 - \frac{4}{3} \times \frac{5}{11}\right) = -\frac{1}{4\alpha} \ln\frac{13}{33}.
$$

Here f_{ij}, $i,j = 1,2,3$, are fractions of mismatches in sequences x^i and x^j aligned without gaps, and α is the parameter of the Jukes–Cantor model.

Now we assume that x^1, x^2, and x^3 are related by the unrooted three-leaf tree $\tilde{T}$ with edge lengths $\tilde{t}_1$, $\tilde{t}_2$, $\tilde{t}_3$ as follows:

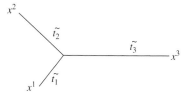

If $\tilde{t}_1$, $\tilde{t}_2$, and $\tilde{t}_3$ satisfy the following equations:

$$\tilde{t}_1 + \tilde{t}_2 = d_{12}^{ML},$$

$$\tilde{t}_1 + \tilde{t}_3 = d_{13}^{ML},$$

$$\tilde{t}_2 + \tilde{t}_3 = d_{23}^{ML},$$

then for the branch lengths we have

$$\tilde{t}_1 = -\frac{1}{8\alpha} \ln \frac{119}{143} = \frac{0.0230}{\alpha},$$

$$\tilde{t}_2 = -\frac{1}{8\alpha} \ln \frac{91}{187} = \frac{0.0900}{\alpha},$$

$$\tilde{t}_3 = -\frac{1}{8\alpha} \ln \frac{221}{693} = \frac{0.1429}{\alpha}.$$

Now we can show that $\tilde{T}$ is not the maximum likelihood tree for x^1, x^2, and x^3.

The likelihood of a tree T with branch lengths t_1, t_2, t_3 and the leaf sequences x^1, x^2, and x^3 is as follows (Problem 8.4):

$$L(T, t_1, t_2, t_3) = P(x^1, x^2, x^3 | T, t_1, t_2, t_3)$$

$$= 4^{-3(n_1+n_2+n_3+n_4)} a(t_1, t_2, t_3)^{n_1} b(t_1, t_2, t_3)^{n_2} b(t_1, t_3, t_2)^{n_3} b(t_3, t_2, t_1)^{n_4}.$$

Here n_1 is the number of sites with CCC or GGG; n_2, n_3, and n_4 are the numbers of sites with CCG or GGC, CGC or GCG, GCC or CGG, respectively, while

$$a(t_1, t_2, t_3) = 1 + 3e^{-4\alpha(t_1+t_2)} + 3e^{-4\alpha(t_1+t_3)} + 3e^{-4\alpha(t_2+t_3)} + 6e^{-4\alpha(t_1+t_2+t_3)}$$

and

$$b(t_1, t_2, t_3) = 1 + 3e^{-4\alpha(t_1+t_2)} - e^{-4\alpha(t_1+t_3)} - 3e^{-4\alpha(t_2+t_3)} - 2e^{-4\alpha(t_1+t_2+t_3)}.$$

In the maximum likelihood tree the branch lengths must maximize the likelihood function L (and its logarithm $\ln L$). At the point of maximum of the function $\ln L$, it is necessary that for $i = 1, 2, 3$

$$\frac{\partial \ln L}{\partial t_i} = 0. \tag{8.10}$$

Direct computations of the left parts of Equation (8.10) using a MATLAB package with the substitution of $\tilde{t}_1, \tilde{t}_2, \tilde{t}_3$ of tree $\tilde{T}$ and $n_1 = 5$, $n_2 = 3$, $n_3 = 2$, and $n_4 = 1$ yield

$$\frac{\partial \ln L}{\partial t_1}(\tilde{t}_1, \tilde{t}_2, \tilde{t}_3)$$

$$= \frac{n_1}{a(\tilde{t}_1, \tilde{t}_2, \tilde{t}_3)} \frac{\partial a(t_1, t_2, t_3)}{\partial t_1}(\tilde{t}_1, \tilde{t}_2, \tilde{t}_3) + \frac{n_2}{b(\tilde{t}_1, \tilde{t}_2, \tilde{t}_3)} \frac{\partial b(t_1, t_2, t_3)}{\partial t_1}(\tilde{t}_1, \tilde{t}_2, \tilde{t}_3)$$

$$+ \frac{n_3}{b(\tilde{t}_1, \tilde{t}_3, \tilde{t}_2)} \frac{\partial b(t_1, t_3, t_2)}{\partial t_1}(\tilde{t}_1, \tilde{t}_2, \tilde{t}_3) + \frac{n_4}{b(\tilde{t}_3, \tilde{t}_2, \tilde{t}_1)} \frac{\partial b(t_3, t_2, t_1)}{\partial t_1}(\tilde{t}_1, \tilde{t}_2, \tilde{t}_3)$$

$$= -206.44\alpha \neq 0. \tag{8.11}$$

Similarly, we compute

$$\frac{\partial \ln L}{\partial t_2}(\tilde{t}_1, \tilde{t}_2, \tilde{t}_3) = 2363.7\alpha \neq 0,$$

$$\frac{\partial \ln L}{\partial t_3}(\tilde{t}_1, \tilde{t}_2, \tilde{t}_3) = 353.71\alpha \neq 0. \tag{8.12}$$

Equations (8.11) and (8.12) show that the necessary conditions in Equation (8.10) for the stationary point are not satisfied. Therefore, a tree $\tilde{T}$ built of pairwise maximum likelihood distances is not the maximum likelihood tree for the sequences x^1, x^2, and x^3. It would be interesting to extend this statement to the general case of n sequences. □

Problem 8.18 (Revisiting Problem 1.18) One theory states that the last common ancestor of birds and crocodiles lived 120 million years ago, while another theory suggests that this time is twice as large. Comparison of fragments of homologous genes of two species (Nile crocodile and Mediterranean seagull) revealed on average 365 differences in 1000 nt long fragments. It was assumed (Problem 1.18) that substitutions at different DNA sites occur independently, and at each site the number of substitutions is described by the Poisson process with parameter p equal to 10^{-9} substitutions per year. Assuming the validity of the Jukes–Cantor model (a) compare the likelihood of the two theories; (b) determine the maximum likelihood distance between the two species.

Solution The Jukes–Cantor matrix of rates of nucleotide substitutions must have the parameter α such that $3\alpha = p = 10^{-9}$, thus $\alpha = 3.33 \times 10^{-10}$.

(a) We assume that DNA sequences x^1 and x^2 of two species and sequence y of their last common ancestor are related by the following tree T with the molecular clock property:

The likelihood of each theory is defined as a likelihood of tree T with branch length t equal to the divergence time proposed by a particular theory ($t = t_1 = 120\,000\,000$ years or $t = t_2 = 240\,000\,000$ years).

Given that substitutions along two edges of the tree are independent, while reversibility and the Markovian property hold for the Jukes–Cantor model, the likelihood L_u at site u is given by

$$L_u(t) = P(x_u^1, x_u^2 | T, t, t) = P(x_u^1, x_u^2 | t) = \sum_{y_u} q_{y_u} P(x_u^1 | y_u, t) P(x_u^2 | y_u, t)$$

$$= \sum_{y_u} q_{x_u^1} P(y_u | x_u^1, t) P(x_u^2 | y_u, t) = q_{x_u^1} P(x_u^2 | x_u^1, t).$$

Hence, for site u with matching nucleotides ($x_u^1 = x_u^2$), the likelihood is given by

$$L_u(t) = P \frac{1}{16} (1 + 3e^{-4\alpha(t+t)}) = \frac{1}{16} (1 + 3e^{-8\alpha t}).$$

Similarly, for site u with a mismatch ($x_u^1 \neq x_u^2$) we have

$$L_u(t) = \frac{1}{16} (1 - e^{-4\alpha(t+t)}) = \frac{1}{16} (1 - e^{-8\alpha t}).$$

Further, assuming independence of substitutions at different sites, we derive the likelihood of the tree T with edge lengths t and leaf sequences x^1 and x^2:

$$L(t) = P(x^1, x^2 | T, t, t) = \prod_{u=1}^{N} L_u(t) = 16^{-N} (1 + 3e^{-8\alpha t})^{N-M} (1 - e^{-8\alpha t})^M.$$

$$(8.13)$$

Here N is the length of both DNA sequences, M is the number of sites with mismatches. For $N = 1000$, $M = 365$, and $\alpha = 3.33 \times 10^{-10}$ the log-likelihood ratio for trees with edge lengths t_1 and t_2 is given by

$$R = \ln \frac{L(t_1)}{L(t_2)} = \ln \frac{(1 + 3e^{-8\alpha t_1})^{635}(1 - e^{-8\alpha t_1})^{365}}{(1 + 3e^{-8\alpha t_2})^{635}(1 - e^{-8\alpha t_2})^{365}}$$

$$= 635(\ln(1 + 3e^{-0.32}) - \ln(1 + 3e^{-0.64}))$$

$$+ 365(\ln(1 - e^{-0.32}) - \ln(1 - e^{-0.64}))$$

$$= 132.00 - 199.23 = -67.23 < 0.$$

Therefore, the Jukes–Cantor model with parameter $\alpha = 3.33 \times 10^{-10}$ suggests that the observed data support the theory that the Nile crocodile and Mediterranean seagull diverged from the common ancestor 240 MYA.

Here we have arrived to the same conclusion as in the solution to Problem 1.18, when the occurrence of substitutions was described by the Poisson process with rate $p = 3\alpha$. An advantage of the Jukes–Cantor model is that it removes a rather unrealistic assumption that no more than one substitution could occur at one site over the time of divergent evolution. For more details on the connection between an evolutionary model of substitutions (the Markov process) and a Poisson process describing the number of substitutions, see Section 8.2.1.

(b) The evolutionary distance between two species connected by a path in a phylogenetic tree is equal to the sum of all branches in the path. We recall the formula for a sum of edge lengths of a two-leaf tree T maximizing the likelihood $L(T)$ of the tree relating the sequences x^1 and x^2 (Problems 8.7 and 8.16):

$$\tau_1 + \tau_2 = -\frac{1}{4\alpha} \ln \left(1 - \frac{4M}{3N} \right).$$

Here τ_1 and τ_2 are edge lengths in the tree. Therefore, the maximum likelihood (evolutionary) distance between sequences x^1 and x^2 of the two species is $\tau_1 + \tau_2 = 500\,000\,000$ (500 million years). The molecular clock property suggests that these two lineages diverged from the last common ancestor 250 MYA. $\qquad\square$

8.2.1 Relationship between sequence evolution models described by the Markov and the Poisson processes

If a stationary Markov process $X(t)$, $t \geq 0$, with a state space $S = \{A, C, G, T\}$ is used as a model for DNA evolution, we assume that nucleotide substitutions at a given site of DNA sequence are considered as transitions between states of process $X(t)$ described by the family of matrices of substitution probabilities, $S(t) = (s_{i,j}(t))$, $i, j \in S$, $t \geq 0$, and matrix $R = (r_{i,j})$, $i, j \in S$, of substitution rates.

We define a random process $N(t)$, $t \geq 0$, as the number of changes of states (nucleotides) of process $X(t)$ over the time interval $[0, t]$. We will show that if all elements of the main diagonal of matrix R are equal, $r_{i,i} = -p$ for any $i \in S$, then the stochastic process $N(t)$ is the Poisson process with rate p (note that all diagonal elements of R are negative, so $p > 0$).

We should verify that $N(t)$, $t \geq 0$, satisfies the definition of the Poisson process with rate p:

(1) $N(0) = 0$;
(2) the process has stationary increments, i.e. $N(t_2) - N(t_1)$ and $N(t_2 + s) - N(t_1 + s)$ have the same distribution for any $0 \leq t_1 \leq t_2$, $s > 0$;
(3) the process has independent increments, i.e. $N(t_4) - N(t_3)$ and $N(t_2) - N(t_1)$ are independent random variables for any $t_1 \leq t_2 \leq t_3 \leq t_4$;
(4) $P(N(t) = n) = e^{-pt}(pt)^n/n!$, for $n \geq 0$.

Obviously, property (1) holds: there is no change over the time interval $[0, 0]$. The stationarity of the process increments follows from the stationarity of the Markov process $X(t)$: the distribution of the number of changes of $X(t)$ states over a time interval depends only on the length of this interval. Since $(t_2 + s) - (t_1 + s) = t_2 - t_1$, $N(t_2) - N(t_1)$ and $N(t_2 + s) - N(t_1 + s)$ are identically distributed.

To prove that properties (3) and (4) hold, we define a random variable $T_j, j \geq 1$, as the time between the $(j-1)$th and jth changes of states of process $X(t)$. It is known from the theory of stochastic processes (Ross, 1996) that if process $X(t)$ is in state i, $i \in S$, after the $(j-1)$th change of state, then T_j is exponentially distributed with parameter $-r_{i,i}$, and random variables $T_j, j \geq 1$, are independent. In our case $-r_{i,i} = p$ for all $i \in S$, and random variables T_j are independent and identically distributed with probability density function $f(t) = pe^{-pt}$. Next we determine the distribution function $F_n(t)$ of random variable $\mathcal{T}_n = \sum_{j=1}^{n} T_j$, the time it takes to observe n changes of states of process $X(t)$. For $n = 1, 2, 3$ we have

$$F_1(t) = P(\mathcal{T}_1 \leq t) = P(T_1 \leq t) = 1 - e^{-pt},$$

$$F_2(t) = P(\mathcal{T}_2 \leq t) = P(T_1 + T_2 \leq t) = \int_0^t f(y)P(T_1 + T_2 \leq t | T_1 = y)dy$$

$$= \int_0^t f(y)P(T_2 \leq t - y | T_1 = y)dy = \int_0^t f(y)P(T_2 \leq t - y)dy$$

$$= \int_0^t pe^{-py}(1 - e^{-p(t-y)})dy = 1 - e^{-pt} - pte^{-pt}, \tag{8.14}$$

$$F_3(t) = P(\mathcal{T}_3 \leq t) = P(T_1 + T_2 + T_3 \leq t) = \int_0^t f(y)P(\mathcal{T}_2 + T_3 \leq t | T_3 = y)dy$$

$$= \int_0^t f(y)F_2(t - y)dy = 1 - e^{-pt} - pte^{-pt} - \frac{(pt)^2}{2}e^{-pt}.$$

The expressions for the distribution functions F_n, $n = 1, 2, 3$, have a pattern that allows us to suggest the general formula for $F_n(t)$ as follows:

$$F_n(t) = P(\mathcal{T}_n \leq t) = 1 - e^{-pt} \sum_{i=0}^{n-1} \frac{(pt)^i}{i!}. \tag{8.15}$$

We prove equality (8.15) by mathematical induction. The basis of this induction was established in Equations (8.14). Now we assume that

$$F_{n-1}(t) = 1 - e^{-pt} \sum_{i=0}^{n-2} \frac{(pt)^i}{i!}$$

and find $F_n(t)$ as follows:

$$F_n(t) = P(\mathcal{T}_{n-1} + T_n \leq t) = \int_0^t f(y) P(\mathcal{T}_{n-1} + T_n \leq t | T_n = y) dy$$

$$= \int_0^t f(y) F_{n-1}(t-y) dy = \int_0^t p e^{-py} \left(1 - e^{-p(t-y)} \sum_{i=0}^{n-2} \frac{(p(t-y))^i}{i!} \right) dy$$

$$= \int_0^t p e^{-py} \, dy - e^{-pt} \sum_{i=0}^{n-2} \int_0^t \frac{p^{i+1}(t-y)^i}{i!} \, dy$$

$$= 1 - e^{-pt} - \sum_{i=0}^{n-2} \frac{(pt)^{i+1}}{(i+1)!} = 1 - e^{-pt} \sum_{i=0}^{n-1} \frac{(pt)^i}{i!}.$$

Thus, we have checked that formula (8.15) is correct. It immediately leads to an analytical expression for the probability density function f_n of $\mathcal{T}_n$:

$$f_n(t) = F'_n(t) = \frac{p^n t^{n-1}}{(n-1)!} e^{-pt}. \tag{8.16}$$

Now we will show that property (4) holds. Indeed, from Equation (8.15) we derive

$$P(N(t) = n) = P(N(t) \leq n) - P(N(t) \leq n-1)$$

$$= P(N(t) < n+1) - P(N(t) < n) = P(\mathcal{T}_{n+1} > t) - P(\mathcal{T}_n > t)$$

$$= (1 - F_{n+1}(t)) - (1 - F_n(t))$$

$$= \left(1 - e^{-pt} \sum_{i=0}^{n-1} \frac{(pt)^i}{i!} \right) - \left(1 - e^{-pt} \sum_{i=0}^{n} \frac{(pt)^i}{i!} \right) = e^{-pt} \frac{(pt)^n}{n!}.$$

Finally, we turn to property (3). Since increments of $N(t)$ are stationary (from property (2)), it is sufficient to prove property (3) for increments

$N(t_1) - N(0) = N(t_1)$ and $N(t_3) - N(t_2)$, $0 \le t_1 \le t_2 \le t_3$. In fact, we will verify the independence of three increments $N(t_1)$, $N(t_2) - N(t_1)$, and $N(t_3) - N(t_2)$:

$$P(N(t_1) = n, N(t_2) - N(t_1) = l, N(t_3) - N(t_2) = m)$$
$$= P(N(t_1) = n)P(N(t_2) - N(t_1) = l)P(N(t_3) - N(t_2) = m) \quad (8.17)$$

For any positive integers n, l, and m. Obviously, if the three increments are mutually independent, then any two of them are independent too. The stationarity of increments and property (4) allows us to rewrite the right-hand side of Equation (8.17) as

$$P(N(t_1) = n)P(N(t_2) - N(t_1) = l)P(N(t_3) - N(t_2) = m)$$
$$= e^{-pt_1}\frac{(pt_1)^n}{n!}e^{-p(t_2-t_1)}\frac{(p(t_2 - t_1))^l}{l!}e^{-p(t_3-t_2)}\frac{(p(t_3 - t_2))^m}{m!}$$
$$= e^{-pt_3}\frac{p^{n+l+m}}{n!l!m!}t_1^n(t_2 - t_1)^l(t_3 - t_2)^m.$$

On the other hand, the left-hand side of Equation (8.17) can be transformed into the same expression as above by the same arguments as in Equations (8.14):

$$P(N(t_1) = n, N(t_2) - N(t_1) = l, N(t_3) - N(t_2) = m)$$
$$= P(\mathcal{T}_n \le t_1, \mathcal{T}_{n+1} > t_1, \mathcal{T}_{n+l} \le t_2, \mathcal{T}_{n+l+1} > t_2, \mathcal{T}_{n+l+m} \le t_3, \mathcal{T}_{n+l+m} > t_3)$$
$$= \int_{y_1=0}^{t_1}\int_{y_2=t_1-y_1}^{t_2-y_1}\int_{y_3=0}^{t_2-y_1-y_2}\int_{y_4=t_2-y_1-y_2}^{t_3-y_1-y_2-y_3}\int_{y_5=0}^{t_3-y_1-y_2-y_3-y_4} (f_n(y_1)$$
$$\times f(y_2)f_{l-1}(y_3)f(y_4)f_{m-1}(y_5)$$
$$\times P(\mathcal{T}_{n+l+m+1} > t_3 - y_1 - y_2 - y_3 - y_4 - y_5))dy_1\,dy_2\,dy_3\,dy_4\,dy_5$$
$$= \int \left(\frac{p^n y_1^{n-1}}{(n-1)!}e^{-py_1}pe^{-py_2}\frac{p^{l-1}y_3^{l-2}}{(l-2)!}e^{-py_3}pe^{-py_4} \right.$$
$$\left. \times \frac{p^{m-1}y_5^{m-2}}{(m-2)!}e^{-py_5}e^{-p(t_3-y_1-y_2-y_3-y_4-y_5)} \right) dy_1\,dy_2\,dy_3\,dy_4\,dy_5$$
$$= e^{-pt_3}\frac{p^{n+l+m}}{n!l!m!}t_1^n(t_2 - t_1)^l(t_3 - t_2)^m.$$

This completes the proof of Equation (8.17). Therefore, we have shown that the process $N(t)$, $t \ge 0$, satisfies the definition of the Poisson process.

One of the implications of this statement is that for both the Jukes–Cantor and Kimura models of DNA evolution, the number $N(t)$ of changes of states (nucleotides) over a time interval $[0, t]$ is the Poisson process with rate 3α for the Jukes–Cantor model and rate $2\beta + \alpha$ for the Kimura model. Thus, the Poisson

Table 8.1. *The set of eigenvalues* $\lambda_1, \ldots, \lambda_{20}$ *of the mutation probability matrix 1PAM = S(1)*

1(Ala)	2(Arg)	3(Asn)	4(Asp)	5(Cys)	6(Gln)	7(Glu)
0.978	0.981	1.000	0.983	0.984	0.985	0.987
8(Gly)	9(His)	10(Ile)	11(Leu)	12(Lys)	13(Met)	14(Phe)
0.987	0.998	0.997	0.997	0.989	0.990	0.996
15(Pro)	16(Ser)	17(Thr)	18(Trp)	19(Tyr)	20(Val)	
0.995	0.991	0.993	0.993	0.992	0.992	

model of substitutions used in Problem 1.18 can be considered as a model associated with the Jukes–Cantor model (Problem 8.18). This associated Poisson model allows us to count the substitutions (as changes of states), but lacks information about the current type of nucleotide at a given site, because the Poisson process $N(t)$ cannot distinguish between nucleotides (states) of the Markov process $X(t)$. In other words, process $X(t)$ unambiguously determines the Poisson process $N(t)$, but the converse does not hold; $X(t)$ cannot be recovered from $N(t)$.

The difference in Equations (1.7) and (8.13) for the likelihood of a tree with two leaves and branch length t (Problems 1.18 and 8.18) stems from the fact that in Problem 1.18 we computed the likelihood assuming that no more than one substitution could occur in one site in the whole lineage. If we include the same condition in the statement of Problem 8.18, the (conditional) likelihood would be the same is in Equation (1.7), and the maximum likelihood estimate of the divergence time would change from 250 MYA to 287.4 MYA. (If no more than one substitution per site is allowed, the greater divergence time is necessary for the observed mismatches M to appear.)

Problem 8.19 In Section 2.2.1 it was shown in detail how to derive PAM matrices under the assumption that amino acid substitutions in proteins could be described by a stationary Markov chain. This method produces a series of mutation probability matrices n PAM corresponding to discrete evolutionary time intervals multiple of the 1 PAM time unit.

Derive the mutation probability matrix t PAM, explicitly depending on continuous time $t \geq 0$, an analog of the substitution matrix $S(t)$ defined for the Jukes–Cantor and Kimura models of DNA evolution. Matrix R of the rates of substitution has to be derived as well.

Solution We will use the designation $S(1)$ for the 1 PAM matrix with elements M_{ij} (Table 2.7).

Table 8.2. *The values of parameters μ_i, $\mu_i = -\ln \lambda_i$, $i = 1, \ldots, 20$, used in Equation (8.20)*

1	2	3	4	5	6	7	8	9	10
0.022	0.020	0.000	0.018	0.017	0.016	0.014	0.013	0.002	0.003
11	12	13	14	15	16	17	18	19	20
0.003	0.011	0.011	0.004	0.005	0.009	0.007	0.007	0.008	0.008

Matrix $S(1)$ could be interpreted as the matrix of a linear operator S in a basis $e_1, \ldots, e_{20}$ of $\mathbf{R}^{20}$. All eigenvalues λ_i, $i = 1, \ldots, 20$, of operator S are distinct (Table 8.1). Therefore, in $\mathbf{R}^{20}$ there exists a basis of eigenvectors $e_1', \ldots, e_{20}'$ of linear operator S. Then, matrix $S(1)$ can be written as $S(1) = VDV^{-1}$, where matrix D is a diagonal matrix with values λ_i, $i = 1, \ldots, 20$, on the diagonal (D is the matrix of operator S in basis $e_1', \ldots, e_{20}'$), V is the matrix of a coordinate transformation from basis $e_1, \ldots, e_{20}$ to basis $e_1', \ldots, e_{20}'$, and V^{-1} is the inverse matrix for V. Matrices V and V^{-1} are shown in Tables 1 and 2 of the Web Supplemental Materials available at opal.biology.gatech.edu/PSBSA.

Since all eigenvalues λ_i, $i = 1, \ldots, 20$, belong to interval $(0, 1]$ (Table 8.1), we can write them in the form $\lambda_i = \exp(-\mu_i)$, $\mu_i \geq 0$. For $S(n)$, the matrix of probabilities of amino acid substitutions at a time equal to n PAM units, we have:

$$(S(1))^n = S(n) = (VDV^{-1})^n = VDV^{-1}VDV^{-1} \cdots VDV^{-1} = VD^nV^{-1}. \quad (8.18)$$

Therefore, from Equation (8.18) for the elements of matrix $S(n)$ we have

$$S_{ij}(n) = \sum_k v_{ik} \lambda_k^n v_{kj}^{-1} = \sum_k v_{ik} \exp(-\mu_k n) v_{kj}^{-1}. \quad (8.19)$$

Here v_{ik}, v_{kj}^{-1} are the elements of matrices V and V^{-1}, respectively. Values of μ_k, $k = 1, \ldots, 20$, are shown in Table 8.2. Replacing n in Equation (8.19) by a continuous variable t, $t \geq 0$, gives the formula for the elements of matrix $S(t)$, $t \geq 0$. Thus, the probability that an amino acid in column j will be substituted by an amino acid in row i after time t is

$$S_{ij}(t) = \sum_{k=1}^{20} v_{ik} \exp(-\mu_k t) v_{kj}^{-1}. \quad (8.20)$$

Note that similarly to the elements of substitution matrices in the Jukes–Cantor and Kimura models, $S_{ij}(t)$ is a linear combination of the exponential functions.

The elements of matrix R which are the rates of amino acid substitution can be derived from Equation (8.20) and $S'(0) = R$ as follows:

$$R_{ij} = \sum_{k=1}^{20} -\mu_k v_{ik} \exp(-\mu_k 0) v_{kj}^{-1} = \sum_{k=1}^{20} -\mu_k v_{ik} v_{kj}^{-1}.$$

Matrix R is shown in Table 8.3.

Another approach to modeling the continuous-time Markov process of amino acid substitution related to the model of Dayhoff *et al.* (1978) was proposed by Wilbur (1985).

The family of continuous-time substitution matrices, (8.20) is useful for applications such as the determination of the maximum likelihood evolutionary distance between two protein sequences, the comparison of the likelihoods of candidate evolutionary trees for a given set of leaf sequences, finding the most probable lengths of branches for a given tree topology, etc. Note that solving such problems is computationally expensive for real sequences.

For the sake of giving a "toy" example, we determine the maximum likelihood distance t^* (in PAM units) between two amino acid sequences $x = KVAD$ and $y = KVFK$. Under the assumption of independence of sites, we have the following product of the t PAM elements defined by Equation (8.20):

$$P(t) = P(x \text{ has been changed to } y \text{ over time } t) = S_{12,12}(t)S_{20,20}(t)S_{14,1}(t)S_{12,4}(t)$$

$$= \sum_{k,l,m,q=1}^{20} v_{12,k} v_{k,12}^{-1} v_{20,l} v_{l,20}^{-1} v_{14,m} v_{m,1}^{-1} v_{12,q} v_{q,4}^{-1} \exp(-(\mu_k + \mu_l + \mu_m + \mu_q)t).$$

The MATLAB package allows us to find the point of maximum of $P(t)$, $t^* \approx 119.75$ PAM, with $P(x$ has been changed to y over time $t^*) = 0.00010279$. Note that t^* is the evolutionary distance between sequences x and y. $\square$

8.2.2 Thorne–Kishino–Felsenstein model of sequence evolution with substitutions, insertions, and deletions

Theoretical introduction to Problem 8.20

A probabilistic model of evolution of DNA or protein sequences as a process of substitutions, insertions, and deletions (the *links* or the TKF model) was proposed by Thorne, Kishino, and Felsenstein (1991) and was later generalized to allow multiple-base insertions and deletions by Thorne *et al.* (1992). The TKF model provides a base for the new alignment algorithms developed by Hein *et al.* (2000) and Holmes and Bruno (2001). In what follows we will use DNA sequence terminology; however, all arguments could be applied to amino acid sequences as well.

The TKF model combines the continuous-time Markov process describing the nucleotide substitutions with the birth–death process with immigration describing

Table 8.3. *Matrix R of rates of substitution of amino acids for the continuous-time version of the Dayhoff model of protein evolution*

The element R_{ij} is the instantaneous rate of substitution of an amino acid in column j to an amino acid in row i. All elements of the matrix have been multiplied by 10^4 and rounded to the closest integer for visualization convenience. Note that the concept of instantaneous change from amino acid a to amino acid b such that more than one nucleotide in a codon should be changed is not entirely realistic; therefore, this model is not suitable for small t.

	Ala	Arg	Asn	Asp	Cys	Gln	Glu	Gly	His	Ile	Leu	Lys	Met	Phe	Pro	Ser	Thr	Trp	Tyr	Val
Ala	−134	2	9	10	3	8	17	21	2	6	4	2	6	2	22	35	32	0	2	18
Arg	1	−88	1	0	1	10	0	0	10	3	1	19	4	1	4	6	1	8	0	1
Asn	4	1	−180	37	0	4	6	6	21	3	1	13	0	1	2	20	9	1	4	1
Asp	6	0	43	−142	0	6	54	6	4	1	0	3	0	0	1	5	3	0	0	1
Cys	1	1	0	0	−27	0	0	0	1	1	0	0	0	0	1	5	1	0	3	2
Gln	3	9	4	5	0	−125	27	1	23	1	3	6	4	0	6	2	2	0	0	1
Glu	10	0	7	57	0	35	−136	4	2	3	1	4	1	0	3	4	2	0	1	2
Gly	21	1	12	11	1	3	7	−65	2	0	1	2	1	1	3	21	3	0	0	5
His	1	8	18	3	1	20	7	0	−88	0	1	1	0	2	3	1	1	1	4	1
Ile	2	2	3	1	2	1	2	0	0	−129	9	2	12	7	0	1	7	0	1	33
Leu	3	1	3	0	0	6	1	1	4	22	−53	2	45	13	3	1	3	4	2	15
Lys	2	37	25	6	0	12	7	2	2	4	1	−75	20	0	3	8	11	0	1	1
Met	1	1	0	0	0	2	0	0	0	5	8	4	−127	1	0	2	2	0	0	4
Phe	1	1	1	0	0	0	0	1	2	8	6	0	4	−54	0	2	1	3	28	0
Pro	13	5	2	1	1	8	3	2	5	1	2	2	1	1	−75	12	4	0	0	2
Ser	28	11	35	7	11	4	6	16	2	2	1	7	4	3	17	−161	38	5	2	2
Thr	22	2	13	4	1	3	2	2	1	11	2	8	6	1	5	32	−130	0	2	9
Trp	0	2	0	0	0	0	0	0	0	0	0	0	0	1	0	1	0	−24	1	0
Tyr	1	0	3	0	3	0	1	0	4	1	0	1	0	21	0	1	1	2	−55	2
Val	13	2	1	1	3	2	2	3	3	58	11	1	17	1	3	2	10	0	2	−100

the insertions and deletions. Similarly to the Jukes–Cantor and the Kimura models, the Markov process is defined by a family of substitution matrices $S(t) = (s_t(a, b))$, $t \geq 0$. An element $s_t(a, b)$ of $S(t)$ determines the probability that a nucleotide a has been substituted by a nucleotide b over time t. At any time the birth–death process allows a single nucleotide to give birth to a child nucleotide, or to die. The child nucleotide is inserted adjacent to the parent on its right, a dead nucleotide is removed from the sequence. To allow insertions at the left end of the sequence, where nucleotides cannot appear as a result of birth events, the immigration events regulate appearances of new nucleotides (insertions) at the left end of the sequence. The rate of such insertions (immigration) is assumed to be equal to the birth rate for an individual nucleotide. Therefore, the insertions at the beginning of the sequence correspond to the immigration events of the birth–death process. The condition $\lambda < \mu$ on the birth and immigration rate λ and the death rate μ prevents both infinite growth and complete disappearance of the sequence and is necessary for the existence of the equilibrium distribution of sequence lengths. The TKF model defines a stationary distribution on DNA sequences:

$$P(s) = \left(1 - \frac{\lambda}{\mu}\right)\left(\frac{\lambda}{\mu}\right)^l \pi_A^{n_A} \pi_C^{n_C} \pi_G^{n_G} \pi_T^{n_T},$$

where l, $l \geq 0$, is the length of the sequence; π_A, π_C, π_G, and π_T are equilibrium frequencies of nucleotides, n_a is the number of occurrences of nucleotide a in sequence s. Let $p_n(t)$ be the probability that a given nucleotide has survived and has given birth to $n - 1$ descendants over time t; let $q_n(t)$ be the probability that by time t a nucleotide has died, leaving n descendants; and let $g_n(t)$ be the probability that n immigrant nucleotides are inserted at the left end of the sequence by time t. These probability functions have to satisfy a system of differential equations (Holmes and Bruno, 2001) with the solution in the following analytical form:

$$
\begin{aligned}
&p_0 = 0, \\
&p_n(t) = \alpha(t)(\beta(t))^{n-1}(1 - \beta(t)), &&n \geq 1, \\
&q_0(t) = (1 - \alpha(t))(1 - \gamma(t)), \\
&q_n(t) = (1 - \alpha(t))(1 - \beta(t))\gamma(t)(\beta(t))^{n-1}, &&n \geq 1, \\
&g_n(t) = (1 - \beta(t))(\beta(t))^n, &&n \geq 0.
\end{aligned}
$$

Here

$$\alpha(t) = e^{-\mu t},$$

$$\beta(t) = \frac{\lambda(1 - e^{(\lambda - \mu)t})}{\mu - \lambda e^{(\lambda - \mu)t}},$$

$$\gamma(t) = 1 - \frac{\mu(1 - e^{(\lambda - \mu)t})}{(1 - e^{-\mu t})(\mu - \lambda e^{(\lambda - \mu)t})}.$$

Given the family of substitution matrices $S(t)$, $t \geq 0$, and parameters λ and μ, one can determine the probabilities of various events in DNA sequence evolution, such as the probability that sequence x has evolved into sequence y over time t, the likelihood of phylogenetic trees with given leaves and edge lengths t_i, etc.

Problem 8.20 Assuming that nucleotide substitutions in the TKF model occur according to the Jukes–Cantor model with parameter $\alpha' = 2 \times 10^{-9}$, while the insertion and deletion rates are $\lambda = 10^{-9}$ and $\mu = 2 \times 10^{-9}$, respectively, find (a) the probability $P(y|x, t = 10^8)$ that sequence $x = CA$ has evolved into sequence $y = AAA$ over time $t = 100$ million years; (b) the probability $P(x, y, z|T, t_1, t_2)$ that sequences x, y, and $z = GA$ are related by tree T with edge lengths $t_1 = t_2 = 10^8$ shown below.

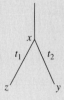

Solution (a) To determine $P(y|x, t = 10^8)$ for $x = (x_1, \ldots, x_n)$ and $y = (y_1, \ldots, y_m)$, we apply the recurrent procedure (Hein *et al.*, 2000), which allows, us to decompose the probability of subsequence $P(y_1, \ldots, y_j | x_1, \ldots, x_i, t)$, $j \leq m$, $i \leq n$, into probabilities of independent events involving either $x_1, \ldots, x_{i-1}$ or x_i:

$$P(y_1, \ldots, y_j | x_1, \ldots, x_i, t) = q_0(t) P(y_1, \ldots, y_j | x_1, \ldots, x_{i-1}, t)$$

$$+ \sum_{1 \leq k \leq j} P(y_1, \ldots, y_{j-k} | x_1, \ldots, x_{i-1}, t)$$

$$\times (p_k(t) s(x_i, y_{j-k+1}) \pi(y_{j-k+2}, \ldots, y_j) + q_k(t) \pi(y_{j-k+1}, \ldots, y_j)), \quad (8.21)$$

$$P(y_1, \ldots, y_j | x_0, t) = g_j(t) \pi(y_1, \ldots, y_j); \quad (8.22)$$

where $\pi(sequence)$ denotes $\prod_{a \in sequence} \pi(a)$. Under the Jukes–Cantor model with substitution rate $\alpha' = 2 \times 10^{-9}$, we have the following probabilities of substitutions over $t = 100$ million years:

$$s(a, a) = \frac{1}{4}(1 + 3e^{-4\alpha' t}) = 0.5870,$$

$$s(a, b) = \frac{1}{4}(1 - e^{-4\alpha' t}) = 0.1377,$$

with equilibrium frequency $\pi_a = 0.25$ for any nucleotide a. For $\lambda = 10^{-9}$, $\mu = 2 \times 10^{-9}$, and $t = 10^8$, computation of several initial values of probabilistic functions $p_n(10^8) = p_n$, $q_n(10^8) = q_n$, and $g_n(10^8) = g_n$ yields

$$
\begin{aligned}
p_0 &= 0.0000, & q_0 &= 0.1738, & g_0 &= 0.9131, \\
p_1 &= 0.7476, & q_1 &= 0.0068, & g_1 &= 0.0793, \\
p_2 &= 0.0650, & q_2 &= 0.0006, & g_2 &= 0.0069, \\
p_3 &= 0.0056, & q_3 &= 0.0001, & g_3 &= 0.0006.
\end{aligned}
$$

From Equations (8.21) and (8.22) for sequences $x = CA$ and $y = AAA$ (x_0, y_0, and z_0 below, the elements of sequences with index zero, all assumed to be the empty sets) we have:

$$
\begin{aligned}
P(y_0|x_0, t = 10^8) &= g_0 = 0.9131, \\
P(y_1|x_0, t = 10^8) &= g_1 \pi_A = 0.0198, \\
P(y_1, y_2|x_0, t = 10^8) &= g_2 \pi_A^2 = 0.0004, \\
P(y_1, y_2, y_3|x_0, t = 10^8) &= g_3 \pi_A^3 = 0.00001.
\end{aligned}
\tag{8.23}
$$

As recursion continues, after several steps we obtain

$$
\begin{aligned}
P(y|x, t = 10^8) &= P(y_1, y_2, y_3|x_1, x_2, t = 10^8) = q_0 P(y|x_1, t = 10^8) \\
&\quad + P(y_1, y_2|x_1, t = 10^8)(p_1 s(A, A) + q_1 \pi_A) \\
&\quad + P(y_1|x_1, t = 10^8)(p_2 s(A, A)\pi_A + q_2 \pi_A^2) \\
&\quad + P(y_0|x_1, t = 10^8)(p_3 s(A, A)\pi_A^2 + q_3 \pi_A^3) = 0.00299.
\end{aligned}
$$

Now we find the probability that sequences x and y are related:

$$
P(x, y|t = 10^8) = P(x)P(y|x, t = 10^8)
$$

$$
= \left(1 - \frac{\lambda}{\mu}\right)\left(\frac{\lambda}{\mu}\right)^2 \pi_C \pi_A P(y|x, t = 10^8) = 2.33 \times 10^{-5}.
$$

The probability $P(x, y|t = 10^8)$ is the sum of probabilities of all possible alignments of sequences x and y. To find the alignment with the largest probability (the optimal alignment of x and y), one needs to modify Equation (8.21) by replacing the summation with the operation of taking maximum and carry out this modified recursion up to the last step, when the probability of the optimal alignment will be determined. The optimal alignment itself can be identified by a traceback procedure.

(b) Since evolution events along different edges of tree T are independent, the likelihood of the tree with edge lengths $t_1 = t_2 = 10^8$, sequence x at the root and leaf sequences y and z is given by

$$
P(x, y, z|T, t_1 = t_2 = 10^8) = P(x)P(y|x, t = 10^8)P(z|x, t = 10^8).
$$

We have already calculated $P(x)P(y|x, t = 10^8)$; to find $P(z|x, t = 10^8)$, we apply recursion Equations (8.21) and (8.22). The first three steps produce the same results as in Equations (8.23). We proceed and obtain

$$P(z|x, t = 10^8) = P(z_1, z_2|x_1, x_2, t = 10^8) = q_0 P(z|x_1, t = 10^8)$$
$$+ P(z_1|x_1, t = 10^8)(p_1 s(A, A) + q_1 0.25)$$
$$+ P(z_0|x_1, t = 10^8)(p_2 s(A, G)0.25 + q_2 0.25^2) = 0.0447.$$

Finally, we find the likelihood of tree T:

$$P(x, y, z|T, t_1 = t_2 = 10^8) = 2.33 \times 10^{-5} \times 0.044701 = 1.05 \times 10^{-6}.$$

Remark Generally speaking, the TKF model allows us to generalize the results obtained for the tree-building problems utilizing ungapped sequence alignments to the problems concerned with tree construction using the gapped alignments. These problems include: finding the optimal edge lengths for a tree with given topology, leaf, and branch nodes; determining optimal ancestral sequences at the branch nodes given tree and leaf sequences; building the maximum likelihood tree for a given set of leaf sequences. Solving these and related problems, however, is a computationally and analytically much more demanding task than solving such problems using the Jukes–Cantor or the Kimura models, since the probabilistic functions of the TKF model are much more complicated. □

8.2.3 *More on the rates of substitution*

The models of DNA and protein evolution described earlier (Jukes–Cantor, Kimura, and Dayhoff) require an assumption that the rates of substitution at all sites of DNA or protein sequence are equal. Several attempts have been made to overcome this limitation. For example, Goldman, Thorne, and Jones (1996) utilized the three distinct reversible stationary Markov processes describing the amino acid substitutions at protein sites that were supposed to be situated within the three different types of secondary structure: α-helix, β-sheet, and loop, interpreted as hidden states of an HMM (with each hidden state characterized by its own matrix of instantaneous rates of substitution rather than by a set of emission probabilities). The authors estimated transition probabilities between hidden states using protein sequences with known secondary structures. The rates of amino acid substitution were derived from evolutionary closely related protein pairs under the assumption that the evolutionary time separating these sequences is small enough and that the probability of substitution is approximately equal to the product of the rate of substitution and the elapsed time.

Nei, Chakraborty, and Fuerst (1976) introduced the distribution of the relative substitution rates among sites in the form of the gamma density ($x \geq 0$):

$$\rho(x) = \frac{\alpha^\alpha}{\Gamma(\alpha)} x^{\alpha-1} e^{-\alpha x}. \tag{8.24}$$

Here Γ is the gamma function and α is the parameter of the gamma distribution.

Under the assumption that the distribution of relative substitution rates among sites does not change with time (but allowing the absolute rates for individual sites to change), Grishin, Wolf, and Koonin (2000) determined approximate intraprotein and interprotein rate distributions. Here the intraprotein rate distribution reflects the site variation of substitution rates within the proteins of one family, while the interprotein rate reflects the variation of substitution rates between proteins of different families. For the intraprotein rate distribution in the form of Equation (8.24) the authors estimated parameter α from multiple alignments of fourteen large protein families. These alignments produced the fraction of unchanged sites u and the average number of substitutions per site $d(u)$. Further, an assumption was made that u, d, and ρ are related by the following formula:

$$u(d) = \int_0^{+\infty} \rho(x) e^{-xd}\, dx.$$

Then $u(d)$ is the moment-generating function of the intraprotein rate distribution with expected value and the variance 1 and $1/\alpha$, respectively. The first estimate of parameter α was derived from $1/\alpha = \frac{d^2 u}{dd^2}(0) - 1$. The second, independent estimate of α was obtained as a minimum point of the chi-square statistics composed of the differences between distributions of the normalized evolutionary distances for all pairs of nineteen complete genomes. Remarkably, both methods have led to the same result, $\alpha \approx 0.31$.

The empirical interprotein rate distribution was generated from pairs of orthologous proteins found as reciprocal best hits of the BLAST program (Altschul *et al.*, 1997) in all-against-all searches in nineteen completely sequenced genomes. The probability density function η determined as the best fit to the data was given by

$$\eta(x) = \frac{b(b+c)}{c}(1 - e^{-cx}) e^{-bx}.$$

Here $x \geq 0$, $b = 1.01$, and $c = 5.88$. The authors argue that the presence of the exponential tail in the interprotein rate distribution makes this distribution compatible with the constant mutation rate. Therefore, the protein change at large evolutionary intervals is reasonably approximated by the molecular clock hypothesis.

8.3 Further reading

Several frequently used models of nucleotide and amino acid substitution were described and compared by Whelan, Liò, and Goldman (2001). The authors also reviewed up-to-date statistical tools available in the molecular phylogenetics: maximum likelihood inference, testing of evolutionary models, testing of tree topologies, etc. The issues of testing models of nucleotide substitution and selecting the best-fit model, given empirical data, were discussed by Posada and Crandall (2001).

An empirical model of protein evolution combining a parsimony-based counting (as in the model by Dayhoff *et al.*, 1978) and the maximum likelihood approach was suggested by Whelan and Goldman (2001). The codon-based model of nucleotide substitution for protein-coding DNA sequences, introduced independently by Goldman and Yang (1994) and by Muse and Gaut (1994), was later modified (Yang, 1998) to accommodate the so-called acceptance rate ω defined as the ratio d_N/d_S of the number of non-synonymous substitutions to the number of synonymous substitutions per site. Under this model the maximum likelihood estimates for ω were obtained and the positive selection (corresponding to the values $\omega > 1$) was detected for some genes (Yang *et al.*, 2000). Statistical methods of estimating d_N and d_S and measuring molecular adaptation were further developed by Yang and Bielawski (2000), and Yang and Nielsen (2000). Using a modification of the method described by Yang and Nielsen (2000), Clark *et al.* (2003) tested hypotheses on the presence of positive selection in human genes. The codon-based models, as well as secondary structure-based models of sequence evolution, were reviewed by Lewis (2001).

Two phylogeny reconstructing methods based on discrete Fourier calculus have been developed: the method of invariants (Cavender and Felsenstein, 1987; Lake, 1987; Evans and Speed, 1993; Székely, Steel, and Erdös, 1993) and the Hadamard conjugation method (Hendy, Penny, and Steel, 1994; Steel, Hendy, and Penny, 1998). These methods are applicable to DNA sequences with substitutions described by the Kimura three-parameter model, the Kimura two-parameter model (the Kimura model in BSA), and the Jukes–Cantor model.

Yang and Rannala (1997) proposed a Bayesian method for estimating phylogenetic trees using DNA sequence data. This method uses the birth–death process to define the prior distribution on the tree space and the Markov chain Monte Carlo method to determine the maximum posterior probability tree. Comparative analysis of the maximum likelihood-based methods versus the Bayesian approach for estimating phylogenetic trees can be found in Suzuki, Glazko, and Nei (2002), Holder and Lewis (2003), Douady *et al.* (2003). Kishino and Hasegawa (1989) proposed a method for estimating the standard error and the confidence intervals for the difference in the log-likelihoods between two distinct topologies (KH test).

The KH test has been often used in practice, even for comparing several topologies. In the latter case the test could lead to overconfidence for a wrong tree. Therefore, Shimodaira and Hasegawa (1999) developed a modification of the KH test to deal with multiple-comparison tests. The KH test and other likelihood-based tests of topologies of phylogenetic trees were reviewed by Goldman, Anderson, and Rodrigo (2000).

CONSEL, the program for assessing the confidence of a tree selection by giving P-values for the trees, was developed by Shimodaira and Hasegawa (2001). Shimodaira (2002) introduced the approximately unbiased (AU) test for testing tree topologies. The AU test provides yet another procedure for assessing the confidence of the tree selection and makes corrections for the selection bias ignored in the bootstrap algorithm and the KH test.

The Molecular Evolutionary Genetics Analysis (MEGA) software developed by Kumar, Tamura, and Nei (1993) focuses on the analysis of the evolutionary changes in DNA and protein sequences. The latest version, MEGA3 (Kumar, Tamura, and Nei, 2004) contains routines for sequence alignment, web-based mining of databases, inference of the phylogenetic trees, estimation of evolutionary distances, and testing evolutionary hypotheses.

9

Transformational grammars

The one-dimensional string is too simple a model to reflect fully the properties of a real biological molecule, which have, after all, been determined by its three-dimensional structure selected in the course of evolution. Physical interactions of amino acids and nucleotides in the three-dimensional folds have to be described by the models that would go beyond the short range correlations which are the typical targets of the Markov chain models. The long range correlations are more important for proteins than for DNA, which has a rather uniform double helix structure. However, the structure of another nucleic acid, RNA, commonly has a significant number of long range interactions of special type, which could be a target for yet another class of probabilistic models.

Chapter 9 introduces the Chomsky hierarchy of deterministic transformational grammars, the models developed originally for natural languages and then applied to computer languages. These grammars could be readily used for the description of a protein (a regular grammar could generate amino acid sequences described as the PROSITE patterns) and RNA (a context-free grammar could generate RNA sequences with a given secondary structure).

Further generalization of these deterministic grammar classes to stochastic ones increases opportunities for sequence modeling. Stochastic regular grammars could be shown to be equivalent to hidden Markov models. Stochastic context-free grammars (SCFGs) are useful for modeling RNA sequences. Major SCFG algorithms solve tasks strikingly similar to those of the major algorithms for hidden Markov models (Chapter 3 of *BSA*): the CYK alignment algorithm finds the optimal alignment of a sequence to an already parameterized SCFG; the inside–outside algorithm finds the probability of a sequence given a parameterized SCFG; and the expectation–maximization algorithm estimates the SCFG parameters from a set of training sequences by using the inside and outside variables.

Chapter 9 includes twelve problems which facilitate understanding of the major concepts introduced in this chapter and their relationships. The problems offer

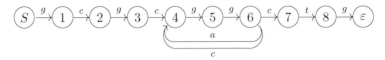

Figure 9.1. This finite state automaton recognizes FMR-1 triplet repeat regions containing any number of *CGG* or *AGG* nucleotide triplets

practice with the use of grammar derivation rules for decoding input sequences, creating grammar rules for particular sequence classes and complete languages, converting the grammar production rules into the Chomsky normal form. Finally, these problems elucidate the relationships between (i) the finite state automaton and the Moore machine, (ii) deterministic and non-deterministic automata, (iii) the push-down automata and the stochastic context-free grammars, and (iv) the HMM and the stochastic regular grammars.

9.1 Original problems

Problem 9.1 Convert the FMR-1 automaton in Figure 9.1 to a Moore machine in which each state accepts a particular symbol, instead of each transition accepting a particular symbol.

Solution To build the Moore machine, each transition-state pair designated as (x, n) in the FMR-1 automaton is converted into the Moore machine's state accepting the symbol associated with transition. We use S as the start non-terminal, G_i, C_i, A_i, and T_i as the non-terminals corresponding to the states of the Moore machine. In the conversion of the FMR-1 automaton, the initial state S remains the same. The transition-state pair $(g, 1)$ in the FMR-1 gives a new state G_1, pair $(c, 2)$ – state C_2, etc. until state 6 in the FMR-1 automaton is reached. Exit from state 6 is possible via one of the three transition-state pairs. In the continuation of the convergence process, $(c, 7)$ produces state C_7 of the Moore machine, $(c, 4)$ produces state C_4, and $(a, 4)$ makes additional state $A_{6'}$. States C_7, C_4, $A_{6'}$ accept symbols c, c, a, respectively. The straightforward addition of the remaining states, T_8 and G_9, completes the Moore machine shown in Figure 9.2. □

Problem 9.2 Convert the FMR-1 automaton to a deterministic automaton.

Solution The FMR-1 automaton is non-deterministic, since state 6 has two possible transitions for the same symbol c of the input language. A non-deterministic finite automaton (NDFA) can be converted into a deterministic one by the following general procedure (Carroll and Long, 1989).

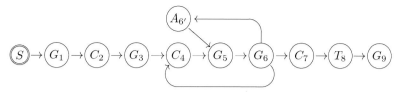

Figure 9.2. The Moore machine that recognizes FMR-1 triplet repeat regions containing any number of *CGG* or *AGG* triplets

If a given non-deterministic finite automaton (NDFA) named N with a set of states S is supposed to be converted to a deterministic finite automaton (DFA) A, we have to keep track of all states that can be reached by some string in the NDFA. As the first step, we define a state space of A as the power set of S, i.e. a set of all subsets of S. Thus, each state in the new machine will be labeled by some subset of S.

Further, we designate an NDFA as $N = (S, \Sigma, \delta, q, F)$ with a set of states S, an alphabet of an input language Σ, a set of starting states q, a set of accepting (final) states F, and a transition function δ, $\delta : S \times \Sigma \to \rho(S)$. Here $\rho(S)$ is a power set of S. Function δ is such that: for any $s \in S, \sigma \in \Sigma : \delta(s, \sigma) = \{s_1, \dots, s_n\}$, where $s_i \in S$, and $\delta(s, \varepsilon) = \{s\}$ for an empty string ε.

The designation for the corresponding DFA is $A = (S', \Sigma, \delta', q', F')$, where $S' = \rho(S)$, $q' = q$, $F' = \{T \in S' | T \cap F \neq \emptyset\}$. To implement our general strategy "to remember" all the states that can be reached by means of some input string, we define a transition function δ', $\delta' : S' \times \Sigma \to S'$, as follows: for any $T \in S', \sigma \in \Sigma$,

$$\delta'(T, \sigma) = \bigcup_{t \in T} \delta(t, \sigma) \quad \text{and} \quad \delta'(T, \varepsilon) = T.$$

For an FMR-1 NDFA with $S = \{S_0, W_1, W_2, W_3, W_4, W_5, W_6, W_7, W_8, W_9\}$, $\Sigma = \{g, c, a, t\}$, the set of starting states $q = \{S_0\}$ and the set of final states $F = \{W_9\}$, the state transition function δ is defined by Table 9.1.

The corresponding DFA has 2^{10} states, but most of them are not reachable by transitions that are permitted by the transition function δ'. Only the reachable states of the DFA are listed in Table 9.2. The DFA with states and transitions listed in Table 9.2 is shown in Figure 9.3. For clarity, the starting state $\{S_0\}$ is shown as S, the final state $\{W_9\}$ is shown as ε, the intermediate states $\{W_1\}, \dots, \{W_8\}$ are shown as numbered circles $1, \dots, 8$, and state $\{W_4, W_7\}$ is shown as circle 7.　　□

Problem 9.3 The PROSITE pattern (Bairoch *et al.*, 1997) for a C2H2 zinc finger, an important DNA binding protein motif, is

$$C - x(2, 4) - C - x(3) - [LIVMFYWC] - x(8) - H - x(3, 5) - H.$$

Draw a finite automaton that accepts this pattern.

Table 9.1. *State transitions of the NDFA FMR-1*

	g	c	t	a
S_0	$\{W_1\}$	$\emptyset$	$\emptyset$	$\emptyset$
W_1	$\emptyset$	$\{W_2\}$	$\emptyset$	$\emptyset$
W_2	$\{W_3\}$	$\emptyset$	$\emptyset$	$\emptyset$
W_3	$\emptyset$	$\{W_4\}$	$\emptyset$	$\emptyset$
W_4	$\{W_5\}$	$\emptyset$	$\emptyset$	$\emptyset$
W_5	$\{W_6\}$	$\emptyset$	$\emptyset$	$\emptyset$
W_6	$\emptyset$	$\{W_4, W_7\}$	$\emptyset$	$\{W_4\}$
W_7	$\emptyset$	$\emptyset$	$\{W_8\}$	$\emptyset$
W_8	$\{W_9\}$	$\emptyset$	$\emptyset$	$\emptyset$
W_9	$\emptyset$	$\emptyset$	$\emptyset$	$\emptyset$

Table 9.2. *State transitions of DFA FMR-1*

	g	c	t	a
$\{S_0\}$	$\{W_1\}$	$\emptyset$	$\emptyset$	$\emptyset$
$\{W_1\}$	$\emptyset$	$\{W_2\}$	$\emptyset$	$\emptyset$
$\{W_2\}$	$\{W_3\}$	$\emptyset$	$\emptyset$	$\emptyset$
$\{W_3\}$	$\emptyset$	$\{W_4\}$	$\emptyset$	$\emptyset$
$\{W_4\}$	$\{W_5\}$	$\emptyset$	$\emptyset$	$\emptyset$
$\{W_5\}$	$\{W_6\}$	$\emptyset$	$\emptyset$	$\emptyset$
$\{W_6\}$	$\emptyset$	$\{W_4, W_7\}$	$\emptyset$	$\{W_4\}$
$\{W_4, W_7\}$	$\{W_5\}$	$\emptyset$	$\{W_8\}$	$\emptyset$
$\{W_8\}$	$\{W_9\}$	$\emptyset$	$\emptyset$	$\emptyset$
$\{W_9\}$	$\emptyset$	$\emptyset$	$\emptyset$	$\emptyset$

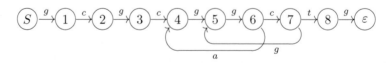

Figure 9.3. The deterministic finite FMR-1 automaton.

Solution It is easy to see that the length of this motif varies from twenty-one to twenty-five residues. The motif includes four positions with highly conserved C and H as well as the stretches of amino acid sequence with low conservation, where any amino acid, x, could occur. It would be cumbersome to draw a finite automaton with

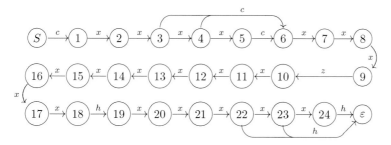

Figure 9.4. Automaton for recognition of the C2H2 zinc finger PROSITE pattern.

	seq1		seq2		seq3	
	A	A	C	A	C	A
	G	A	G	A	G	A
	$G \bullet C$		$U \bullet A$		$U \times C$	
	$A \bullet U$		$C \bullet G$		$C \times U$	
	$C \bullet G$		$G \bullet C$		$G \times G$	

$C \ A \ G \ G \ A \ A \ A \ C \ U \ G$ seq1

$G \ C \ U \ G \ C \ A \ A \ A \ G \ C$ seq2

$G \ C \ U \ G \ C \ A \ A \ C \ U \ G$ seq3

Figure 9.5. Sequences 1 and 2 are forming a hairpin structure with stem and loop.

twenty amino acid transitions shown individually, hence we use one transition with x instead. Similarly, we designate as z a symbol from a set $\{L, I, V, M, F, Y, W, C\}$.

The diagram of the finite automaton for a C2H2 zinc finger PROSITE pattern is shown in Figure 9.4. Note that this is a deterministic automaton. □

Problem 9.4 Write derivations for *seq1* and *seq2* shown in Figure 9.5 using the context-free grammar (CFG) described below:

$$S \rightarrow aW_1u|cW_1g|gW_1c|uW_1a,$$

$$W_1 \rightarrow aW_2u|cW_2g|gW_2c|uW_2a,$$

$$W_2 \rightarrow aW_3u|cW_3g|gW_3c|uW_3a,$$

$$W_3 \rightarrow gaaa|gcaa.$$

Solution The derivation for *seq*1 is

$$S \Rightarrow cSg \Rightarrow caSug \Rightarrow cagScug \Rightarrow caggaaacug;$$

while the derivation for *seq*2 is

$$S \Rightarrow gSc \Rightarrow gcSgc \Rightarrow gcuSagc \Rightarrow gcugcaaagc.$$

□

Problem 9.5 Write a regular grammar that generates *seq*1 and *seq*2 but not *seq*3 in the example given in Figure 9.5.

Solution The problem does not specify if a regular grammar should generate only *seq*1 and *seq*2, or if it may generate numerous sequences, *seq*1 and *seq*2 among them, but should not generate *seq*3. We will show solutions for both interpretations.

First, we will write a regular grammar that generates *seq*1 and *seq*2 only. It forks at the first symbol and then simply proceeds with generation of either *seq*1 or *seq*2:

$$S \rightarrow cW_1 | gW_2,$$

$$W_1 \rightarrow aW_3, \qquad\qquad W_2 \quad \rightarrow cW_4,$$

$$W_3 \rightarrow gW_5, \qquad\qquad W_4 \quad \rightarrow uW_6,$$

$$W_5 \rightarrow gW_7, \qquad\qquad W_6 \quad \rightarrow gW_8,$$

$$W_7 \rightarrow aW_9, \qquad\qquad W_8 \quad \rightarrow cW_{10},$$

$$W_9 \rightarrow aW_{11}, \qquad\qquad W_{10} \quad \rightarrow aW_{12},$$

$$W_{11} \rightarrow aW_{13}, \qquad\qquad W_{12} \quad \rightarrow aW_{14},$$

$$W_{13} \rightarrow cW_{15}, \qquad\qquad W_{14} \quad \rightarrow aW_{16},$$

$$W_{15} \rightarrow uW_{17}, \qquad\qquad W_{16} \quad \rightarrow gW_{18},$$

$$W_{17} \rightarrow g, \qquad\qquad W_{18} \quad \rightarrow c.$$

Now we will write a regular grammar that generates *seq*1, *seq*2, and a few others made of the same alphabet, such as *ccggaaacgg*, but not *seq*3:

$$S \rightarrow cW_1 | gW_1,$$

$$W_1 \rightarrow aW_2 | cW_2, \qquad\qquad W_7 \rightarrow cW_{11},$$

$$W_2 \rightarrow gW_3 | uW_3, \qquad\qquad W_8 \rightarrow aW_9,$$

$$W_3 \rightarrow gW_4, \qquad\qquad\qquad W_9 \rightarrow aW_{10},$$

$$W_4 \rightarrow aW_5 | cW_8, \qquad\qquad W_{10} \rightarrow aW_{11},$$

$$W_5 \rightarrow aW_6, \qquad\qquad\qquad W_{11} \rightarrow uW_{12} | gW_{12},$$

$$W_6 \rightarrow aW_7, \qquad\qquad\qquad W_{12} \rightarrow g | c.$$

A fork after state W_4 leads to the generation of fragments *aaac* and *caaa* from sequences *seq*1 and *seq*2, respectively. Neither of the fragments appear in *seq*3; thus, the grammar does not generate fragment *caac* of *seq*3 and therefore is not able to generate *seq*3. $\qquad\qquad\qquad\qquad\qquad\qquad\qquad\qquad\qquad\qquad\qquad\square$

Problem 9.6 Consider the complete language generated by the CFG in Problem 9.4. Describe a regular grammar that generates exactly the same language. Does describing this sequence family with a regular grammar seem like a good idea?

Solution The language generated by the grammar consists of ten-symbol sequences. In each sequence the first three symbols complement the last three symbols, as defined by rules S, W_1, and W_2 of CFG (Problem 9.4). A regular grammar should "remember" the first three symbols of a sequence to generate a correct ending. Therefore, a regular grammar will consist of:

- four rules for generation of the first symbol;
- sixteen rules for the second symbol;
- sixty-four rules for the third symbol;
- sixty-four rules for the fourth symbol, which is g for all of them;
- 128 rules for the fifth symbol, which is either a or c;
- sixty-four rules for the sixth symbol, which is a;
- sixty-four rules for the seventh symbol, which is a;
- sixty-four rules for the eighth symbol;
- sixtgeen rules for the ninth symbol;
- four rules for the tenth symbol.

Thus, the regular grammar consists of 488 production rules instead of fourteen in CFG. So it is probably not a good idea to use a regular grammar to describe this sequence family. □

Problem 9.7 Modify the push-down automaton parsing algorithm so that it randomly *generates* one of the possible valid sequences in a context-free grammar's language.

Solution The modified push-down automaton will generate a sequence from left to right. The automaton's stack is initialized by pushing the start non-terminal onto it. The following steps are then made in iteration until the stack is empty.

Pop a symbol off the stack.
If the popped symbol is a non-terminal:

- choose a production at random from the set of allowed production rules for this non-terminal.
- push the right-hand side of the expression for the chosen production rule onto the stack, rightmost symbols first.

If the popped symbol is a terminal:

- produce this terminal symbol as a part of a sequence.   □

Problem 9.8 $G - U$ pairs are accepted in base paired RNA stems but occur with lower frequency than $G - C$ and $A - U$ Watson–Crick pairs. Make the RNA stem loop context-free grammar from Problem 9.4 into a stochastic context-free grammar, allowing $G - U$ pairs in the stem with half the probability of a Watson–Crick pair.

Solution To allow $G - U$ pairs, the original CFG must be supplemented with production rules allowing transitions $W_i \rightarrow gW_{i+1}u$, $W_i \rightarrow uW_{i+1}g$, $i = 0, 1, 2$, $W_0 = S$. To convert the CFG into a stochastic one, we have to assign probabilities to the production rules of the original CFG. Finally, the production rules of SCFG and their probabilities should be defined as follows:

$S \rightarrow aW_1u,$	$S \rightarrow cW_1g,$	$S \rightarrow gW_1c,$
(0.2)	(0.2)	(0.2)
$S \rightarrow uW_1a,$	$S \rightarrow gW_1u,$	$S \rightarrow uW_1g,$
(0.2)	(0.1)	(0.1)

$W_1 \rightarrow aW_2u,$	$W_1 \rightarrow cW_2g,$	$W_1 \rightarrow gW_2c,$
(0.2)	(0.2)	(0.2)
$W_1 \rightarrow uW_2a,$	$W_1 \rightarrow gW_2u,$	$W_1 \rightarrow uW_2g,$
(0.2)	(0.1)	(0.1)

$W_2 \rightarrow aW_3u,$	$W_2 \rightarrow cW_3g,$	$W_2 \rightarrow gW_3c,$
(0.2)	(0.2)	(0.2)
$W_2 \rightarrow uW_3a,$	$W_2 \rightarrow gW_3u,$	$W_2 \rightarrow uW_3g,$
(0.2)	(0.1)	(0.1)

$W_3 \rightarrow gaaa,$	$W_3 \rightarrow gcaa.$
(0.5)	(0.5)

□

Problem 9.9 Extend the push-down automaton algorithm from Problem 9.7 to generate sequences from a stochastic context-free grammar according to their probability. (Note: this gives an efficient algorithm for sampling sequences from any SCFG, including the more complex RNA SCFGs in Chapter 10.)

Solution The modified push-down automaton generates a sequence from left to right. The automaton's stack is initialized by pushing the start non-terminal onto it. Then the following steps are iterated until the stack is empty.

Pop a symbol off the stack.
If the popped symbol is a non-terminal:

• choose a production rule for the non-terminal according to its probability;
• push the right-hand side of the expression for the chosen production rule onto the stack, rightmost symbols first.

If the popped symbol is a terminal:

• produce this terminal symbol as a part of a sequence. □

Problem 9.10 Consider a simple HMM that models two kinds of base composition in DNA. The model has two states fully interconnected by four state transitions. State 1 emits $C + G$-rich sequences with probabilities of symbols $(p_a, p_c, p_g, p_t) = (0.1, 0.4, 0.4, 0.1)$ and state 2 emits $A + T$-rich sequences with probabilities of symbols $(p_a, p_c, p_g, p_t) = (0.3, 0.2, 0.2, 0.3)$.
 (a) Draw this HMM.
 (b) Set the transition probabilities so that the expected length of a run of state 1s is 1000 bases, and the expected length of a run of state 2s is 100 bases.
 (c) Give the same model in stochastic regular grammar form with terminals, non-terminals, and production rules with their associated probabilities.

Solution (a) The HMM diagram is as follows:

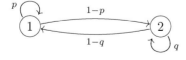

The transition probabilities between states are supposed to be such that $p_{11} = p$, $p_{12} = 1 - p$, $p_{22} = q$, $p_{21} = 1 - q$, where p and q are from interval $(0, 1)$. States 1 and 2 emit four symbols A, C, G, T with probabilities $e_1(A) = 0.1$, $e_1(C) = 0.4$, $e_1(G) = 0.4$, $e_1(T) = 0.1$, and $e_2(A) = 0.3$, $e_2(C) = 0.2$, $e_2(G) = 0.2$, $e_2(T) = 0.3$, respectively.

(b) The transition probabilities are determined as follows. If a random variable L_1 is the length of a sequence of symbols generated by state 1, then L_1 has values $n = 1, 2, \ldots$ with the following probabilities: $P(L_1 = n) = qp^{n-1}(1 - p)$ (the first transition must be from state 2 to state 1, followed by $n - 1$ transitions from state 1 to itself, and the last transition must be from state 1 to state 2). Similarly, we define a random variable L_2 as the length of a run of state 2, with values $k = 1, 2, \ldots$ and probabilities $P(L_2 = k) = pq^{k-1}(1 - q)$. The expected values of L_1 and L_2 are given by

$$EL_1 = \sum_{n=1}^{+\infty} nqp^{n-1}(1 - p) = \frac{q(1 - p)}{(1 - p)^2} = \frac{q}{1 - p};$$

$$EL_2 = \sum_{k=1}^{+\infty} kpq^{k-1}(1 - q) = \frac{p(1 - q)}{(1 - q)^2} = \frac{p}{1 - q}.$$

Conditions $EL_1 = 1000$, and $EL_2 = 100$ lead to two equations for parameters p and q. Solving the system gives $p = 0.999$, $q = 0.99$.

(c) The regular grammar form of the model in question consists of non-terminals W_1 and W_2 and the following production rules:

$$W_1 \rightarrow aW_1|cW_1|gW_1|tW_1|aW_2|cW_2|gW_2|tW_2, \tag{9.1}$$

$$W_2 \rightarrow aW_1|cW_1|gW_1|tW_1|aW_2|cW_2|gW_2|tW_2. \tag{9.2}$$

To convert a regular grammar into a stochastic regular grammar, we need to assign to the eight production rules (9.1) for W_1 the following transition probabilities: $e_1(A)p$, $e_1(C)p$, $e_1(G)p$, $e_1(T)p$, $e_1(A)(1 - p)$, $e_1(C)(1 - p)$, $e_1(G)(1 - p)$, $e_1(T)(1 - p)$, in the order established by Equation (9.1); and to the eight production rules (9.2) for W_2 the transition probabilities $e_2(A)(1 - q)$, $e_2(C)(1 - q)$, $e_2(G)(1 - q)$, $e_2(T)(1 - q)$, $e_2(A)q$, $e_2(C)q$, $e_2(G)q$, $e_2(T)q$, in the order established by (9.2). Therefore, we have the following list of stochastic regular grammar production rules and their probabilities:

$W_1 \rightarrow aW_1$,	$W_1 \rightarrow cW_1$,	$W_1 \rightarrow gW_1$,	$W_1 \rightarrow tW_1$,
(0.0999)	(0.3996)	(0.3996)	(0.0999)
$W_1 \rightarrow aW_2$,	$W_1 \rightarrow cW_2$,	$W_1 \rightarrow gW_2$,	$W_1 \rightarrow tW_2$,
(0.0001)	(0.0004)	(0.0004)	(0.0001)

$$W_2 \to aW_1, \qquad W_2 \to cW_1, \qquad W_2 \to gW_1, \qquad W_2 \to tW_1,$$

$$(0.003) \qquad\qquad (0.002) \qquad\qquad (0.002) \qquad\qquad (0.003)$$

$$W_2 \to aW_2, \qquad W_2 \to cW_2, \qquad W_2 \to gW_2, \qquad W_2 \to tW_2.$$

$$(0.297) \qquad\qquad (0.198) \qquad\qquad (0.198) \qquad\qquad (0.297) \qquad \square$$

Problem 9.11 Convert the production rule $W \to aWbW$ to Chomsky normal form. If the probability of the original production is p, show the probabilities for the productions in your normal form version.

Solution Since the Chomsky normal form requires that all production rules are of the form $W_v \to W_y W_z$ or $W_v \to a$, the rule $W \to aWbW$ is replaced with the following sequence of rules:

$$W \to \hat{W}_1 \hat{W}_2, \quad \hat{W}_1 \to W_a W, \quad \hat{W}_2 \to W_b W, \quad W_a \to a, \quad W_b \to b.$$

Only one production rule exists for each of the non-terminals $\hat{W}_1, \hat{W}_2, W_a, W_b$; hence, probability 1 should be assigned to each production rule. The rule $W \to \hat{W}_1 \hat{W}_2$ inherits the probability of the original production rule. Therefore, for the Chomsky normal form we have

$$W \to \hat{W}_1 \hat{W}_2,$$

$$(p)$$

$$\hat{W}_1 \to W_a W, \qquad \hat{W}_2 \to W_b W, \qquad W_a \to a, \qquad W_b \to b.$$

$$(1) \qquad\qquad (1) \qquad\qquad (1) \qquad\qquad (1) \qquad \square$$

Problem 9.12 Convert the production rules $W_3 \to gaaa|gcaa$ from the RNA stem model grammar in Problem 9.4 to Chomsky normal form. Assuming that $W_3 \to gaaa$ has probability p_1 and $W_3 \to gcaa$ has probability $p_2 = 1 - p_1$, assign probabilities to your normal form productions. Show that your normal form version correctly assigns probabilities p_1 and p_2 for *GAAA* and *GCAA* loops, respectively.

Solution Similarly to the previous problem, we convert the context-free grammar production rule to the following Chomsky normal form:

$$W_3 \to W_g \hat{W}_1,$$

$$(1)$$

$$\hat{W}_1 \rightarrow W_a\hat{W}_2, \qquad\qquad \hat{W}_1 \rightarrow W_c\hat{W}_2,$$

$$(p_1) \qquad\qquad\qquad (p_2)$$

$$\hat{W}_2 \rightarrow W_aW_a, \qquad\qquad W_g \rightarrow g,$$

$$(1) \qquad\qquad\qquad (1)$$

$$W_a \rightarrow a, \qquad\qquad W_c \rightarrow c.$$

$$(1) \qquad\qquad\qquad (1)$$

For the probabilities of loops *GAAA* and *GCAA*, we have

$$P(GAAA) = P(W_3 \rightarrow W_g\hat{W}_1)P(\hat{W}_1 \rightarrow W_a\hat{W}_2)P(\hat{W}_2 \rightarrow W_aW_a)$$
$$\times P(W_g \rightarrow g)P(W_a \rightarrow a)P(W_a \rightarrow a)P(W_a \rightarrow a) = p_1,$$

$$P(GCAA) = P(W_3 \rightarrow W_g\hat{W}_1)P(\hat{W}_1 \rightarrow W_c\hat{W}_2)P(\hat{W}_2 \rightarrow W_aW_a)$$
$$\times P(W_g \rightarrow g)P(W_c \rightarrow c)P(W_a \rightarrow a)P(W_a \rightarrow a) = p_2.$$

The results confirm that the normal form correctly assigns probabilities to *GAAA* and *GCAA* loops. $\qquad\qquad\qquad\qquad\qquad\qquad\qquad\qquad\qquad\qquad$ □

9.2 Further reading

In addition to the texts mentioned in Section 9.7 of Durbin *et al.* (1998), the following textbooks can be useful for the reader as systematic and detailed descriptions of the theory of formal languages, transformational grammars, and the automata theory: Carroll and Long (1989); Kozen (1999); Simon (1999); Khoussainov and Nerode (2001); Motwani, Ullman, and Hopcroft (2003); Gopalakrishnan (2006). Recent advances in the automata theory are described in Salomaa, Wood, and Yu (2001); and Ito (2004).

10

RNA structure analysis

Stochastic transformational grammars, particularly stochastic context-free grammars, turned out to be effective modeling tools for RNA sequence analysis. Two biologically interesting problems are the prediction of RNA secondary structure and the construction of multiple alignments of RNA families. Non-stochastic algorithms for the RNA secondary structure prediction were developed more than twenty years ago (by Nussinov *et al.* (1978) and by Zuker and Stiegler (1981)). Notably, the Nussinov algorithm could be immediately rewritten in SCFG terms as a version of the Cocke–Younger–Kasami (CYK) algorithm. The SCFG interpretation provides an insight into the probabilistic meaning of parameters of the original Nussinov algorithm and also suggests statistical procedures for parameter estimation. A similar translation into SCFG terms is possible for the Zuker algorithm.

Interestingly, equivalence between the non-probabilistic dynamic programming sequence alignment algorithm and the Viterbi algorithm for a pair HMM is analogous to equivalence between the non-probabilistic algorithm of RNA structure prediction and the CYK algorithm for a SCFG. There is also an analogy between the use of the profile HMM for alignment of multiple DNA or protein sequences and the use of the SCFG-based RNA structure profiles, called covariance models (CMs), for constructing structurally sound alignments of multiple RNAs. Furthermore, parameters of the covariance models could be derived by the inside–outside expectation maximization algorithm (compare with the simultaneous profile HMM parameter estimation and construction of multiple sequence alignment). Finally, the CYK algorithm for local alignment of genomic DNA to a given covariance model could be used to search a nucleotide sequence database for a particular type of RNA genes (compare with searching for homologs in a database of proteins using the algorithm of local alignment of a protein sequence to a profile HMM representing the protein domain).

The problems in Chapter 10 provide useful illustrations of application of both deterministic and probabilistic algorithms of the RNA sequence analysis. They also emphasize the relationships between the algorithms, for example the relationship between the deterministic algorithm of the RNA secondary structure prediction and the CYK algorithm that finds the maximum probability secondary structure for a given RNA. The use of the CYK, the inside and the outside algorithms, their elements, and modifications is the subject of several problems.

10.1 Original problems

Problem 10.1 Calculation of the mutual information by the following formula:

$$M_{ij} = \sum_{x_i, x_j} f_{x_i x_j} \log_2 \frac{f_{x_i x_j}}{f_{x_i} f_{x_j}}$$

requires counting frequencies of all sixteen different base pairs. This has the advantage that it makes no assumptions about Watson–Crick base pairing, so mutual information can be detected between covarying non-canonical pairs, such as $A - A$ and $G - G$. On the other hand, the calculation requires a large number of aligned sequences to obtain reasonable frequencies for sixteen probabilities. Write down an alternative information theoretic measure of base-pairing correlation that considers only two classes of i, j identities instead of all sixteen: Watson–Crick and $G - U$ pairs grouped in one class, and all other pairs grouped in the other. Compare the properties of this calculation to the M_{ij} calculation both for small numbers of sequences and in the limit of infinite data.

Solution For two columns, i and j, of the structurally correct multiple alignment of N RNA sequences $x^1, x^2, \ldots, x^N$ the mutual information M_{ij} is defined as the sum of sixteen terms, each carrying information about a frequency of a particular pair of ribonucleotides (in the specified order). For instance, for the pair (A, G) such a term is given by

$$f_{ij}(AG) \log_2 \frac{f_{ij}(AG)}{f_i(A) f_j(G)}. \tag{10.1}$$

Here $f_{ij}(AG)$ is the frequency of (A, G) in columns i and j, and $f_i(A)$ ($f_j(G)$) is the frequency of base A (G) in column i (j).

Now we define two classes of base pairs. The first class **CP** (complementary pairs) consists of Watson–Crick pairs (G, C), (C, G), (A, U), (U, A), along with (G, U), (U, G) pairs. The second class **NCP** consists of another ten base pairs. We define the distributions μ_{ij} and ν_{ij} on the two-point space $\{$**CP**, **NCP**$\}$ as follows: $\mu_{ij}($**CP**$)$ is the frequency of **CP**-pairs observed in columns i and j, $\mu_{ij}($**NCP**$)$ is the frequency of **NCP**-pairs observed in columns i and j, $\nu_{ij}($**CP**$)$ is the probability of observing a **CP**-pair in two independent columns with the same base frequencies as in columns i and j, $v_{ij}(\mathbf{NCP})$ is the probability of observing an **NCP**-pair in two independent columns with the same base frequencies as in columns i and j. Obviously,

$$\mu_{ij}(\mathbf{CP}) = \sum_{(a,b)\in\mathbf{CP}} f_{ij}(ab),$$

$$\mu_{ij}(\mathbf{NCP}) = \sum_{(a,b)\in\mathbf{NCP}} f_{ij}(ab),$$

$$v_{ij}(\mathbf{CP}) = \sum_{(a,b)\in\mathbf{CP}} f_i(a)f_j(b),$$

$$v_{ij}(\mathbf{NCP}) = \sum_{(a,b)\in\mathbf{NCP}} f_i(a)f_j(b).$$

We define the new information measure $\mathcal{M}_{ij}$ between the columns i and j as follows:

$$\mathcal{M}_{ij} = \mu_{ij}(\mathbf{CP})\log_2\frac{\mu_{ij}(\mathbf{CP})}{v_{ij}(\mathbf{CP})} + \mu_{ij}(\mathbf{NCP})\log_2\frac{\mu_{ij}(\mathbf{NCP})}{v_{ij}(\mathbf{NCP})}.$$

According to this definition, $\mathcal{M}_{ij}$ is in fact the Kullback-Leibler distance between the distributions μ_{ij} and v_{ij}.

Let us compare the properties of M_{ij} and $\mathcal{M}_{ij}$. First of all, unlike M_{ij}, the information measure $\mathcal{M}_{ij}$ is sensitive to complementary pairs. For example, for two pairs of columns:

i	j		i	k
A	C		A	U
A	C		A	U
C	G		C	G
C	G		C	G

$M_{ij} = M_{ik} = 1$, while $\mathcal{M}_{ij} < \mathcal{M}_{ik}$ ($\mathcal{M}_{ij} = 1 - \frac{1}{2}\log_2 3$, $\mathcal{M}_{ik} = 1$).

For a small number of sequences we can expect that many of the frequencies $f_{ij}(ab)$ are equal to zero, making the corresponding terms in the sum for M_{ij} zeroes, too. For $\mathcal{M}_{ij}$, however, we expect that both terms contribute positive values to the sum, even for small N. When the number of sequences N grows and the frequencies

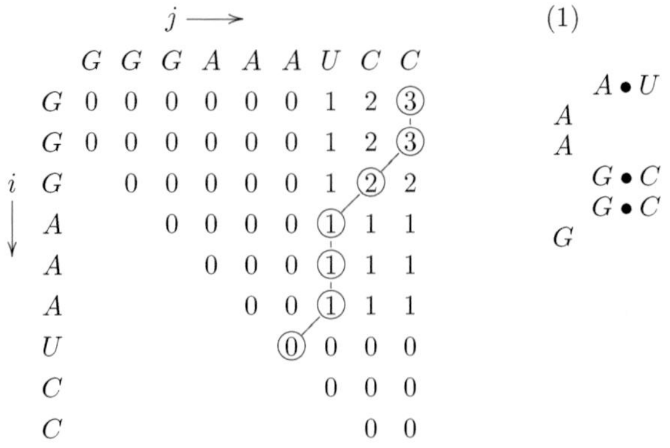

RNA structure analysis

$$j \longrightarrow \qquad\qquad\qquad (1)$$

	G	G	G	A	A	A	U	C	C
G	0	0	0	0	0	0	1	2	③
G	0	0	0	0	0	0	1	2	③
G		0	0	0	0	0	1	②	2
A			0	0	0	0	①	1	1
A				0	0	0	①	1	1
A					0	0	①	1	1
U						⓪	0	0	0
C							0	0	0
C								0	0

$i \downarrow$

$$A \bullet U$$
$$A$$
$$A$$
$$G \bullet C$$
$$G \bullet C$$
$$G$$

Figure 10.1. The traceback path produced by the Nussinov folding algorithm for the sequence *GGGAAAUCC*. The scores on the optimal path are indicated circles. The starting point of the path is located at the top right-hand corner. The secondary structure (1) associated with the optimal path is shown on the right.

converge to the values of probabilities, we have

$$\lim_{N \to +\infty} M_{ij} = \sum_{a,b} p_{ij}(ab) \log_2 \frac{p_{ij}(ab)}{p_i(a)p_j(b)},$$

$$\lim_{N \to +\infty} M_{ij} = \left(\sum_{(a,b)\in\mathbf{CP}} p_{ij}(ab) \right) \log_2 \frac{\sum_{(a,b)\in\mathbf{CP}} p_{ij}(ab)}{\sum_{(a,b)\in\mathbf{CP}} p_i(a)p_j(b)}$$

$$+ \left(\sum_{(a,b)\in\mathbf{NCP}} p_{ij}(ab) \right) \log_2 \frac{\sum_{(a,b)\in\mathbf{NCP}} p_{ij}(ab)}{\sum_{(a,b)\in\mathbf{NCP}} p_i(a)p_j(b)}.$$

Here $p_{ij}(ab)$ is the probability of occurrence of the base pair (a, b) (in the specified order) in columns i and j, $p_i(a)$ is the probability of occurrence of base a in column i, and $p_j(b)$ is the probability of occurrence of base b in column j. $\qquad\square$

Problem 10.2 Use the traceback stage of the Nussinov RNA folding algorithm (Durbin *et al.*, 1998, p. 272) to find two more optimal structures for the sequence *GGGAAAUCC* with three base pairs besides the one in Figure 10.1. Modify the traceback algorithm so it finds one of new structures instead of the one obtained in Figure 10.1.

Solution The traceback stage of the Nussinov RNA folding algorithm (Nussinov *et al.*, 1978) is as follows.

Initialization: push $(1, L)$ onto stack.

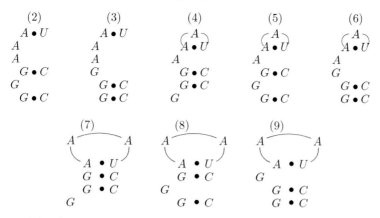

Figure 10.2. Eight alternative secondary structures (with three base pairs) of sequence *GGGAAAUCC*.

Recursion: repeat until stack is empty:

- pop (i,j);
- if $i >= j$ continue;
- else if $\gamma(i+1,j) = \gamma(i,j)$ push $(i+1,j)$;
- else if $\gamma(i,j-1) = \gamma(i,j)$ push $(i,j-1)$;
- else if $\gamma(i+1,j-1) + \delta(i,j) = \gamma(i,j)$

  - record i,j base pair
  - push $(i+1,j-1)$;

- else for $k = i+1$ to $j-1$: if $\gamma(i,k) + \gamma(k+1,j) = \gamma(i,j)$:

  - push $(k+1,j)$
  - push (i,k)
  - break.

The traceback algorithm recovers the path through the semi-matrix shown in Figure 10.1 along with the corresponding RNA secondary structure. Although the structure has three Watson–Crick pairs (the maximum possible number of pairs for this sequence), it is topologically unlikely to make a zero length hairpin loop by forming hydrogen bonds between adjacent ribonucleotides. Eight other folding structures could have three base pairs (Figure 10.2). Each of the eight structures can be recovered by a proper modification of the traceback algorithm. For instance, structure (7) can be found by the following algorithm.

Initialization: push $(1, L)$ onto stack.

Recursion: repeat until stack is empty:

- pop (i,j);
- if $i >= j$ continue;

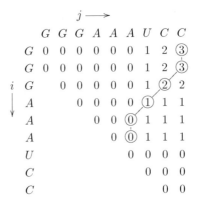

Figure 10.3. The same matrix of base pair scores as in Figure 10.2. The circles show the traceback path corresponding to secondary structure (7) in Figure 10.2.

- else if $\delta(i,j) = \gamma(i,j)$ and $\delta(i,j) = 1$

  - record i,j base pair
  - push $(i+1, j-1)$;

- else if $\gamma(i+1,j) = \gamma(i,j)$ push $(i+1,j)$;
- else if $\gamma(i,j-1) = \gamma(i,j)$ push $(i,j-1)$;
- else if $\gamma(i+1,j-1) + \delta(i,j) = \gamma(i,j)$

  - record i,j base pair
  - push $(i+1, j-1)$;

- else for $k = i+1$ to $j-1$: if $\gamma(i,k) + \gamma(k+1,j) = \gamma(i,j)$:

  - push $(k+1, j)$
  - push (i, k)
  - break.

The path through the semi-matrix for the given sequence is shown in Figure 10.3.

                                                        □

Problem 10.3 The Nussinov algorithm, as it was defined, can produce non-sensical 'base pairs' between adjacent complementary residues (for example, several structures admitted in the preceding problem contains such an *AU* base pair). Modify the Nussinov folding algorithm so that hairpin loops must have a minimum length of h. Give the new recursion equations for the fill and traceback.

Solution To guarantee that the secondary structure prediction algorithm generates no hairpin loops with a length below h, the semi-matrix produced by the Nussinov algorithm should have zeroes in $h+1$ sub-diagonals. This type of algorithm initialization precludes pairing and forming a hairpin loop within at least h consecutive bases. The modified Nussinov RNA folding algorithm operates as follows. For a given RNA sequence $x = (x_1, \ldots, x_L)$ we define $\delta(i,j) = 1$ if x_i and x_j make a complementary base pair; otherwise, $\delta(i,j) = 0$. For each pair (i,j), score $\gamma(i,j)$ denotes the maximal number of pairs that can be formed by subsequence $x_i, \ldots, x_j$ with the hairpin loop length at least h. Therefore, at the initialization step of the algorithm, we assign

$$\gamma(i, i-1) = 0 \quad \text{for } i = 2 \text{ to } L;$$
$$\gamma(i, i+k) = 0 \quad \text{for } i = 1 \text{ to } L - k, \quad k = 0 \text{ to } h.$$

At the recursion step for $j - i + 1 = h + 1, \ldots, L$ we calculate

$$\gamma(i,j) = \max \begin{cases} \gamma(i+1,j), \\ \gamma(i,j-1), \\ \gamma(i+1,j-1) + \delta(i,j), \\ \max_{i<k<j}(\gamma(i,k) + \gamma(k+1,j)). \end{cases}$$

The traceback rules are the same as for the Nussinov algorithm with $h = 0$.   □

Problem 10.4 Show that the Nussinov folding algorithm can be trivially extended to find a maximally *scoring* structure wherea base pair between residues a and b gets a score $s(a, b)$.

Solution The only part of the algorithm that should be changed is the definition of $\delta(i,j)$. Now we define $\delta(i,j) = s(x_i, x_j)$. There is no change in the matrix fill stage of the algorithm.

Initialization:

$$\gamma(i, i-1) = 0 \quad \text{for } i = 2 \text{ to } L;$$
$$\gamma(i, i) = 0 \quad \quad \text{for } i = 1 \text{ to } L.$$

Recursion: starting with all subsequences of length 2, to length L:

$$\gamma(i,j) = \max \begin{cases} \gamma(i+1,j), \\ \gamma(i,j-1), \\ \gamma(i+1,j-1) + s(i,j), \\ \max_{i<k<j}(\gamma(i,k) + \gamma(k+1,j)). \end{cases}$$

Having completed the matrix fill stage for a given RNA sequence, we obtain the maximum score located in the top-right cell of the semi-matrix.   □

Problem 10.5 The Cocke–Younger–Kasami (CYK) algorithm (Kasami, 1965; Younger, 1967; Cocke and Schwartz, 1970), for Nussinov-style RNA SCFA that finds the maximum probability secondary structure is defined as follows. Let the probability parameters of the SCFG be denoted by $p(aS)$, $p(aSu)$, etc.

Initialization:

$$\gamma(i, i - 1) = -\infty \qquad \text{for } i = 2 \text{ to } L;$$

$$\gamma(i, i) = \max \begin{cases} \log p(x_i S) \\ \log p(S x_j) \end{cases} \qquad \text{for } i = 1 \text{ to } L.$$

Recursion: for $i = 1$ to $L - 1, j = i + 1$ to L:

$$\gamma(i, j) = \max \begin{cases} \gamma(i + 1, j) + \log p(x_i S); \\ \gamma(i, j - 1) + \log p(S x_j); \\ \gamma(i + 1, j - 1) + \log p(x_i S x_j); \\ \max_{i < k < j}(\gamma(i, k) + \gamma(k + 1, j) + \log p(SS)). \end{cases}$$

Upon completion, the algorithm finds $\gamma(1, L)$, the log-likelihood of the optimal secondary structure under the SCFG model. Write down a traceback algorithm for determining the optimal RNA secondary structure.

Solution The traceback stage of the RNA folding SCFG is similar to the traceback stage of the Nussinov algorithm. It operates as follows.

Initialization: push $(1, L)$ onto stack.

Recursion: repeat until stack is empty:

- pop (i, j).
- if $i >= j$ continue;
- else if $\gamma(i + 1, j) + \log p(x_i S) = \gamma(i, j)$ push $(i + 1, j)$;
- else if $\gamma(i, j - 1) + \log p(S x_j) = \gamma(i, j)$ push $(i, j - 1)$;
- else if $\gamma(i + 1, j - 1) + \log p(x_i S x_j) = \gamma(i, j)$

  - record i, j base pair
  - push $(i + 1, j - 1)$.

- else for $k = i + 1$ to $j - 1$: if $\gamma(i, k) + \gamma(k + 1, j) + \log p(SS) = \gamma(i, j)$:

  - push $(k + 1, j)$
  - push (i, k)
  - break.   □

Problem 10.6 Devise an SCFG which uses different non-terminals to model bulge loops, hairpin loops, and single strands.

Solution We consider the SCFG containing five non-terminals S, $W_1,\ldots, W_5$, with W_1 and W_2 to model single strands, W_3 to model hairpin loops, W_4 to model bulge loops, and W_5 to model multifurcation loops. The SCFG allows us to determine the probabilities of different parts of RNA secondary structure, as well as probabilities of individual bases. The grammar rules (without the probability parameters) are as follows:

$$S \rightarrow W_1S \mid SW_2 \qquad\qquad\qquad \text{(1) single strands,}$$
$$S \rightarrow aW_3 \mid gW_3 \mid cW_3 \mid uW_3 \qquad \text{(2) hairpin loops,}$$
$$S \rightarrow W_4S \mid SW_4 \qquad\qquad\qquad \text{(3) bulge loops,}$$
$$S \rightarrow W_5S \qquad\qquad\qquad\qquad \text{(4) multifurcation,}$$
$$S \rightarrow aSu \mid cSg \mid gSc \mid uSa \qquad \text{(5) the Watson–Crick pairs;}$$

$$W_1 \rightarrow aW_1 \mid cW_1 \mid gW_1 \mid uW_1 \mid a \mid c \mid g \mid u \quad \text{(6)},$$
$$W_2 \rightarrow W_2a \mid W_2c \mid W_2g \mid W_2u \mid a \mid c \mid g \mid u \quad \text{(7)},$$
$$W_3 \rightarrow aW_3 \mid cW_3 \mid gW_3 \mid uW_3 \mid a \mid c \mid g \mid u \quad \text{(8)},$$
$$W_4 \rightarrow aW_4 \mid cW_4 \mid gW_4 \mid uW_4 \mid a \mid c \mid g \mid u \quad \text{(9)},$$
$$W_5 \rightarrow W_5S \mid S.$$

As one can see, the sets of production rules (6), (8), and (9) appear to be identical, and the sets (1) and (3) define essentially the same language. However, those production rules are different in the semantic meaning and the actual probability parameters involved. □

Problem 10.7 Write down the inside algorithm, outside algorithm, and inside–outside re-estimation equations for the Nussinov-style RNA folding SCFG given below:

$$
\begin{array}{lll}
S \rightarrow aS \mid cS \mid gS \mid uS & (i \text{ unpaired}), & \\
S \rightarrow Sa \mid Sc \mid Sg \mid Su & (j \text{ unpaired}), & \\
S \rightarrow aSu \mid cSg \mid gSc \mid uSa & (i,j \text{ pair}), & (10.2) \\
S \rightarrow SS & \text{bifurcation.} &
\end{array}
$$

Solution The inside algorithm, outside algorithm, and inside–outside re-estimation equations can be defined for any SCFG in the Chomsky normal form. To convert the given SCFG (10.2), to the Chomsky normal form, we use W_1 as the start non-terminal along with the nine non-terminals $W_1, \ldots, W_9$:

$$
\begin{aligned}
W_1 &\to W_6 W_1 \mid W_7 W_1 \mid W_8 W_1 \mid W_9 W_1 & \text{(i unpaired)}, \\
W_1 &\to W_1 W_6 \mid W_1 W_7 \mid W_1 W_8 \mid W_1 W_9 & \text{(j unpaired)}, \\
W_1 &\to W_6 W_2 \mid W_7 W_3 \mid W_8 W_4 \mid W_9 W_5 & \text{(i,j pair)}, \\
W_1 &\to W_1 W_1 & \text{bifurcation},
\end{aligned}
$$

$$
\begin{aligned}
W_2 &\to W_1 W_9, & W_6 &\to a, \\
W_3 &\to W_1 W_8, & W_7 &\to c, \\
W_4 &\to W_1 W_7, & W_8 &\to g, \\
W_5 &\to W_1 W_6, & W_9 &\to u.
\end{aligned}
$$

We assume that the probability parameters of the SCFG are denoted by $p(aS)$, $p(aSu)$, etc. Then the probabilities associated with the SCFG production rules are as follows:

$$
\begin{aligned}
&t_1(6,1) = p(aS), && t_1(7,1) = p(cS), && t_1(8,1) = p(gS), && t_1(9,1) = p(uS), \\
&t_1(1,6) = p(Sa), && t_1(1,7) = p(Sc), && t_1(1,8) = p(Sg), && t_1(1,9) = p(Su), \\
&t_1(6,2) = p(aSu), && t_1(7,3) = p(cSg), && t_1(8,4) = p(gSc), && t_1(9,5) = p(uSa), \\
&t_1(1,1) = p(SS), \\
&t_2(1,9) = 1, && t_3(1,8) = 1, && t_4(1,7) = 1, && t_5(1,6) = 1, \\
&e_6(a) = 1, && e_7(c) = 1, && e_8(g) = 1, && e_9(u) = 1.
\end{aligned}
$$

All other values of $t_v(y,z)$, $e_v(a)$, $e_v(c)$, $e_v(g)$, $e_v(u)$, for $v, y, z = 1, \ldots, 9$ are equal to zero.

For an RNA sequence $x = (x_1, \ldots, x_L)$, $x_i \in \{a, c, g, u\}$, the inside algorithm (Lari and Young, 1990) determines the probability $\alpha(i, j, v)$ of a parse subtree rooted at the non-terminal W_v for the subsequence $x_i, \ldots, x_j$ for all possible i, j, and v. The algorithm includes the following steps.

Initialization: for $i = 1$ to L, $v = 1$ to 9,

$$
\alpha(i, i, v) = e_v(x_i).
$$

Iteration: for $i = 1$ to $L - 1$, $j = i + 1$ to L, $v = 1$ to 9,

$$
\alpha(i, j, v) = \sum_{y=1}^{9} \sum_{z=1}^{9} \sum_{k=i}^{j-1} \alpha(i, k, y) \alpha(k+1, j, z) t_v(y, z).
$$

Termination: $P(x|\theta) = \alpha(1, L, 1)$.

Thus, the inside algorithm determines the probability (score) of a sequence x generated by the SCFG, Equation (10.2), with the set of parameters designated as θ.

The outside algorithm (Lari and Young, 1990) finds the probability $\beta(i, j, v)$ of a parse tree rooted at the start non-terminal W_1 for the complete sequence x, excluding the parse subtree for the subsequence $x_i, \ldots, x_j$ rooted at the non-terminals W_v for all possible i, j, and v. The outside algorithm assumes that the 'inside' probabilities $\alpha(i, j, v)$ are known. The algorithm proceeds as follows.

Initialization: $\beta(1, L, 1) = 1$; for $v = 2$ to 9: $\beta(1, L, v) = 0$.

Iteration: for $i = 1$ to $L, j = L$ to $i, v = 1$ to 9,

$$\beta(i, j, v) = \sum_{y,z} \sum_{k=1}^{i-1} \alpha(k, i-1, z)\beta(k, j, y)t_y(z, v)$$

$$+ \sum_{y,z} \sum_{k=j+1}^{L} \alpha(j+1, k, z)\beta(i, k, y)t_y(v, z).$$

Termination:

$$P(x|\theta) = \sum_{v=1}^{9} \beta(i, i, v)e_v(x_i) \qquad \text{for any } i.$$

The probabilities of production rules for the non-terminals W_i, $i = 1, \ldots, 9$ could be estimated by the expectation maximization (EM) re-estimation equations (Dempster, Laird, and Rubin, 1977). For the non-terminal W_1 we have

$$\hat{t}_1(y, z) = \frac{\sum_{i=1}^{L-1} \sum_{j=i+1}^{L} \sum_{k=i}^{j-1} \beta(i, j, 1)\alpha(i, k, y)\alpha(k+1, j, z)t_1(y, z)}{\sum_{i=1}^{L} \sum_{j=i}^{L} \alpha(i, j, 1)\beta(i, j, 1)},$$

where pairs (y, z) assume values of (1,1), (1,6), (1,7), (1,8), (1,9), (6,1), (6,2), (7,1), (7,3), (8,1), (8,4), (9,1), and (9,5). All other values of $\hat{t}_1(y, z)$ are equal to zero.

The EM re-estimation equation for the probabilities of production rules for the non-terminals $W_2, \ldots, W_5$ is given by

$$\hat{t}_v(1, 11-v) = \frac{\sum_{i=1}^{L-1} \sum_{j=i+1}^{L} \sum_{k=i}^{j-1} \beta(i, j, v)\alpha(i, k, 1)\alpha(k+1, j, 11-v)}{\sum_{i=1}^{L} \sum_{j=i}^{L} \alpha(i, j, v)\beta(i, j, v)},$$

where $v = 2, \ldots, 5$. For all other pairs of (y, z), $\hat{t}_2(y, z), \ldots, \hat{t}_9(y, z)$ are equal to zero.

Finally, for the non-terminals $W_6, \ldots, W_9$ we have

$$\hat{e}_6(a) = \frac{\sum_{i:x_i=a} \beta(i,i,6)}{\sum_{i=1}^L \sum_{j=i}^L \alpha(i,j,6)\beta(i,j,6)},$$

$$\hat{e}_7(c) = \frac{\sum_{i:x_i=c} \beta(i,i,7)}{\sum_{i=1}^L \sum_{j=i}^L \alpha(i,j,7)\beta(i,j,7)},$$

$$\hat{e}_8(g) = \frac{\sum_{i:x_i=g} \beta(i,i,8)}{\sum_{i=1}^L \sum_{j=i}^L \alpha(i,j,8)\beta(i,j,8)},$$

$$\hat{e}_9(u) = \frac{\sum_{i:x_i=u} \beta(i,i,9)}{\sum_{i=1}^L \sum_{j=i}^L \alpha(i,j,9)\beta(i,j,9)}.$$

The values of all other $\hat{e}_v(a)$, $\hat{e}_v(c)$, $\hat{e}_v(g)$, and $\hat{e}_v(u)$ are equal to zero.   □

Problem 10.8 By analogy to profile HMM suboptimal alignment sampling, give an algorithm for sampling structures probabilistically from your inside matrix.

Solution Suppose we have completed the inside algorithm and filled out the cells of the three-dimensional dynamic programming matrix $\alpha(i,j,v)$ of the order $L \times L \times M$ (where L is the length of sequence x and M is the number of non-terminals in the SCFG). The value $\alpha(1,L,1)$ of the total probability of sequence x given the SCFG, is the sum of probabilities of all possible parses of x, given SCFG (the sum of probabilities of all possible folding structures of x). We have to define the probability distribution over all possible structures such that the probability of each structure is proportional to the contribution of this structure to $\alpha(1,L,1)$. To generate a sample structure, we trace back through the three-dimensional matrix $\alpha(i,j,v)$ and assign the probabilities to the parse subtrees as follows. Suppose we reach a cell (i,j,v) associated with a parse subtree rooted at the non-terminal W_v for the subsequence $x_i, \ldots, x_j$. The probability $\alpha(i,j,v)$, $i = 1, \ldots, L-1$, $j = i+1, \ldots, L$, $v = 1, \ldots, M$, of this parse subtree is given by the inside algorithm as follows:

$$\alpha(i,j,v) = \sum_{y=1}^M \sum_{z=1}^M \sum_{k=i}^{j-1} \alpha(i,k,y)\alpha(k+1,j,z)t_v(y,z).$$

For a pair of states y, z and index k, $k = i, \ldots, j-1$, we consider a parse subtree $T_{i,j,v}(y,z,k)$ rooted at state v that includes a pair of parse subtrees for y and z and shorter subsequences $x_i, \ldots, x_k$ and $x_{k+1}, \ldots, x_j$:

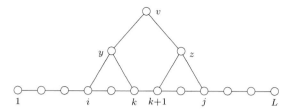

To subtree $T_{i,j,v}(y, z, k)$ we assign the following probability:

$$P(T_{i,j,v}(y, z, k)) = \frac{\alpha(i, k, y)\alpha(k + 1, j, z)t_v(y, z)}{\alpha(i, j, v)}.$$

Then the smallest trees $T_{i,i+1,v}(y, z, i)$, $i = 1, \ldots, L - 1$, have the following probabilities:

$$P(T_{i,i+1,v}(y, z, i)) = \frac{\alpha(i, i, y)\alpha(i + 1, i + 1, z)t_v(y, z)}{\alpha(i, i + 1, v)} = \frac{e_y(x_i)e_z(x_{i+1})t_v(y, z)}{\alpha(i, i + 1, v)}.$$

Upon completing the traceback procedure we determine the probability of any folding structure of x as follows. For each structure we find the consistent sequence of subtrees as described above (starting with a tree $T_{1,L,1}(y, z, k)$ for states y, z, and index k and ending with the smallest subtrees $T_{i,i+1,v}(p, q, i)$, for all i, $i = 1, \ldots, L - 1$, and some states p and q). Then the probability of the structure is the product of the probabilities of participating subtrees. Therefore, we have defined the probability distribution over the set of all possible secondary structures of the sequence x given the SCFG. $\qquad\square$

Problem 10.9 Show how to use your inside and outside variables to calculate the probability that positions i, j are base paired and summed over all structures. The functional form of the answer will be analogous to your inside–outside re-estimation equations.

Solution The RNA folding SCFG in the Chomsky normal form (Problem 10.7) implies that positions i, j are base paired when the productions

$$W_1 \rightarrow W_6 W_2 \mid W_7 W_3 \mid W_8 W_4 \mid W_9 W_5$$

are executed. Let us assume that a particular production, say $W_1 \rightarrow W_6 W_2$, has the transition probability $t_1(6, 2)$. The probability of the event that the subsequence $x_i, \ldots, x_j$ is parsed using this production rule is equal to the probability of the parse of the whole sequence, excluding subsequence $x_i, \ldots, x_j$. This value is defined as the product of the value $\beta(i, j, 1)$ determined by the outside algorithm and the

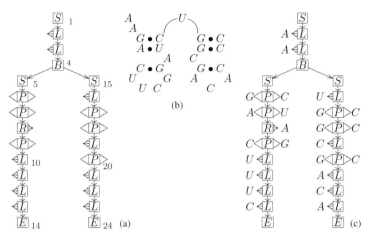

Figure 10.4. (a) Graphical representation of the ungapped RNA SCFG example. Boxes labeled P represent sixteen pairwise production rules; boxes labeled L and R represent four leftwise and four rightwise production rules, respectively; boxes labeled S, B, and E represent start, bifurcation, and end terminals, respectively. (b) The RNA consensus structure is redrawn to correspond more closely to the tree structure of the SCFG. (c) A parse tree for this RNA structure with the ribonucleotides assigned to states in the SCFG.

probability of the parse subtrees rooted at the chosen production rule, which is given by

$$\frac{t_1(6,2)\sum_{k=i}^{j-1}\alpha(i,k,6)\alpha(k+1,j,2)}{\sum_{i'=1}^{L}\sum_{j'=i'}^{L}\alpha(i',j',1)\beta(i',j',1)}.$$

Then, the probability $p(i,j)$ that positions i,j are base paired is equal to the sum of such probabilities for all four production rules:

$$p(i,j) = \frac{\beta(i,j,1)\sum_{(y,z)\in\Pi}t_1(y,z)\sum_{k=i}^{j-1}\alpha(i,k,y)\alpha(k+1,j,z)}{\sum_{i'=1}^{L}\sum_{j'=i'}^{L}\alpha(i',j',1)\beta(i',j',1)},$$

where $\Pi = \{(6,2),(7,3),(8,4),(9,5)\}$. $\Box$

Problem 10.10 Rewrite the list of production rules from the ungapped RNA model such that symbols are emitted independently of the previous state, as in an HMM. This is the formal stochastic transformational grammar that corresponds to the graphical SCFG representation in Figure 10.4.

Solution Following the logic used in Problem 9.1, we use the non-terminals $W_2, \ldots, W_{23}$ to model state transitions. The non-terminals involved in symbols emission, P_i, L_i, and R_i, will retain the same meaning as in the original grammar.

	Stem 1	**Stem 2**
$S_1 \rightarrow W_2$	$S_5 \rightarrow W_6$	$S_{15} \rightarrow W_{16}$
$W_2 \rightarrow aL_2 \mid \ldots$	$W_6 \rightarrow gP_6c \mid \ldots$	$W_{16} \rightarrow uL_{16} \mid \ldots$
$L_2 \rightarrow W_3$	$P_6 \rightarrow W_7$	$L_{16} \rightarrow W_{17}$
$W_3 \rightarrow aL_3 \mid \ldots$	$W_7 \rightarrow aP_7u \mid \ldots$	$W_{17} \rightarrow gP_{17}c \mid \ldots$
$L_3 \rightarrow B_4$	$P_7 \rightarrow W_8$	$P_{17} \rightarrow W_{18}$
$B_4 \rightarrow S_5 S_{15}$	$W_8 \rightarrow R_8a \mid \ldots$	$W_{18} \rightarrow gP_{18}c \mid \ldots$
	$R_8 \rightarrow W_9$	$P_{18} \rightarrow W_{19}$
	$W_9 \rightarrow cP_9g \mid \ldots$	$W_{19} \rightarrow cL_{19} \mid \ldots$
	$P_9 \rightarrow W_{10}$	$L_{19} \rightarrow W_{20}$
	$W_{10} \rightarrow uL_{10} \mid \ldots$	$W_{20} \rightarrow gP_{20}c \mid \ldots$
	$L_{10} \rightarrow W_{11}$	$P_{20} \rightarrow W_{21}$
	$W_{11} \rightarrow uL_{11} \mid \ldots$	$W_{21} \rightarrow aL_{21} \mid \ldots$
	$L_{11} \rightarrow W_{12}$	$L_{21} \rightarrow W_{22}$
	$W_{12} \rightarrow cL_{12} \mid \ldots$	$W_{22} \rightarrow cL_{22} \mid \ldots$
	$L_{12} \rightarrow W_{13}$	$L_{22} \rightarrow W_{23}$
	$W_{13} \rightarrow gL_{13} \mid \ldots$	$W_{23} \rightarrow aL_{23} \mid \ldots$
	$L_{13} \rightarrow E_{14}$	$L_{23} \rightarrow E_{24}$
	$E_{14} \rightarrow \varepsilon$	$E_{24} \rightarrow \varepsilon$

Figure 10.5. Production rules of the stochastic transformational grammar of the ungapped RNA model.

Each of the sixteen pairwise productions will be associated with one of the sixteen production probabilities (for all possible pairs); a single nucleotide emission by the four leftwise and the four rightwise productions will be defined by the two groups of four probabilities. Other productions (bifurcation, start, end) will have a probability of one.

The whole stochastic transformational grammar is then described by the productions shown in Figure 10.5 (for brevity, only one of the possible productions is shown for each non-terminal). □

Problem 10.11 Suppose that we are given a long nucleotide sequence and our task is to find one or more subsequences that match the RNA covariance model. By employing a transformation of the dynamic programming matrix coordinate system, we can implement an efficient algorithm (CYK, inside, or outside algorithm) for the local subsequence database search.

Compared to CYK, the inside algorithm has the advantage that it sums over the probabilities of all possible structures and alignments for the subsequences, yet it is no more computationally complex than the CYK version. Give the inside algorithm for searching for local subsequence matches no greater than length D.

Solution We consider a covariance model (CM) with M states (non-terminals) denoted by $W_1, \ldots, W_M$. Seven types of states are labeled as P, L, R, D, S, B, and E, for pairwise emitting, leftwise ($5'$) emitting, rightwise ($3'$) emitting, delete, start, bifurcation, and end, respectively; W_1 is the start (the root) state for the whole CM. The seven state types are associated with the symbol emission and the state transition probabilities (Table 10.1). The dynamic programming matrix is indexed by v, j, d (instead of v, i, j for SCFG), where d is the length of the subsequence $x_i, \ldots, x_j$ ($d = j - i + 1$) and $d \leq D$. For convenience, we will use the notation $e_v(x_i, x_j)$ for all emission probabilities. For L states $e_v(x_i, x_j) = e_v(x_i)$, for R states $e_v(x_i, x_j) = e_v(x_j)$, and for non-emitting states $e_v(x_i, x_j) = 1$. Numbers Δ_v^L and Δ_v^R are the numbers of symbols emitted by state v to the left and to the right, respectively. Let s_v be the *state type*, $s \in \{P, L, R, D, S, B, E\}$, indicating one of the seven possible types of production rule. Let $\mathcal{C}_v$ be the set of possible *children* of state v, represented by the list of one or more indices y of states W_y that state W_v can make a transition to. The convention of numbering states here implies that $y > v$ for all $y \in \mathcal{C}_v$, except for the insert states, where $y \geq v$ for all $y \in \mathcal{C}_v$.

The inside algorithm searching for local subsequence matches proceeds as follows.

Initialization: for $j = 0$ to L, $v = M$ to 1,

$$
\alpha_v(j, 0) = \begin{cases}
\text{for } s_v = \mathrm{E} : 1; \\
\text{for } s_v \in \mathrm{D,S} : \sum_{y \in \mathcal{C}_v} t_v(y)\alpha_y(j, 0); \\
\text{for } s_v = \mathrm{B}, \mathcal{C}_v = (y, z) : \alpha_y(j, 0)\alpha_z(j, 0); \\
\text{for } s_v \in \mathrm{P,L,R} : 0.
\end{cases}
$$

Recursion: for $j = 1$ to L, $d = 1$ to $\min(D, j)$, $v = M$ to 1,

$$
\alpha_v(j, d) = \begin{cases}
\text{for } s_v = \mathrm{E} : 0; \\
\text{for } s_v = \mathrm{P, and } d < 2 : 0; \\
\text{for } s_v = \mathrm{B}, \mathcal{C}_v = (y, z) : \sum_{k=0}^{d} \alpha_y(j - k, d - k)\alpha_z(j, k); \\
\text{otherwise:} \\
\quad e_v(x_{j-d+1}, x_j) \sum_{y \in \mathcal{C}_v} t_v(y)\alpha_y(j - \Delta_v^R, d - \Delta_v^L - \Delta_v^R).
\end{cases}
$$

When the recursion is completed, cells $\alpha_1(j, d)$ should contain the probabilities of matches for subsequences $x_{j-d+1}, \ldots, x_j$ with lengths no greater than D. $\square$

Table 10.1. *The seven types of states of a covariance model*

State (s_v)	Production	Δ_v^L	Δ_v^R	Emission	Transition
P	$W_v \rightarrow x_i W_y x_j$	1	1	$e_v(x_i, x_j)$	$t_v(y)$
L	$W_v \rightarrow x_i W_y$	1	0	$e_v(x_i)$	$t_v(y)$
R	$W_v \rightarrow W_y x_j$	0	1	$e_v(x_j)$	$t_v(y)$
D	$W_v \rightarrow W_y$	0	0	1	$t_v(y)$
S	$W_v \rightarrow W_y$	0	0	1	$t_v(y)$
B	$W_v \rightarrow W_y W_z$	0	0	1	1
E	$W_v \rightarrow \varepsilon$	0	0	1	1

Problem 10.12 Modify the CYK algorithm so that it keeps traceback information in each cell to assist in recovering the optimal parse tree. What is the
minimum information that needs to be kept for tracing back from the bifurcation
state? What is the minimum information that needs to be kept for tracing back
from any other state?

Solution We will modify the CYK algorithm with the index system described in
Problem 10.11. For every value $\gamma_v(j, d)$, we will keep the corresponding traceback
value $\tau_v(j, d)$. The modified CYK algorithm works as follows.
  Initialization: for $v = M, \ldots, 1$,

- if $s_v = E$, then $\gamma_v(0, 0) = 0$, $\quad \tau_v(0, 0) = \emptyset$,
- else if $s_v \in D, S$, then

$$\gamma_v(0, 0) = \max_{y \in C_v}(\gamma_y(0, 0) + \log t_v(y)),$$

$$\tau_v(0, 0) = \underset{y \in C_v}{\mathrm{argmax}}(\gamma_y(0, 0) + \log t_v(y)),$$

- else if $s_v = B$, with $C_v = (y, z)$, then

$$\gamma_v(0, 0) = \gamma_y(0, 0) + \gamma_z(0, 0), \quad \tau_v(0, 0) = 0,$$

- else $\gamma_v(0, 0) = -\infty$, $\quad \tau_v(0, 0) = \emptyset$;
- for $j = 1, \ldots, L$, $\gamma_v(j, 0) = \gamma_v(0, 0)$, $\quad \tau_v(j, 0) = \tau_v(0, 0)$.

  Recursion: for $j = 1, \ldots, L$, $\quad d = 1, \ldots, \min(j, D)$, $\quad v = M, \ldots, 1$,

- if $s_v = E$, then $\gamma_v(j, d) = -\infty$, $\quad \tau_v(j, d) = \emptyset$,
- else if $s_v = P$ and $d < 2$, then $\gamma_v(j, d) = -\infty$, $\quad \tau_v(j, d) = \emptyset$,

- else if $s_v = B$, with $C_v = (y, z)$, then

$$\gamma_v(j, d) = \max_{k=0,\dots,d} (\gamma_y(j - k, d - k) + \gamma_z(j, k)),$$

$$\tau_v(j, d) = \operatorname{argmax}_{k=0,\dots,d}(\gamma_y(j - k, d - k) + \gamma_z(j, k)),$$

- else

$$\gamma_v(j, d) = \max_{y \in C_v}(\gamma_y(j - \Delta_v^R, d - \Delta_v^L - \Delta_v^R) + \log t_v(y))$$

$$+ \log \hat{e}_v(x_{j-d+1}, x_j),$$

$$\tau_v(j, d) = \operatorname{argmax}_{y \in C_v}(\gamma_y(j - \Delta_v^R, d - \Delta_v^L - \Delta_v^R) + \log t_v(y)).$$

Here $\hat{e}_v(a, b) = \log(e_v(a, b)/f_a f_b)$ for $s_v = P$; $\log(e_v(a)/f_a)$ for $s_v = L$; and $\log(e_v(b)/f_b)$ for $s_v = R$; f_a, f_b are the frequencies of individual bases.

The traceback starts from $\gamma_1(j, d)$ for the highest-scoring subsequence of length d ending at position j and works back. Note that for a global rather than a local alignment, the traceback starts from $\gamma_1(L, L)$. The traceback algorithm works as follows:

For score $\gamma_v(j, d)$,

- if $\tau_v(j, d) = \emptyset$, then stop;
- if $s_v = B$, with $C_v = (y, z)$, then let $k = \tau_v(j, d)$, make bifurcation to $\gamma_y(j - k, d - k)$ and $\gamma_z(j, k)$;
- else let $y = \tau_v(j, d)$, go to $\gamma_y(j - \Delta_v^R, d - \Delta_v^L - \Delta_v^R)$.

The minimum information that needs to be kept for tracing back from a bifurcation state is the subsequence fork point k that gives the maximum score. The minimum information that needs to be kept for tracing back from any other state is the non-terminal state index y that gives the maximum score. □

10.2 Further reading

The stochastic context-free grammars continue to be employed as the efficient modeling tools for the algorithms of RNA secondary structure prediction. A new algorithm using SCFGs and evolutionary history for finding the maximum *a posteriori* probability (MAP) secondary structure for a set of RNA sequences was proposed by Knudsen and Hein (1999). Later this algorithm was implemented as the software program Pfold able to construct an alignment of up to fourty sequences of 500 bases long (Knudsen and Hein, 2003). Holmes and Rubin (2002) proposed several dynamic programming algorithms (the inside algorithm, the CYK algorithm, the inside–outside algorithm) for pair SCFGs generalizing a concept of (single) SCFGs. Dowell and Eddy (2004) evaluated the accuracy of different SCFG designs for the single-sequence RNA secondary structure prediction on the benchmark of RNA sequences with verified structures. A comparison with the accuracy of predictions made by the energy minimization methods was also provided.

A dynamic programming algorithm for aligning a target RNA sequence with unknown structure to a query sequence with a known structure was proposed by Eddy (2002). This algorithm uses the covariance model of RNA secondary structure and employs a divide-and-conquer strategy that significantly reduces the memory requirement when compared with the SCFG-based algorithms, which makes possible the computation of optimal structural alignments for large RNAs.

Several algorithms have been developed to predict RNA secondary structure with pseudoknots. The approach proposed by Tabaska *et al.* (1998) uses the maximum weighted matching algorithm of Gabow (1973) to find an optimal set of base-pairing interactions, including pseudoknots and other ternary pairs. This algorithm works in polynomial time and memory. Rivas and Eddy (1999) have developed a dynamic programming algorithm that generates the minimal energy structure for a single RNA sequence. This algorithm uses the standard RNA folding thermodynamic parameters including ones describing the thermodynamic stability of pseudoknots.

Mathews *et al.* (1999) augmented the conventional energy minimization dynamic programming algorithm for RNA secondary structure prediction with new thermodynamic parameters that define the secondary structure motif stabilities with expanded sequence dependence. Further refinement of these parameters along with the experimental determination of the chemical modification constraints allowed for additional improvement of this algorithm (Mathews *et al.*, 2004). The free energy minimization algorithms were reviewed by Zuker (2000).

Juan and Wilson (1999) developed a computational method which uses a combination of free energy and covariational analysis to identify an evolutionary conserved secondary structure in a set of aligned homologous RNA sequences. Ding and Lawrence (2003) proposed a statistical RNA secondary structure prediction algorithm which works by sampling from the Boltzmann equilibrium probability distribution defined on a set of possible secondary structures for a given RNA. New RNA secondary structure prediction algorithms were also developed by Gorodkin, Stricklin, and Stormo (2001) and by Perriquet, Touzet, and Dauchet (2003).

Several RNA sequence databases and RNA processing servers are currently available: the Ribonuclease P database (Brown, 1999a) is a compilation of RNase P sequences, RNA sequence alignments, secondary structures, and three-dimensional models; the ribosomal database project (Maidak *et al.*, 2000) includes both sequence data and the software for aligning rRNA sequences, building phylogenetic trees, and predicting rRNA secondary structure; the European Large Subunit Ribosomal RNA database (Wuyts *et al.*, 2001) contains the complete list of LSU rRNA sequences, the RNA multiple alignments, and the secondary structures predicted from comparative sequence analysis; the Pfam database (Griffiths-Jones *et al.*, 2003) includes a collection of multiple sequence alignments and covariance models representing non-coding RNA families; the tmRNA database (Zwieb *et al.*, 2003) provides the regularly updated list of tmRNAs, the multiple alignments, the secondary and ternary structure of each tmRNA molecule. Cannone *et al.* (2002) developed the comparative RNA web site (CRW), a database of comparative sequence and structure information for ribosomal, intron, and other RNAs. The Vienna RNA secondary structure server (Hofacker, 2003), offers prediction of RNA secondary structure from a single sequence, prediction of the consensus secondary structure for a set of aligned sequences, and designs RNA sequences that fold into a predefined structure.

11

Background on probability

The last chapter of *BSA* gives a brief review of some important probabilistic concepts and analytical results, including the properties of frequently used probability distributions, a discussion of the notions of entropy and mutual entropy, the maximum likelihood principle, and principles of rational sampling and parameter estimation. The chapter ends with the description of the general expectation maximization (EM) algorithm, which underlies several important and perhaps the most complex algorithms discussed in *BSA*, such as estimation of parameters of a profile HMM in parallel with the construction of multiple sequence alignment, and estimation of parameters of a covariance model in parallel with the construction of multiple alignment of RNA sequences. The problems offered in this chapter further illustrate the major mathematical concepts used in biological sequence analysis.

11.1 Original problems

Problem 11.1 Calculate the mean and variance of the binomial distribution.

Solution For a random variable X binomially distributed with parameters p and N, thus $P(X = k) = \binom{N}{k} p^k (1 - p)^{N-k}$, $k = 0, 1, \ldots, N$, $0 \leq p \leq 1$, the formulas for the mean value m and variance $\mathbf{Var}X$ may be derived in three different ways.

(1) By definition,

$$
m = \mathbf{E}X = \sum_{k=0}^{N} k \binom{N}{k} p^k (1 - p)^{N-k} = \sum_{k=1}^{N} \frac{kN!}{k!(N-k)!} p^k (1 - p)^{N-k}
$$

$$
= pN \sum_{k=1}^{N} \frac{(N-1)!}{(k-1)!(N-k)!} p^{k-1} (1 - p)^{N-k}.
$$

311

After changing the summation index to $l = k - 1$ and using the binomial expansion, we have

$$m = pN \sum_{l=0}^{N-1} \frac{(N-1)!}{l!(N-1-l)!} p^l (1-p)^{N-1-l} = pN(p + (1-p))^{N-1} = pN. \quad (11.1)$$

Since

$$\mathbf{Var}X = \mathbf{E}X^2 - (\mathbf{E}X)^2 = \mathbf{E}X^2 - m^2,$$

we only need the expression for the second moment $\mathbf{E}X^2$. We introduce new summation indexes $l = k - 1$ and $j = k - 2$ and proceed as follows (note that $0! = 1$):

$$\mathbf{E}X^2 = \sum_{k=0}^{N} k^2 \binom{N}{k} p^k (1-p)^{N-k} = \sum_{k=1}^{N} \frac{kN!}{(k-1)!(N-k)!} p^k (1-p)^{N-k}$$

$$= \sum_{k=1}^{N} \frac{(k-1)N!}{(k-1)!(N-k)!} p^k (1-p)^{N-k} + \sum_{k=1}^{N} \frac{N!}{(k-1)!(N-k)!} p^k (1-p)^{N-k}$$

$$= \sum_{k=2}^{N} \frac{N!}{(k-2)!(N-k)!} p^k (1-p)^{N-k} + pN \sum_{l=0}^{N-1} \frac{(N-1)!}{l!(N-1-l)!} p^l (1-p)^{N-1-l}$$

$$= p^2 N(N-1) \sum_{j=0}^{N-2} \frac{(N-2)!}{j!(N-2-j)!} p^j (1-p)^{N-2-j} + pN(p + (1-p))^{N-1}$$

$$= p^2 N(N-1)(p + (1-p))^{N-2} + pN = p^2 N^2 - p^2 N + pN.$$

Thus, we obtain

$$\mathbf{Var}X = p^2 N^2 - p^2 N + pN - p^2 N^2 = Np(1-p). \quad (11.2)$$

(2) The random variable X can be considered as a sum of N independent Bernoulli variables $Y_1, \ldots, Y_N$. This representation can be used to derive Equations (11.1) and (11.2) in a rather easy way. Each of Y_i, $i = 1, \ldots, N$, has the probability distribution $P(Y_i = 1) = p$, $P(Y_i = 0) = 1 - p$. Therefore,

$$\mathbf{E}Y_i = 1 \times p + 0 \times (1-p) = p,$$

$$\mathbf{Var}Y_i = \mathbf{E}Y_i^2 - (\mathbf{E}Y_i)^2 = 1^2 \times p + 0^2 \times (1-p) - p^2 = p(1-p).$$

For $X = \sum_{i=1}^{N} Y_i$ we have

$$m = \mathbf{E}X = \mathbf{E}\left(\sum_{i=1}^{N} Y_i\right) = \sum_{i=1}^{N} \mathbf{E}Y_i = Np,$$

$$\mathbf{Var}X = \mathbf{Var}\left(\sum_{i=1}^{N} Y_i\right) = \sum_{i=1}^{N} \mathbf{Var}Y_i = Np(1-p).$$

(3) Finally, we show an approach that illustrates an application of one important analytical technique. We consider the binomial expansion for the two variables p and q:

$$(p+q)^N = \sum_{k=0}^{N} \binom{N}{k} p^k q^{N-k} \tag{11.3}$$

and differentiate both parts of Equation (11.3) with respect to variable p:

$$N(p+q)^{N-1} = \sum_{k=1}^{N} k \binom{N}{k} p^{k-1} q^{N-k}. \tag{11.4}$$

By multiplying both parts of Equation (11.4) by p we have

$$Np(p+q)^{N-1} = \sum_{k=1}^{N} k \binom{N}{k} p^k q^{N-k}. \tag{11.5}$$

Now, if $q = 1 - p$, Equation (11.5) becomes

$$Np = \sum_{k=1}^{N} k \binom{N}{k} p^k (1-p)^{N-k} = \sum_{k=0}^{N} k \binom{N}{k} p^k (1-p)^{N-k} = m.$$

Similarly, by differentiating the binomial expansion in Equation (11.3), twice with respect to variable p we have

$$N(N-1)(p+q)^{N-2} = \sum_{k=2}^{N} k(k-1) \binom{N}{k} p^{k-2} q^{N-k} = \sum_{k=1}^{N} k(k-1) \binom{N}{k} p^{k-2} q^{N-k}$$

$$= \sum_{k=1}^{N} k^2 \binom{N}{k} p^{k-2} q^{N-k} - \sum_{k=1}^{N} k \binom{N}{k} p^{k-2} q^{N-k}$$

$$= \sum_{k=0}^{N} k^2 \binom{N}{k} p^{k-2} q^{N-k} - \sum_{k=0}^{N} k \binom{N}{k} p^{k-2} q^{N-k}.$$

We multiply both parts of the last equality by p^2, and for $q = 1 - p$ we obtain

$$N(N-1)p^2 = EX^2 - EX.$$

Therefore,

$$EX^2 = N^2 p^2 - Np^2 + EX = N^2 p^2 - Np^2 + Np.$$

Finally,

$$\mathbf{Var}X = EX^2 - (EX)^2 = N^2 p^2 - Np^2 + Np - N^2 p^2 = Np - Np^2 = Np(1-p). \quad \square$$

Problem 11.2 Assume a model in which $p_i(a)$ is the probability of amino acid a occurring in the ith position of a sequence of length l. The amino acids are considered independent. What is the probability $P(x)$ of a particular sequence x = x_1, \ldots, x_l? Show that the average of the log of the probability is the negative entropy $\sum P(x) \log P(x)$, where the sum is over all possible sequences x of length l.

Solution Assuming independence of the amino acids at different positions of the sequence, the probability of sequence x is given by

$$P(x) = P(x_1, \ldots, x_l) = p_1(x_1) \times \cdots \times p_l(x_l) = \prod_{i=1}^{l} p_i(x_i). \tag{11.6}$$

Equation (11.6) defines the probability distribution P over all amino acid sequences of length l. The entropy of this distribution is given by

$$H(P) = -\sum_x P(x) \log P(x).$$

On the other hand, the average of the logarithm of the probability of sequence x is given by

$$E(\log P(x)) = \sum_x P(x) \log P(x) = -H(P). \qquad \square$$

Problem 11.3 Prove the equivalence of information content and relative entropy when Q is uniform.

Solution The relative entropy (or the Kullback–Leibler distance) for two discrete distributions P and Q taking non-negative values on one and the same set $\{x_i\}$ is defined as follows:

$$H(P||Q) = \sum_i P(x_i) \log \frac{P(x_i)}{Q(x_i)},$$

If the distribution Q is uniform over $x_1, \ldots, x_N$, then the relative entropy becomes

$$H(P||Q) = \sum_i P(x_i) \log \frac{P(x_i)}{1/N} = \sum_i P(x_i) \log P(x_i) - \sum_i P(x_i) \log(1/N)$$

$$= -H(P) - \log(1/N) = -H(P) - \sum_i 1/N \log(1/N)$$

$$= -H(P) + H(Q)$$

Here $H(P)$ and $H(Q)$ are the entropies of the distributions P and Q, respectively. We can interpret the value $H(Q) = \log N$ as the entropy of distribution on $x_1, \ldots, x_N$ defined *a priori* (the prior distribution is supposed to be uniform), while $H(P)$ is the entropy of the empirical distribution (with the probabilities of outcomes observed in the experiment defined as the outcome frequencies). Therefore, for the information content I we have

$$I = H_{before} - H_{after} = H(Q) - H(P) = H(P||Q). \qquad \square$$

Problem 11.4 Show that $M(X; Y) = M(Y; X)$.

Solution We consider the random variables X and Y that take values $\{x_i\}$ and $\{y_j\}$, respectively, with probabilities $P(x_i)$ and $P(y_j)$. The random vector (X, Y) takes values (x_i, y_j) with probabilities $P(x_i, y_j)$. The mutual information $M(X; Y)$ is defined by the following formula:

$$M(X; Y) = \sum_{i,j} P(x_i, y_j) \log \frac{P(x_i, y_j)}{P(x_i)P(y_j)}.$$

This formula is symmetric with respect to X and Y; thus, $M(X; Y) = M(Y; X)$. $\quad\square$

Problem 11.5 Show that

$$M(X; Y) = H(X) + H(Y) - H(Y, X), \qquad (11.7)$$

where $H(Y, X)$ is the entropy of the joint distribution $P(X, Y)$.

Solution Definitions of the mutual information and the entropy allow us to derive the chain of equalities as follows:

$$M(X; Y) = \sum_{i,j} P(x_i, y_j) \log \frac{P(x_i, y_j)}{P(x_i)P(y_j)}$$

$$= \sum_{i,j} P(x_i, y_j) \log P(x_i, y_j) - \sum_{i,j} P(x_i, y_j) \log P(x_i)P(y_j)$$

$$= -H(Y; X) - \sum_{i,j} P(x_i, y_j) \log P(x_i) - \sum_{i,j} P(x_i, y_j) \log P(y_j).$$

We have to show that the last two sums define the entropies $H(X)$ and $H(Y)$. Indeed,

$$-\sum_{i,j} P(x_i, y_j) \log P(x_i) = -\sum_i \left(\sum_j P(x_i, y_j) \log P(x_i) \right)$$

$$= -\sum_i P(x_i) \log P(x_i) = H(X),$$

$$-\sum_{i,j} P(x_i, y_j) \log P(y_j) = -\sum_j \left(\sum_i P(x_i, y_j) \log P(y_j) \right)$$

$$= -\sum_j P(y_j) \log P(y_j) = H(Y).$$

Thus, the proof of formula (11.7) is complete.   □

Problem 11.6 The *weak law of large numbers* says that the mean of a sample of size N differs from the true mean by an amount d or more with probability $\sigma^2/(Nd^2)$, where σ^2 is the variance of the distribution. Show that this implies that $n_i / \sum n_k$ tends to $P(\omega_i)$ as $\sum n_k \to \infty$, where n_i is the frequency of occurrence of ω_i.

Solution The *weak law of large numbers* is a corollary of Chebyshov's inequality. We prove below the statements we need.

(1) If a random variable ξ is non-negative *almost surely*, then for any $\varepsilon > 0$ the following Chebyshov's inequality holds:

$$P(\xi \geq \varepsilon) \leq \frac{\mathbf{E}\xi}{\varepsilon}. \tag{11.8}$$

We note that

$$\xi = \mathbf{I}\{\xi \geq \varepsilon\}\xi + \mathbf{I}\{\xi < \varepsilon\}\xi \geq \mathbf{I}\{\xi \geq \varepsilon\}\xi \geq \mathbf{I}\{\xi \geq \varepsilon\}\varepsilon, \tag{11.9}$$

where $\mathbf{I}\{A\}$ designates the indicator of event A. By taking expectations of both sides of Equation (11.9), we obtain Chebyshov's inequality in the form of (11.8). Similarly, for any integer n,

$$\mathbf{E}\xi^n \geq P(\xi \geq \varepsilon)\varepsilon^n.$$

(2) As a consequence of (11.8) we can state that for a random variable ξ, a positive ε and an integer n the following inequalities hold:

$$P(|\xi| \geq \varepsilon) \leq \frac{\mathbf{E}|\xi|^n}{\varepsilon^n},$$

$$P(|\xi - \mathbf{E}\xi| \geq \varepsilon) \leq \frac{\mathbf{E}|\xi - \mathbf{E}\xi|^n}{\varepsilon^n}.$$

The last inequality is especially well known for $n = 2$:

$$P(|\xi - \mathbf{E}\xi| \geq \varepsilon) \leq \frac{\mathbf{Var}\xi}{\varepsilon^2}. \tag{11.10}$$

Now, we consider a sample $x = (x_1, \ldots, x_N)$ from a distribution with expectation $\mathbf{E}x_i = d$ and variance $\mathbf{Var}x_i = \sigma^2$, $i = 1, \ldots, N$. The sample mean $\bar{x} = (1/N)\sum_i x_i$ has the expectation $\mathbf{E}\bar{x} = (1/N)\sum_i \mathbf{E}x_i = d$, and the variance

Var$\bar{x} = (1/N^2) \sum_i$ **Var**$x_i = (\sigma^2/N)$. For the random variable $\bar{x}$ and a positive ε, Chebyshov's inequality, Equation (11.10), implies

$$P(|\bar{x} - d| \geq \varepsilon) \leq \frac{\sigma^2}{N\varepsilon^2}.$$

If the sample size N tends to infinity while the variance $\sigma^2 < +\infty$, the right-hand side of the inequality converges to zero and

$$P(|\bar{x} - d| \geq \varepsilon) \rightarrow 0.$$

The last statement is equivalent to the definition of convergence in probability. Thus, when $N \rightarrow +\infty$,

$$\bar{x} \rightarrow^P \mathbf{E}\bar{x} = d.$$

In fact, since $x_1, \ldots, x_n$ are independent and σ^2 is finite, the sample mean obeys the *strong law of large numbers* by Kolmogorov, namely

$$\bar{x} \rightarrow \mathbf{E}\bar{x} = d$$

almost surely.

In particular, if an elementary event w has occurred n times among N independent repetitions of a random experiment, then for x_i, $i = 1, \ldots, N$, defined as the indicator of occurrence of w in the ith experiment, we have: $\bar{x} = (n/N)$, $d = \mathbf{EI}(w) = P(w)$, $\sigma^2 = P(w)(1 - P(w)) < +\infty$. Now the law of large numbers yields that for the elementary event w

$$\frac{n}{N} \rightarrow P(w). \qquad \square$$

Problem 11.7 Let $f(x) = 2(1 - x)$ be a density on $[0, 1]$. Show how this transforms to a density on y under $x = y^2$. Show that the peak and the *posterior mean estimator* (PME) of the density both shift under this transformation.

Solution If a random variable X has the probability density function f, and a random variable Y is defined as $X = Y^2$, several methods to determine the probability density function g of Y are available. Here we use two basic methods.

(1) First we use the definition of the cumulative distribution function (CDF) and the relationship between the CDF and the density function. For the CDF of X we have

$$F(x) = \int_{-\infty}^{x} f(\tau)\, d\tau = \begin{cases} 0, & x \leq 0, \\ \int_0^x (2 - 2\tau)\, d\tau = 2x - x^2, & 0 < x \leq 1, \\ \int_0^1 (2 - 2\tau)\, d\tau = 1, & x > 1. \end{cases}$$

Then, the formula for the cumulative distribution function G of Y comes straight from the definition of CDFs:

$$G(x) = P(Y \leq x) = P(\sqrt{X} \leq x) = \begin{cases} 0, & x < 0, \\ P(X \leq x^2) = F(x^2), & x \geq 0; \end{cases}$$

$$= \begin{cases} 0, & x < 0, \\ 2x^2 - x^4, & 0 \leq x \leq 1, \\ 1, & x > 1. \end{cases}$$

Then, the probability density function g of Y is given by

$$g(x) = G'(x) = \begin{cases} 4x - 4x^3, & 0 \leq x \leq 1, \\ 0, & \text{elsewhere.} \end{cases}$$

(2) Next, we use the transformation rule: if $X = \phi(Y)$, then

$$g(y) = f(\phi(y))|\phi'(y)|.$$

In our case $\phi(y) = y^2$, and

$$g(y) = f(y^2)2y = 2(1 - y^2)2y = 4y - 4y^3.$$

Since the random variable X takes values on interval $[0, 1]$, $Y = X^2$ takes values on $[0, 1]$, too. Hence,

$$g(y) = \begin{cases} 4y - 4y^3, & 0 \leq y \leq 1, \\ 0, & \text{elsewhere.} \end{cases}$$

To verify that g is a density function, we check the normalization condition:

$$\int_{-\infty}^{+\infty} g(x)\, dx = \int_0^1 (4x - 4x^3)\, dx = \left(\frac{4x^2}{2} - \frac{4x^4}{4} \right)\Big|_{x=0}^1 = 2 - 1 = 1.$$

The density function $f(x)$ has its maximum at $x = 0$, $\max_x f(x) = f(0) = 2$. However, the density function $g(x)$ attains its maximum at the point $x = 1/\sqrt{3}$, $\max_x g(x) = g(1/\sqrt{3}) = 8/3\sqrt{3}$. The transformation from X to Y also changes the expectation:

$$\mathbf{E}X = \int_{-\infty}^{+\infty} xf(x)\, dx = \int_0^1 2x(1 - x)\, dx = \frac{1}{3},$$

$$\mathbf{E}Y = \int_{-\infty}^{+\infty} xg(x)\, dx = \int_0^1 x(4x - 4x^3)\, dx = \frac{8}{15}. \qquad \square$$

Problem 11.8 Show that the function

$$g(y) = \frac{\alpha^\lambda \lambda y^{\lambda-1}}{(\alpha^\lambda + y^\lambda)^2}$$

is a density on $0 \le y < \infty$. Show that picking x uniformly from $(0, 1)$ and mapping x to $y = \alpha(x/(1-x))^{1/\lambda}$ samples from $g(y)$.

Solution The function $g(y)$ is positive for all $y \ge 0$ and we need to verify the normalization condition:

$$I = \int_{-\infty}^{+\infty} g(y)\,dy = \int_0^{+\infty} \frac{\alpha^\lambda \lambda y^{\lambda-1}}{(\alpha^\lambda + y^\lambda)^2} = 1.$$

Indeed, the use of a new variable $z = \alpha^\lambda + y^\lambda$, $dz = \lambda y^{\lambda-1}\,dy$, with z taking values on the interval $[\alpha^\lambda, +\infty)$ leads to the following equality:

$$I = \alpha^\lambda \int_{\alpha^\lambda}^{+\infty} \frac{dz}{z^2} = -\frac{\alpha^\lambda}{z}\Big|_{\alpha^\lambda}^{+\infty} = \frac{\alpha^\lambda}{\alpha^\lambda} = 1.$$

It means that g is a probability density function.

Let X be a uniform random variable on $[0, 1]$, and $Y = \alpha(X/(1-X))^{1/\lambda}$. We want to prove that the function g is the probability density function of the random variable Y. The CDF of random variable X is given by

$$F(x) = \begin{cases} 0, & x \le 0, \\ x, & 0 < x \le 1, \\ 1, & x > 1. \end{cases}$$

Then, the cumulative distribution function G of Y is defined as follows:

$$G(y) = P(Y \le y) = P\left(\alpha\left(\frac{X}{1-X}\right)^{1/\lambda} \le y\right)$$

$$= \begin{cases} 0, & y < 0, \\ P\left(X \le \dfrac{y^\lambda}{y^\lambda + \alpha^\lambda}\right) = F\left(\dfrac{y^\lambda}{y^\lambda + \alpha^\lambda}\right), & y \ge 0. \end{cases}$$

The value $y^\lambda/(y^\lambda + \alpha^\lambda)$ belongs to interval $[0, 1]$ for any positive y; therefore, we have that

$$G(y) = \begin{cases} 0, & y < 0, \\ \dfrac{y^\lambda}{y^\lambda + \alpha^\lambda}, & y \ge 0. \end{cases}$$

Finally, the probability density function of random variable Y is given by

$$G'(y) = \begin{cases} 0, & y < 0, \\ \dfrac{\alpha^\lambda \lambda y^{\lambda-1}}{(\alpha^\lambda + y^\lambda)^2}, & y \geq 0, \end{cases}$$

or, $G'(y) = g(y)$. The proof has been completed. □

Problem 11.9 Define a mapping ϕ from the variables (x, y) to (u, w) by $x = uw$, $y = (1 - u)w$. Show that $J(\phi) = w$, where J is the Jacobean.

Solution The Jacobean of the multivariable mapping ϕ : $y \rightarrow x$, $x_i = \phi_i(y_1, \ldots, y_n)$, $i = 1, \ldots, n$, is the determinant $J(\phi) = |a_{ij}|$ of the order n with elements defined by partial derivatives: $a_{ij} = \partial \phi_i / \partial y_j$. Therefore, we have

$$J = \begin{vmatrix} \partial x / \partial u & \partial x / \partial w \\ \partial y / \partial u & \partial y / \partial w \end{vmatrix} = \begin{vmatrix} w & u \\ -w & 1 - u \end{vmatrix} = w(1 - u) + uw = w. \qquad \square$$

Problem 11.10 Suppose we pick two random variables X_1 and X_2 in the range $[0, 1]$ and map (x_1, x_2) to the sample point $\cos(2\pi x_1) \sqrt{\ln(1/x_2^2)}$. Prove that this samples correctly from a Gaussian. This is called the Box–Muller method.

Solution The Box–Muller method (Box and Muller, 1958) was designed to generate two independent standard Gaussian random variables from two independent uniformly distributed random variables. It works as follows. Let X_1 and X_2 be independent uniform random variables on $[0, 1]$. We define Y_1 and Y_2 as follows:

$$Y_1 = \sqrt{\ln \frac{1}{(X_2)^2}} \cos(2\pi X_1),$$

$$Y_2 = \sqrt{\ln \frac{1}{(X_2)^2}} \sin(2\pi X_1).$$

To show that Y_1 and Y_2 are independent standard Gaussian random variables, we start with finding the CDF F_R of random variable $R = \sqrt{\ln(1/(X_2)^2)}$. If $x < 0$, then $F_R(x) = P(R \leq x) = 0$. If x is non-negative, then

$$F_R(x) = P(R \leq x) = P(R^2 \leq x^2) = P\left(\ln \frac{1}{(X_2)^2} \leq x^2\right)$$

$$= P\left((X_2)^2 \geq e^{-x^2}\right) = P\left(X_2 \geq e^{-x^2/2}\right) = 1 - F_{X_2}\left(e^{-x^2/2}\right).$$

Here F_{X_2} is the CDF of the uniform distribution on $[0, 1]$. As $e^{-x^2/2}$ belongs to $[0, 1]$, F_R becomes

$$F_R(x) = \begin{cases} 0, & x < 0, \\ 1 - e^{-x^2/2}, & x \geq 0. \end{cases}$$

The associated probability density function is given by

$$f_R(x) = F'_R(x) = \begin{cases} 0, & x < 0, \\ xe^{-x^2/2}, & x \geq 0. \end{cases}$$

To compute the joint density function of a random vector (Y_1, Y_2), we have to determine the map ϕ: $X_1 = \phi_1(Y_1, Y_2)$, $R = \phi_2(Y_1, Y_2)$, which appears to be as follows:

$$X_1 = \frac{1}{2\pi} \arctan \frac{Y_2}{Y_1},$$

$$R = \sqrt{Y_1^2 + Y_2^2}.$$

Then, for the Jacobean of the map ϕ we have

$$J(\phi) = \begin{vmatrix} \dfrac{\partial \phi_1}{\partial y_1} & \dfrac{\partial \phi_1}{\partial y_2} \\[2ex] \dfrac{\partial \phi_2}{\partial y_1} & \dfrac{\partial \phi_2}{\partial y_2} \end{vmatrix} = \begin{vmatrix} -\dfrac{y_2}{2\pi(y_1^2 + y_2^2)} & \dfrac{y_1}{2\pi(y_1^2 + y_2^2)} \\[2ex] \dfrac{y_1}{\sqrt{y_1^2 + y_2^2}} & \dfrac{y_2}{\sqrt{y_1^2 + y_2^2}} \end{vmatrix} = -\frac{1}{2\pi\sqrt{y_1^2 + y_2^2}}.$$

Finally, the joint probability density function of the random vector (Y_1, Y_2) is given by the multivariate version of the transformation formula:

$$g(y_1, y_2) = f_{X_1}(\phi_1(y_1, y_2)) f_R(\phi_2(y_1, y_2)) |J(\phi)|$$
$$= \sqrt{y_1^2 + y_2^2} e^{-(y_1^2 + y_2^2)/2} \frac{1}{2\pi\sqrt{y_1^2 + y_2^2}} = \frac{1}{\sqrt{2\pi}} e^{-y_1^2/2} \frac{1}{\sqrt{2\pi}} e^{-y_2^2/2}$$
$$= g_1(y_1) g_2(y_2).$$

Here g_1, g_2 are the probability density functions of the standard normal distribution. Thus, we have established that Y_1 and Y_2 are independent standard Gaussian random variables. □

Problem 11.11 Show that the formula

$$D(u, v) = \frac{\int_0^\infty \delta(u + v - 1) e^{-uw} (uw)^{\alpha_1 - 1} e^{vw} (vw)^{\alpha_2 - 1} w \, dw}{\Gamma(\alpha_1)\Gamma(\alpha_2)}$$

$$= \frac{u^{\alpha_1 - 1} v^{\alpha_2 - 1} \delta(u + v - 1)}{\Gamma(\alpha_1)\Gamma(\alpha_2)} \int_0^\infty e^{-w} w^{\alpha_1 + \alpha_2 - 1} \, dw$$

$$= u^{\alpha_1 - 1} v^{\alpha_2 - 1} \delta(u + v - 1) \frac{\Gamma(\alpha_1 + \alpha_2)}{\Gamma(\alpha_1)\Gamma(\alpha_2)} = \mathcal{D}(u, v | \alpha_1, \alpha_2)$$

can be extended to the case of K gamma distributions, i.e. that sampling from $g(x, \alpha_i, 1)$, for $i = 1, \ldots, K$, then averaging, is equivalent to sampling from the Dirichlet $\mathcal{D}(\Theta_1, \ldots, \Theta_K | \alpha_1, \ldots, \alpha_2)$.

Solution Suppose we sample the values $x_1, \ldots, x_K$ from the K gamma distributions with parameters $\alpha_1, \ldots, \alpha_K$, respectively, and define K values $u_1, \ldots, u_K$ as $u_i = x_i / \Sigma_i x_i$. Equivalently, we can set $x_i = \phi_i(u_1, \ldots, u_{K-1}, w) = u_i w, i = 1, \ldots, K-1$, $x_K = \phi_K(u_1, \ldots, u_{K-1}, w) = (1 - \sum_{i=1}^{K-1} u_i)w$ and integrate over variable w. The Jacobean of the map $\phi : (u_1, \ldots, u_{K-1}, w) \to (x_1, \ldots, x_K)$ is given by

$$
J(\phi) = \begin{vmatrix}
\dfrac{\partial \phi_1}{\partial u_1} & \dfrac{\partial \phi_1}{\partial u_2} & \dfrac{\partial \phi_1}{\partial u_3} & \cdots & \dfrac{\partial \phi_1}{\partial w} \\[2mm]
\dfrac{\partial \phi_2}{\partial u_1} & \dfrac{\partial \phi_2}{\partial u_2} & \dfrac{\partial \phi_2}{\partial u_3} & \cdots & \dfrac{\partial \phi_2}{\partial w} \\[2mm]
\dfrac{\partial \phi_3}{\partial u_1} & \dfrac{\partial \phi_3}{\partial u_2} & \dfrac{\partial \phi_3}{\partial u_3} & \cdots & \dfrac{\partial \phi_3}{\partial w} \\[2mm]
\cdots & \cdots & \cdots & \cdots & \cdots \\[2mm]
\dfrac{\partial \phi_K}{\partial u_1} & \dfrac{\partial \phi_K}{\partial u_2} & \dfrac{\partial \phi_K}{\partial u_3} & \cdots & \dfrac{\partial \phi_K}{\partial w}
\end{vmatrix}
$$

$$
= \begin{vmatrix}
w & 0 & 0 & \cdots & u_1 \\
0 & w & 0 & \cdots & u_2 \\
0 & 0 & w & \cdots & u_3 \\
\cdots & \cdots & \cdots & \cdots & \cdots \\
-w & -w & -w & \cdots & 1 - \sum_{i=1}^{K} u_i
\end{vmatrix}.
$$

To compute $J(\phi)$, we add to the last row of the determinant all other rows and obtain $J(\phi) = w^{K-1}$. The joint density D of $u_1, \ldots, u_K$ becomes

$$
D(u_1, u_2, \ldots, u_K) = \frac{\int_0^{\infty} \delta(\sum_{i=1}^{K} u_i - 1) \left(\prod_{i=1}^{K} e^{-u_i w}(u_i w)^{\alpha_i - 1} \right) w^{K-1} \, dw}{\prod_{i=1}^{K} \Gamma(\alpha_i)}
$$

$$
= \frac{\delta(\sum_{i=1}^{K} u_i - 1) \prod_{i=1}^{K} u_i^{\alpha_i - 1}}{\prod_{i=1}^{K} \Gamma(\alpha_i)} \int_0^{\infty} e^{-w} w^{\Sigma_{i=1}^{K} \alpha_i - 1} \, dw
$$

$$
= \delta \left(\sum_{i=1}^{K} u_i - 1 \right) \prod_{i=1}^{K} u_i^{\alpha_i - 1} \frac{\Gamma(\sum_{i=1}^{K} \alpha_i)}{\prod_{i=1}^{K} \Gamma(\alpha_i)}.
$$

Since the last expression is the density $\mathcal{D}(u_1, u_2, \ldots, u_K | \alpha_1, \alpha_2, \ldots, \alpha_K)$ of the Dirichlet distribution with parameters $\alpha_1, \ldots, \alpha_K$, the problem has been solved.   □

Problem 11.12 Prove that for the probability density function $g(x, \alpha, 1)$ of a gamma-distribution, $g(x, \alpha, 1) = e^{-x}x^{\alpha-1}/\Gamma(\alpha)$, $x > 0$, and a function

$$f(x) = \frac{4e^{-\alpha}\alpha^{\lambda+\alpha}x^{\lambda-1}}{\Gamma(\alpha)(\alpha^\lambda + x^\lambda)^2},$$

the inequality $g(x, \alpha, 1) \leq f(x)$ holds for all $x > 0$, $\alpha > 1$, and $1 \leq \lambda < \sqrt{2\alpha - 1}$. What happens when $\lambda \geq \sqrt{2\alpha - 1}$?

Solution To prove that $g(x, \alpha, 1) \leq f(x)$ for all positive x and given values of parameters α and λ, we take the ratio $R(x) = f(x)/g(x)$ and check that $R(x) \geq 1$ for all positive x. We write $R(x)$ in the following form:

$$R(x) = \frac{f(x)}{g(x)} = \frac{4}{(\alpha^\lambda + x^\lambda)^2}e^{x-\alpha}\alpha^{\lambda+\alpha}x^{\lambda-\alpha}.$$

Note that if $\lambda - \alpha > 0$, $R(x) \to 0$ as $x \to 0$. Thus, the condition $\lambda \leq \alpha$ is necessary for $R(x) \geq 1$.

To study a behavior of R, we consider

$$\ln R(x) = \ln 4 - 2\ln(\alpha^\lambda + x^\lambda) + (x - \alpha) + (\alpha + \lambda)\ln\alpha + (\lambda - \alpha)\ln x.$$

We calculate the derivative of $\ln R(x)$ as follows:

$$\frac{d\ln R(x)}{dx} = \frac{x^{\lambda+1} + (-\alpha - \lambda)x^\lambda + \alpha^\lambda x + (\lambda - \alpha)\alpha^\lambda}{x(\alpha^\lambda + x^\lambda)}, \tag{11.11}$$

then introduce a new variable t, $x = \alpha t$, and denote (11.11) as $Q(t)$; thus,

$$Q(t) = \frac{\alpha t^{\lambda+1} + (-\alpha - \lambda)t^\lambda + \alpha t + (\lambda - \alpha)}{\alpha t^{\lambda+1} + \alpha t}.$$

The derivative,

$$\frac{dQ(t)}{dt} = \frac{(\alpha + \lambda)t^{2\lambda} + 2(\alpha - \lambda^2)t^\lambda + (\alpha - \lambda)}{\alpha t^2(t^\lambda + 1)^2}, \tag{11.12}$$

is positive for all $t > 0$ if and only if the numerator of (11.12) is positive. Therefore, if $4(\alpha - \lambda^2)^2 - 4(\alpha + \lambda)(\alpha - \lambda) < 0$ (which holds if $\lambda < \sqrt{2\alpha - 1}$), then $dQ(t)/dt > 0$. Since both $Q(t)$ and $t = x/\alpha$ are monotonic increasing functions (for $t > 0$, $x > 0$), their composition $d\ln R(x)/dx$ is also a monotonic increasing function on the interval $(0, +\infty)$. It grows from large negative values (in the vicinity of $x = 0$) to 1 (as x tends to $+\infty$) and therefore has only one root $x = \alpha$. Hence, both $\ln R(x)$ and $R(x)$ have one and the same critical point, $x = \alpha$, which is the common minimum point of these two functions. We have

$$R(x) = \frac{f(x)}{g(x)} \geq \min_{x>0} R(x) = R(\alpha) = 1,$$

and conclude that $f(x) \geq g(x)$ for all positive x, $\alpha > 1$, and $1 \leq \lambda < \sqrt{2\alpha - 1}$.

If $\lambda \geq \sqrt{2\alpha - 1}$, then $dQ(t)/dt = 0$ has either one or two positive solutions. Therefore, there might be not one but several intervals of monotonicity of $Q(t)$ and, consequently, $d \ln R(x)/dx = 0$ might have more than one solution and $R(x) = f(x)/g(x)$ might have more than one minimum point. For example, if $\lambda = \alpha = 1$, then $d \ln R(x)/dx = (x - 1)/(x + 1)$ and $R(x)$ has only one minimum point, $x = 1$. If $\lambda = 2$ and $\alpha = 2.1$, then

$$\frac{d \ln R(x)}{dx} = \frac{x^3 - 4.1x^2 + 4.41x - 0.441}{x(x^2 + 4.41)},$$

and $R(x)$ has three positive critical points: the maximum point $x_1 = (2 + \sqrt{3.16})/2$ and the two minimum points, $x_2 = (2 - \sqrt{3.16})/2$ and $x_3 = 2.1$. Therefore, the arguments considered above for $1 \leq \lambda < \sqrt{2\alpha - 1}$ cannot be applied if $\lambda \geq \sqrt{2\alpha - 1}$; there are many subcases which must be treated separately. $\square$

Problem 11.13 A random vector (X, Y) has the following density function:

$$f(x, y) = 0.5\mathbf{I}_{[0,1] \times [0,1]}(x, y) + 0.5\mathbf{I}_{[0.99,1.99] \times [0.99,1.99]}(x, y),$$

where I_A is the indicator of set A. The Gibbs sampling chooses points from the conditional distributions $P(X|Y), P(Y|X), P(X|Y), P(Y|X), \ldots$. What is the expected number of samples within one region, before a crossover occurs into the other?

Solution First we find the probability density functions of X and Y, and the conditional density functions $f(x|y)$ and $f(y|x)$. Let $g(x)$ designate the pdf of random variable X:

$$g(x) = \int_{-\infty}^{+\infty} f(x, y) \, dy$$

$$= \int_{-\infty}^{+\infty} (0.5\mathbf{I}_{[0,1] \times [0,1]}(x, y) + 0.5\mathbf{I}_{[0.99,1.99] \times [0.99,1.99]}(x, y)) \, dy$$

$$= 0.5\mathbf{I}_{[0,1]}(x) + 0.5\mathbf{I}_{[0.99,1.99]}(x),$$

$$f(y|x) = \frac{f(x, y)}{g(x)} = \frac{\mathbf{I}_{[0,1] \times [0,1]}(x, y) + \mathbf{I}_{[0.99,1.99] \times [0.99,1.99]}(x, y)}{\mathbf{I}_{[0,1]}(x) + \mathbf{I}_{[0.99,1.99]}(x)}.$$

Due to symmetry, the density function of Y is $g(y)$, and the conditional density is given by

$$f(x|y) = \frac{f(x, y)}{g(y)} = \frac{\mathbf{I}_{[0,1] \times [0,1]}(x, y) + \mathbf{I}_{[0.99,1.99] \times [0.99,1.99]}(x, y)}{\mathbf{I}_{[0,1]}(y) + \mathbf{I}_{[0.99,1.99]}(y)}.$$

We designate two squares, the supports of the density $f(x, y)$, as $S_1 = [0, 1] \times [0, 1]$ and $S_2 = [0.99, 1.99] \times [0.99, 1.99]$. Following Casella and George (1992), we generate the 'Gibbs sequence' of random variables $Y_0', X_0', Y_1', X_1', Y_2', X_2', \ldots$ by the following rule. For an initial value $Y_0' = y_0'$ the rest of the sequence is obtained iteratively by generating values from X_j' with distribution $f(x|Y_j' = y_j')$ and subsequent generation of Y_j' with the distribution $f(y|X_j' = x_j')$. Now we define a random sequence $Z = (Z_i)$ as follows:

$$
\begin{aligned}
Z_{2i+1} &= 1, &&\text{if } Y_i' \in [0, 1], \\
Z_{2i+1} &= 2, &&\text{if } Y_i' \in [1, 1.99], \\
Z_{2i} &= 1, &&\text{if } X_i' \in [0, 1], \\
Z_{2i} &= 2, &&\text{if } X_i' \in [1, 1.99].
\end{aligned}
$$

We check that (Z_i) is the Markov chain with two states $\{1, 2\}$ and determine its initial distribution and transition probabilities. For any i,

$$P(Z_{2i+1} = k_{2i+1}|Z_1 = k_1, Z_2 = k_2, \ldots, Z_{2i} = k_{2i})$$

$$= P(Y_i' \in A(k_{2i+1})|Y_0' \in A(k_1), X_0' \in A(k_2), \ldots, X_i' \in A(k_{2i}))$$

$$= P(Y_i' \in A(k_{2i+1})|X_i' \in A(k_{2i})) = P(Z_{2i+1} = k_{2i+1}|Z_{2i} = k_{2i}).$$

Here k_j is either state 1 or state 2, while $A(1) = [0, 1]$, $A(2) = [1, 1.99]$. Similarly, the Markov property holds for all members Z_{2i}. For the initial distribution of the Markov chain Z, we have

$$p_1 = P(Z_1 = 1) = P(Y_0' \in [0, 1]) = \int_0^1 g(y)\,dy$$

$$= \int_0^1 (0.5\mathbf{I}_{[0,1]}(y) + 0.5\mathbf{I}_{[0.99,1.99]}(y))\,dy = 0.5 + 0.5 \times 0.01 = 0.505,$$

$$p_2 = P(Z_1 = 2) = P(Y_0' \in [1, 1.99]) = 1 - p_1 = 0.495;$$

and for transition probabilities

$$p_{11} = P(Z_2 = 1|Z_1 = 1) = P(X_0' \in [0, 1]|Y_0' \in [0, 1])$$

$$= \int_0^1 P(X_0' \in [0, 1]|Y_0' = y_0)\,dy_0 = \int_0^1 \int_0^1 f(x|y)\,dx\,dy$$

$$= \int_0^1 \int_0^1 \frac{\mathbf{I}_{[0,1]\times[0,1]}(x, y) + \mathbf{I}_{[0.99,1.99]\times[0.99,1.99]}(x, y)}{\mathbf{I}_{[0,1]}(y) + \mathbf{I}_{[0.99,1.99]}(y)}\,dx\,dy$$

$$= \int_0^1 \frac{1}{\mathbf{I}_{[0,1]}(y) + \mathbf{I}_{[0.99,1.99]}(y)}(\mathbf{I}_{[0,1]}(y) + 0.01\mathbf{I}_{[0.99,1.99]}(y))\,dy$$

$$= 0.99 + \frac{1.01}{2}0.01 = 0.99505,$$

$$p_{12} = P(Z_2 = 2|Z_1 = 1) = P(X_0' \in [0.99, 1.99]|Y_0' \in [0, 1])$$

$$= 1 - p_{11} = 0.00495.$$

Let ξ be the number of points (X'_j, Y'_j) of the 'Gibbs sequence,' such that $(X'_j, Y'_j) \in S_1$, before the first transition to region S_2/S_1 occurs $(S_2/S_1 = S_2 - (S_2 \cap S_1))$. To find the distribution of random variable ξ with values $0, 1, 2, \ldots$, we use the Markov chain Z:

$$P(\xi = n)$$
$$= P(Y'_0 \in [0,1], X'_0 \in [0,1], \ldots, Y'_n \in [0,1], X'_n \in [0,1], (Y'_{n+1}, X'_{n+1}) \in S_2/S_1)$$
$$= P(Z_1 = 1, Z_2 = 1, \ldots, Z_{2n-1} = 1, Z_{2n} = 1, Z_{2n+1} = 2)$$
$$\quad + P(Z_1 = 1, Z_2 = 1, \ldots, Z_{2n-1} = 1, Z_{2n} = 1, Z_{2n+1} = 1, Z_{2n+2} = 2)$$
$$= p_1 p_{11}^{2n-1} p_{12} + p_1 p_{11}^{2n} p_{12} = p_1 p_{11}^{2n-1} p_{12}(1 + p_{11}).$$

Then we find the expected value of ξ as follows:

$$E\xi = \sum_{n=0}^{+\infty} n p_1 p_{11}^{2n-1} p_{12}(1 + p_{11}) = p_1 p_{12} p_{11}(1 + p_{11}) \sum_{n=1}^{+\infty} n (p_{11}^2)^{n-1}$$
$$= \frac{p_1 p_{12} p_{11}(1 + p_{11})}{(1 - p_{11}^2)^2} = \frac{p_1 p_{11}}{p_{12}(1 + p_{11})} = \frac{0.505 \times 0.99505}{0.00495 \times 1.99505} \approx 50.8835.$$

Due to symmetry, the expected number of points sampled within square S_2 before a transition to S_1 is also equal to $E\xi \approx 50.8835$. Hence, on average, fifty-one sample points in a row are situated within one of the squares S_1 or S_2 before the first transition to the other square. □

11.2 Additional problem

Problem 11.14 is a combinatorial exercise. We want to prove an auxiliary statement used in Remark 2 (Problem 2.5) to derive for sequences of lengths n and m the number $\mathcal{A}_{n,m}$ of pairwise alignments in which the x-gap does not follow the y-gap and vice versa.

Problem 11.14 It was proved (Shiryaev, 1996, Sect. 1, Ex. 6) that the number of ways to place k undistinguishable items into i boxes is equal to the binomial coefficient $\binom{i+k-1}{k}$. Show that this number becomes $\binom{k-1}{k-i}$ under the additional restriction that no box remains empty.

Solution Let $N(k, i)$ be the number of ways to place k undistinguishable items into i boxes provided that each box contains at least one item. Obviously, if $k < i$, then $N(k, i) = 0$.

For $k \geq i$ we will prove the formula $N(k, i) = \binom{k-1}{k-i}$ by using mathematical induction. It is easy to see that $N(m, 1) = 1 = \binom{m-1}{m-1}$ and $N(m, 2) = m - 1 = \binom{m-1}{m-2}$

for all m, $m \le k$. Next, we suppose that $N(m, i) = \binom{m-1}{m-i}$, $m \le k$, and check that $N(k, i + 1) = \binom{k-1}{k-(i+1)}$. The number $N(k, i + 1)$ is equal to the number of vectors $x = (x_1, \ldots, x_{i+1})$ of length $i + 1$ with positive integer components, such that $x_1 + x_2 + \cdots + x_{i+1} = k$ (component x_l denotes the number of items in the lth box, $l = 1, \ldots, i + 1$). Obviously, x_{i+1} can take integer values from 1 to $k - (i - 1) = k - i + 1$. If $x_{i+1} = 1$, then the sum of the first i components of x must be equal to $k - 1$; thus, the number of possible vectors x is $N(k - 1, i)$. If $x_{i+1} = 2$, the sum of the first i components of x must be equal to $k - 2$ and the number of possible vectors x is $N(k - 2, i)$. Finally, for the maximal possible number of items $x_{i+1} = k - i + 1$ in the $(i + 1)$th box there is the only one way, $N(i, i) = 1$, to choose the remaining components: $x_1 = 1$, $x_2 = 1, \ldots, x_i = 1$. Hence, we arrive at the recurrent formula:

$$N(k, i + 1) = N(k - 1, i) + N(k - 2, i) + \cdots + N(i + 1, i) + N(i, i). \quad (11.13)$$

Next, to every term but the last one on the right-hand side of Equation (11.13) we apply the induction assumption. By using the property of the binomial coefficients, $\binom{m}{l} = \binom{m+1}{l} - \binom{m}{l-1}$, we derive

$$
\begin{aligned}
N(k, i + 1) &= \binom{k-2}{k-1-i} + \binom{k-3}{k-2-i} + \cdots + \binom{i}{1} + \binom{i-1}{0} \\
&= \left(\binom{k-1}{k-1-i} - \binom{k-2}{k-2-i} \right) + \left(\binom{k-2}{k-2-i} - \binom{k-3}{k-3-i} \right) \\
&\quad + \cdots + \left(\binom{i+1}{1} - \binom{i}{0} \right) + \binom{i-1}{0}.
\end{aligned}
\quad (11.14)
$$

On the right-hand side of Equation (11.14) the sums of the second and the third binomial coefficients, the fourth and the fifth, etc., and the two last ones are all equal to zero. Therefore, we obtain the equality

$$N(k, i + 1) = \binom{k-1}{k-1-i} = \binom{k-1}{k-(i+1)},$$

which is equivalent to the required one. The proof is complete. $\qquad \square$

11.3 Further reading

Several references to the popular textbooks on the theory of probability and statistics have already been provided in the introduction to Chapter 1. The following are references on information theory: Shannon and Weaver (1963), Ash (1990), Cover and Thomas (1991), Reza (1994), MacKay (2003), and Goldman (2005).

References

Almagor, H. (1983). A Markov analysis of DNA sequences. *Journal of Theoretical Biology* **104**, 633–645.

Altschul, S. F. (1991). Amino acid substitution matrices from an information theoretic perspective. *Journal of Molecular Biology* **219**, 555–565.

Altschul, S. F. and Gish, W. (1996). Local alignment statistics. *Methods in Enzymology* **266**, 460–480.

Altschul, S. F. and Koonin, E. V. (1998). Iterated profile searches with PSI-BLAST – a tool for discovery in protein databases. *Trends in Biochemical Sciences* **23**, 444–447.

Altschul, S. F., Carroll, R. J., and Lipman, D. J. (1989). Weights for data related by a tree. *Journal of Molecular Biology* **207**, 647–653.

Altschul, S. F., Gish, W., Miller, W., Myers, E. W., and Lipman, D. J. (1990), Basic local alignment search tool. *Journal of Molecular Biology* **215**, 403–410.

Altschul, S. F., Madden, T. L., Schäffer, A. A., Zhang, J., Zhang, Z., Miller, W., and Lipman, D. J. (1997). Gapped BLAST and PSI-BLAST: A new generation of protein database search programs. *Nucleic Acids Research* **25**, 3389–3402.

Altschul, S. F., Bundschuh, R., Olsen, R., and Hwa, T. (2001). The estimation of statistical parameters for local alignment score distributions. *Nucleic Acids Research* **29**, 351–361.

Arbogast, B. S., Edwards, S. V., Wakeley, J., Beerli, P., and Slowinski, J. B. (2002). Estimating divergence times from molecular data on phylogenetic and population genetic time scales. *Annual Review of Ecology and Systematics* **33**, 707–740.

Arnason, U., Gullberg, A., and Janke, A. (1998). Molecular timing of primate divergences as estimated by two nonprimate calibration points. *Journal of Molecular Evolution* **47**, 718–727.

Arratia, R. and Waterman, M. S. (1985). Critical phenomena in sequence matching. *The Annals of Probability* **13**, 1236–1249.

Arratia, R., Gordon, L., and Waterman, M. (1986). An extreme value theory for sequence matching. *The Annals of Statistics* **14**, 971–993.

Ash, R. B. (1990). *Information Theory* (New York: Dover Publications).

Ayala, F. J., Rzhetsky, A., and Ayala, F. J. (1998). Origin of the metazoan phyla: Molecular clocks confirm paleontological estimates. *Proceedings of the National Academy of Sciences of the USA* **95**, 606–611.

Bailey, T. L. and Gribskov, M. (2002). Estimating and evaluating the statistics of gapped local-alignment scores. *Journal of Computational Biology* **9**, 575–593.

Bairoch, A. and Apweiler, R. (1999). The SWISS-PROT protein sequence data bank and its supplement TrEMBL in 1999. *Nucleic Acids Research* **27**, 49–54.

Bairoch, A., Bucher, P., and Hofmann, K. (1997). The PROSITE database, its status in 1997. *Nucleic Acids Research* **25**, 217–221.

Baldi, P. and Brunak, S. (2001). *Bioinformatics: The Machine Learning Approach*, 2nd edn (Cambridge, MA: The MIT Press).

Bateman, A., Birney, E., Cerruti, L. *et al.* (2002). The Pfam protein families database. *Nucleic Acids Research* **30**, 276–280.

Baum, L. E. (1972). An equality and associated maximization technique in statistical estimation for probabilistic functions of Markov processes. *Inequalities* **3**, 1–8.

Berman, A. and Plemmons, R. J. (1979). *Nonnegative Matrices in the Mathematical Sciences* (New York: Academic Press).

Billingsley, P. (1961a). *Statistical Inference in Markov Processes* (Chicago, IL: The University of Chicago Press).

Billingsley, P. (1961b). Statistical methods in Markov chains. *The Annals of Mathematical Statistics* **32**, 12–40.

Borodovsky, M. and McIninch, J. (1993). GeneMark: Parallel gene recognition for both DNA strands. *Computers & Chemistry* **17**, 123–133.

Borodovsky, M. and Peresetsky, A. (1994). Deriving non-homogeneous DNA Markov chain models by cluster analysis algorithm minimizing multiple alignment entropy. *Computers & Chemistry* **18**, 259–267.

Borodovsky, M. Y., Sprizhitsky, Y. A., Golovanov, E. I., and Alexandrov, A. A. (1986a). Statistical patterns in the primary structure of the functional regions of the Escherichia coli genome. I. Frequency characteristics. *Molecularnaya Biologia* **20**, 826–833 (English translation).

Borodovsky, M. Y., Sprizhitsky, Y. A., Golovanov, E. I., and Alexandrov, A. A. (1986b). Statistical patterns in the primary structure of the functional regions of the Escherichia coli genome. II. Nonuniform Markov models. *Molecularnaya Biologia* **20**, 833–840 (English translation).

Borodovsky, M. Y., Sprizhitsky, Y. A., Golovanov, E. I., and Alexandrov, A. A. (1986c). Statistical patterns in the primary structure of the functional regions of the Escherichia coli genome. III. Computer recognition of coding regions. *Molecularnaya Biologia* **20**, 1144–1150 (English translation).

Box, G. E. P. and Muller, M. E. (1958). A note on the generation of random normal deviates. *The Annals of Mathematical Statistics* **29**, 610–611.

Braun, J. V. and Müller, H-G. (1998). Statistical methods for DNA sequence segmentation. *Statistical Science* **13**, 142–162.

Brenner, S. E., Chothia, C. and Hubbard, T. J. P. (1998). Assessing sequence comparison methods with reliable structurally identified distant evolutionary relationships. *Proceedings of the National Academy of Sciences of the USA* **95**, 6073–6078.

Brocchieri, L. and Karlin, S. (1998). A symmetric-iterated multiple alignment of protein sequences. *Journal of Molecular Biology* **276**, 249–264.

Bromham, L. and Penny, D. (2003). The modern molecular clock. *Nature Reviews Genetics* **4**, 216–224.

Brown, J. R., Douady, C. J., Italia, M. J., Marshall, W. E., and Stanhope, M. J. (2001). Universal trees based on large combined protein sequence data sets. *Nature Genetics* **28**, 281–285.

Brown, J. W. (1999a). The ribonuclease P database. *Nucleic Acids Research* **27**, 314.

Brown, T. A. (1999b). *Genomes* (New York: John Wiley & Sons, Inc.).

Brudno, M., Do, C. B., Cooper, G. M. *et al.* (2003). LAGAN and Multi-LAGAN: Efficient tools for large-scale multiple alignment of genomic DNA. *Genome Research* **13**, 721–731.

Bruno, W. J., Socci, N. D., and Halpern, A. L. (2000). Weighted neighbor joining: A likelihood-based approach to distance-based phylogeny reconstruction. *Molecular Biology and Evolution* **17**, 189–197.

Burge, C. and Karlin, S. (1997). Prediction of complete gene structures in human genomic DNA. *Journal of Molecular Biology* **268**, 78–94.

Bystroff, C., Thorsson, V., and Baker, D. (2000). HMMSTR: A hidden Markov model for local sequence-structure correlations in proteins. *Journal of Molecular Biology* **301**, 173–190.

Cannone, J. J., Subramanian, S., Schnare, M. N. *et al.* (2002). The comparative RNA web (CRW) site: An online database of comparative sequence and structure information for ribosomal, intron, and other RNAs. *BMC Bioinformatics* **3**, 2–33.

Carrillo, H. and Lipman, D. (1988). The multiple sequence alignment problem in biology. *SIAM Journal of Applied Mathematics* **48**, 1073–1082.

Carroll, J. and Long, D. (1989). *Theory of Finite Automata: With an Introduction to Formal Languages* (Englewood Cliffs, N.J.: Prentice Hall).

Casella, G. and Berger, R. L. (2001). *Statistical Inference*, 2nd edn (Pacific Grove, CA: Duxbury Press).

Casella, G. and George, E. I. (1992). Explaining the Gibbs sampler. *The American Statistician* **46**, 167–174.

Cavender, J. A. and Felsenstein, J. (1987). Invariants of phylogenies in a simple case with discrete states. *Journal of Classification* **4**, 57–71.

Cayley, A. (1889). A theorem on trees. *Quarterly Journal of Mathematics* **23**, 376–378.

Chenna, R., Sugawara, H., Koike, T. *et al.* (2003). Multiple sequence alignment with the Clustal series of programs. *Nucleic Acids Research* **31**, 3497–3500.

Churchill, G. A. (1989). Stochastic models for heterogeneous DNA sequences. *Bulletin of Mathematical Biology* **51**, 79–94.

Clamp, M., Andrews, D., Barker, D. *et al.* (2003). Ensembl 2002: Accommodating comparative genomics. *Nucleic Acids Research* **31**, 38–42.

Clark, A. G., Glanowski, S., Nielson, R. *et al.* (2003). Inferring nonneutral evolution from human-chimp-mouse orthologous gene trios. *Science* **302**, 1960–1963.

Cocke, J. and Schwartz, J. T. (1970). *Programming Languages and their Compilers: Preliminary Notes.* Technical report, Courant Institute of Mathematical Sciences, New York University.

Coin, L. and Durbin, R. (2004). Improved techniques for the identification of pseudogenes. *Bioinformatics* **20** (suppl. 1), i94–i100.

Cover, T. M. and Thomas, J. A. (1991). *Elements of Information Theory* (New York: Wiley).

Cowan, R. (1991). Expected frequencies of DNA patterns using Whittle's formula. *Journal of Applied Probability* **28**, 886–892.

Cox, D. R. and Hinkley, D. V. (1974). *Theoretical Statistics* (London: Chapman and Hall).

Dayhoff, M. O., Schwartz, R. M., and Orcutt, B. C. (1978). A model of evolutionary change in proteins. In Dayhoff, M. O., ed., *Atlas of Protein Sequence and Structure*, vol. 5, supplement 3 (Washington, D.C.: National Biomedical Research Foundation), pp. 345–352.

Delcher, A. L., Kasif, S., Fleischmann, R. D., Peterson, J., White, O., and Salzberg, S. L. (1999). Alignment of whole genomes. *Nucleic Acids Research* **27**, 2369–2376.

Dembo, A., Karlin, S., and Zeitouni, O. (1994a). Critical phenomena for sequence matching with scoring. *The Annals of Probability* **22**, 1993–2021.

Dembo, A., Karlin, S., and Zeitouni, O. (1994b). Limit distribution of maximal non-aligned two-sequence segmental score. *The Annals of Probability* **22**, 2022–2039.

Dempster, A. P., Laird, N. M., and Rubin, D. B. (1977). Maximum likelihood from incomplete data via the EM Algorithm. *Journal of the Royal Statistical Society* **39**, 1–38.

Deonier, R., Tavaré, S., and Waterman, M. (2005). *Computational Genome Analysis: An Introduction* (New York: Springer-Verlag).

Ding, Y. and Lawrence, C. E. (2003). A statistical sampling algorithm for RNA secondary structure prediction. *Nucleic Acids Research* **31**, 7280–7301.

Do, C. B., Mahabhashyam, M. S. P., Brudno, M., and Batzoglou, S. (2005). ProbCons: Probabilistic consistency-based multiple sequence alignment. *Genome Research* **15**, 330–340.

Douady, C. J., Delsuc, F., Boucher, Y., Doolittle, W. F., and Douzery, E. J. P. (2003). Comparison of Bayesian and maximum likelihood bootstrap measures of phylogenetic reliability. *Molecular Biology and Evolution* **20**, 248–254.

Dowell, R. D. and Eddy, S. R. (2004). Evaluation of several lightweight stochastic context-free grammars for RNA secondary structure prediction. *BMC Bioinformatics* **5**, 71–85.

Durbin, R., Eddy, S., Krogh, A., and Mitchison, G. (1998). *Biological Sequence Analysis: Probabilistic Models of Proteins and Nucleic Acids* (Cambridge: Cambridge University Press).

Eddy, S. R. (1998). Profile hidden Markov models. *Bioinformatics* **14**, 755–763.

Eddy, S. R. (2002). A memory-efficient dynamic programming algorithm for optimal alignment of a sequence to an RNA secondary structure. *BMC Bioinformatics* **3**, 18–34.

Eddy, S. R., Mitchison, G., and Durbin, R. (1995). Maximum discrimination hidden Markov models of sequence consensus. *Journal of Computational Biology* **2**, 9–23.

Edwards, A. W. F. (1970). Estimation of the branch points of a branching diffusion process. *Journal of the Royal Statistical Society B* **32**, 155–174.

Ekisheva, S. and Borodovsky, M. (2006). Probabilistic models for biological sequences: Selection and maximum likelihood estimation. *International Journal of Bioinformatics Research and Applications* **2** (3).

Enright, A. J., Iliopoulos, I., Kyrpides, N. C., and Ouzounis, C. A. (1999). Protein interaction maps for complete genomes based on gene fusion events. *Nature* **402**, 86–90.

Erdös, P. and Revesz, P. (1975). On the length of the longest head-run; Topics in Information Theory. *Colloqia Math. Soc. J. Bolyai* **16**, 219–228.

Evans, S. N. and Speed, T. P. (1993). Invariants of some probability models used in phylogenetic inference. *The Annals of Statistics* **21**, 355–377.

Ewens, W. J. and Grant, G. R. (2001). *Statistical Methods in Bioinformatics: An Introduction* (New York: Springer-Verlag).

Feller, W. (1971). *An Introduction to Probability Theory and its Applications,* 2nd edn, Vols I, II (New York: John Wiley & Sons).

Felsenstein, J. (1981). Evolutionary trees from DNA sequences: A maximum likelihood approach. *Journal of Molecular Evolution* **17**, 368–376.

Felsenstein, J. (2004). *Inferring Phylogenies* (Sunderland, MA: Sinauer Associates, Inc.).

Feng, D-F. and Doolittle, R. F. (1987). Progressive sequence alignment as a prerequisite to correct phylogenetic trees. *Journal of Molecular Evolution* **25**, 351–360.

Feng, D-F. and Doolittle, R. F. (1996). Progressive alignment of amino acid sequences and construction of phylogenetic trees from them. *Methods in Enzymology* **266**, 368–382.

Feng, D-F., Cho, G., and Doolittle, R. F. (1997). Determining divergence times with a protein clock: Update and reevaluation. *Proceedings of the National Academy of Sciences of the USA* **94**, 13 028–13 033.

Fitch, W. M. (1971). Toward defining the course of evolution: Minimum change for specific tree topology. *Systematic Zoology* **20**, 406–416.

Fitch, W. M. (1983). Calculating the expected frequencies of potential secondary structure in nucleic acids as a function of stem length, loop size, base composition and nearest-neighbor frequencies. *Nucleic Acids Research* **11**, 4655–4663.

Fitch, W. M. and Margoliash, E. (1967). Construction of phylogenetic trees. *Science* **155**, 279–284.

Fitz-Gibbon, S. T. and House, C. H. (1999). Whole genome-based phylogenetic analysis of free-living microorganisms. *Nucleic Acids Research* **27**, 4218–4222.

Florea, L., Hartzell, G., Zhang, Z., Rubin, G. M., and Miller, W. (1998). A computer program for aligning a cDNA sequence with a genomic DNA sequence. *Genome Research* **8**, 967–974.

Freedman, D. (1983). *Markov Chains* (New York: Springer-Verlag).

Frith, M. C., Hansen, U., and Weng, Z. (2001). Detection of *cis*-element clusters in higher eukaryotic DNA. *Bioinformatics* **17**, 878–889.

Gabow, H. W. (1973). Implementations of algorithms for maximum matching on nonbipartite graphs. Ph.D. Dissertation, Department of Computer Science, Stanford University.

Gascuel, O. (1997). BIONJ: An improved version of the NJ algorithm based on a simple model of sequence data. *Molecular Biology and Evolution* **14**, 685–695.

Gatlin, L. L. (1972). *Information Theory and the Living System* (New York: Columbia University Press).

George, D. G., Barker, W. C., and Hunt, L. T. (1990). Mutation data matrices and its uses. *Methods in Enzymology* **183**, 333–353.

Gerstein, M., Sonnhammer, E. L. L., and Chothia, C. (1994). Volume changes in protein evolution. *Journal of Molecular Biology* **236**, 1067–1078.

Glazko, G. V. and Nei, M. (2003). Estimation of divergence times for major lineages of primate species. *Journal of Molecular Evolution* **20**, 424–434.

Goldman, N. and Yang, Z. (1994). A codon-based model of nucleotide substitution for protein-coding DNA sequences. *Molecular Biology and Evolution* **11**, 725–736.

Goldman, N., Thorne, J. L., and Jones, D. T. (1996). Using evolutionary trees in protein secondary structure prediction and other comparative sequence analyses. *Journal of Molecular Biology* **263**, 196–208.

Goldman, N., Anderson, J. P., and Rodrigo, A. G. (2000). Likelihood-based tests of topologies in phylogenetics. *Systematic Biology* **49**, 652–670.

Goldman, S. (2005). *Information Theory* (New York: Dover Publications).

Goodman, L. A. (1959). On some statistical tests for M-th order Markov chains. *The Annals of Mathematical Statistics* **30**, 154–164.

Gopalakrishnan, G. (2006). *Computational Engineering: Applied Automata Theory and Logic* (New York: Springer).

Gorodkin, J., Stricklin, S. L., and Stormo, G. D. (2001). Discovering common stem-loop motifs in unaligned RNA sequences. *Nucleic Acids Research* **29**, 2135–2144.

Gough, J., Karplus, K., Hughey, R., and Chothia, C. (2001). Assignment of homology: to genome sequences using a library of hidden Markov models that represent all proteins of known structure. *Journal of Molecular Biology* **313**, 903–919.

Griffiths-Jones, S., Bateman, A., Marshall, M., Khanna, A., and Eddy, S. R. (2003). Rfam: An RNA family database. *Nucleic Acids Research* **31**, 439–441.

Grishin, N. V. (1995). Estimation of the number of amino acid substitutions per site when the substitution rate varies among sites. *Journal of Molecular Evolution* **41**, 675–679.

Grishin, N. V. (1999). A novel approach to phylogeny reconstruction from protein sequences. *Journal of Molecular Evolution* **48**, 264–273.

Grishin, N. V., Wolf, Y. I., and Koonin, E. V. (2000). From complete genomes to measures of substitution rate variability within and between proteins. *Genome Research* **10**, 991–1000.

Grishin, V. N. and Grishin, N. V. (2002). Euclidian space and grouping of biological objects. *Bioinformatics* **18**, 1523–1533.

Grossman, S. and Yakir, B. (2004). Large deviations for global maxima of independent superadditive processes with negative drift and an application to optimal sequence alignments. *Bernoulli* **10**, 829–845.

Gubser, C., Hué, S., Kellam, P., and Smith, G. L. (2004). Poxvirus genomes: A phylogenetic analysis. *Journal of General Virology* **85**, 105–117.

Guindon, S. and Gascuel, O. (2003). A simple, fast, and accurate algorithm to estimate large phylogenies by maximum likelihood. *Systematic Biology* **52**, 696–704.

Hein, J. (1989). A new method that simultaneously aligns and reconstructs ancestral sequences for any number of homologous sequences, when the phylogeny is given. *Molecular Biology and Evolution* **6**, 649–668.

Hein, J., Wiuf, C., Knudsen, B., Møller, M. B., and Wibling, G. (2000). Statistical alignment: Computational properties, homology testing and goodness-of-fit. *Journal of Molecular Biology* **302**, 265–279.

Hendy, M. D., Penny, D., and Steel, M. A. (1994). A discrete Fourier analysis for evolutionary trees. *Proceedings of the National Academy of Sciences of the USA* **91**, 3339–3343.

Henikoff, J. G. and Henikoff, S. (1996). Using substitution probabilities to improve position-specific scoring matrices. *Computer Applications in the Biosciences* **12**, 135–143.

Henikoff, S. and Henikoff, J. G. (1992). Amino acid substitution matrices from protein blocks. *Proceedings of the National Academy of Sciences of the USA* **89**, 10 915–10 919.

Henikoff, S. and Henikoff, J. G. (1994). Position-based sequence weights. *Journal of Molecular Biology* **243**, 574–578.

Hirschberg, D. S. (1975). A linear space algorithm for computing maximal common subsequences. *Communications of the ACM* **18**, 341–343.

Hofacker, I. L. (2003). Vienna RNA secondary structure server. *Nucleic Acids Research* **31**, 3429–3431.

Hogg, R. V. and Craig, A. T. (1994). *Introduction to Mathematical Statistics*, 5th edn (Upper Saddle River, N.J.: Prentice Hall).

Hogg, R. V. and Tanis, E. A. (2005). *Probability and Statistical Inference*, 7th edn (Upper Saddle River, N.J.: Prentice Hall).

Holder, M. and Lewis, P. O. (2003). Phylogeny estimation: Traditional and Bayesian approaches. *Nature Reviews Genetics* **4**, 275–284.

Holmes, I. and Bruno, W. (2001). Evolutionary HMMs: A Bayesian approach to multiple alignment. *Bioinformatics* **17**, 803–820.

Holmes, I. and Rubin, G. M. (2002). Pairwise RNA structure comparison with stochastic context-free grammars. *Pacific Symposium on Biocomputing 2002* (Singapore: World Scientific), pp. 163–174.

Hourai, Y., Akutsu, T., and Akiyama, Y. (2004). Optimizing substitution matrices by separating score distributions. *Bioinformatics* **20**, 863–873.

Hubbard, T., Barker, D., Birney, E. *et al.* (2002). Ensembl genome database project. *Nucleic Acids Research* **30**, 38–41.

Hulo, N., Sigrist, C. J. A., Le Saux, V. *et al.* (2004). Recent improvements to the PROSITE database. *Nucleic Acids Research* **32** (Database issue), D134–D137.

Huynen, M. A. and Bork, P. (1998). Measuring genome evolution. *Proceedings of the National Academy of Sciences of the USA* **95**, 5849–5856.

Iglehart, D. L. (1972). Extreme values in the GI/G/1 queue. *Annals of Mathematical Statistics* **43**, 627–635.

Ito, M. (2004). *Algebraic Theory of Automata & Languages* (Singapore: World Scientific).

Jones, D. T., Taylor, W. R., and Thornton, J. M. (1992). The rapid generation of mutation data matrices from protein sequences. *Computer Applications in Biosciences* **8**, 275–282.

Jones, N. C. and Pevzner, P. A. (2004). *An Introduction to Bioinformatics Algorithms* (Cambridge, MA: The MIT Press).

Juan, V. and Wilson. C. (1999). RNA secondary structure prediction based on free energy and phylogenetic analysis. *Journal of Molecular Biology* **289**, 935–947.

Jukes, T. H. and Cantor, C. (1969). Evolution of protein molecules. In Munro, H. N. and Allison, J. B., eds, *Mammalian Protein Metabolism* (New York: Academic Press), pp. 21–132.

Kanehisa, M., Goto, S., Kawashima, S., and Nakaya, A. (2002). The KEGG databases at GenomeNet. *Nucleic Acids Research* **30**, 42–46.

Kann, M., Qian, B., and Goldstein, R. A. (2000). Optimization of a new score function for the detection of remote homologs. *Proteins: Structure, Function, and Genetics* **41**, 498–503.

Karlin, S. (2005). Statistical signals in bioinformatics. *Proceedings of the National Academy of Sciences of the USA* **102**, 13 355–13 362.

Karlin, S. and Altschul, S. F. (1990). Methods for assessing the statistical significance of molecular sequence features by using general scoring schemes. *Proceedings of the National Academy of Sciences of the USA* **87**, 2264–2268.

Karlin, S. and Altschul, S. F. (1993). Applications and statistics for multiple high-scoring segments in molecular sequences. *Proceedings of the National Academy of Sciences of the USA* **90**, 5873–5877.

Karlin, S. and Brendel, V. (1992). Chance and statistical significance in protein and DNA sequence analysis. *Science* **257**, 39–49.

Karlin, S. and Dembo, A. (1992). Limit distributions of maximal segmental score among Markov-dependent partial sums. *Advances in Applied Probability* **24**, 113–140.

Karlin, S. and Ghandour, G. (1985). Comparative statistics for DNA and protein sequences: Single sequence analysis. *Proceedings of the National Academy of Sciences of the USA* **82**, 5800–5804.

Karlin, S. and Macken, C. (1991). Assessment of inhomogeneities in an *E.Coli* physical map. *Nucleic Acids Research* **19**, 4241–4246.

Karlin, S. and Ost, F. (1987). Counts of long aligned word matches among random letter sequences. *Advances in Applied Probability* **19**, 293–351.

Karlin, S. and Ost, F. (1988). Maximal length of common words among random letter sequences. *The Annals of Probability* **16**, 535–563.

Karlin, S., Dembo, A., and Kawabata, T. (1990). Statistical composition of high-scoring segments from molecular sequences. *The Annals of Statistics* **18**, 571–581.

Karlin, S., Burge, C., and Campbell, A. M. (1992). Statistical analyses of counts and distributions of restriction sites in DNA sequences. *Nucleic Acids Research* **20**, 1363–1370.

Karplus, K., Barrett, C., and Hughey, R. (1998). Hidden Markov models for detecting remote protein homologies. *Bioinformatics* **14**, 846–856.

Kasami, T. (1965). An efficient recognition and syntax algorithm for context-free algorithms. Technical Report AFCRL-65-758, Air Force Cambridge Research Laboratory Bedford, MA.

Kelley, L. A., MacCallum, R. M., and Sternberg, M. J. E. (2000). Enhanced genome annotation using structural profiles in the program 3D-PSSM. *Journal of Molecular Biology* **299**, 499–520.

Kent, W. J. (2002). BLAT – the BLAST-like alignment tool. *Genome Research* **12**, 656–664.

Khoussainov, B. and Nerode, A. (2001). *Automata Theory and its Applications* (Boston, MA: Birkhauser).

Kimura, M. (1980). A simple method for estimating evolutionary rates of base substitutions through comparative studies of nucleotide sequences. *Journal of Molecular Evolution* **16**, 111–120.

Kimura, M. (1983). *The Neutral Theory of Molecular Evolution* (Cambridge: Cambridge University Press).

Kishino, H. and Hasegawa, M. (1989). Evaluation of the maximum likelihood estimate of the evolutionary tree topologies from DNA sequence data, and the branching order in Hominoidea. *Journal of Molecular Evolution* **29**, 170–179.

Kleffe, J. and Borodovsky, M. (1992). First and second moment of counts of words in random texts generated by Markov chains. *Computer Applications in Biosciences* **8**, 433–441.

Knudsen, B. (2003). Optimal multiple parsimony alignment with affine gap cost using a phylogenetic tree. In Benson, G. and Page, R., eds, *Proceedings of Algorithms in Bioinformatics, Third International Workshop,* Lecture Notes in Computer Science 2812 (Berlin: Springer), pp. 433–446.

Knudsen, B. and Hein, J. (1999). RNA secondary structure prediction using stochastic context-free grammars and evolutionary history. *Bioinformatics* **15**, 446–454.

Knudsen, B. and Hein, J. (2003). Pfold: RNA secondary structure prediction using stochastic context-free grammars. *Nucleic Acids Research* **31**, 3423–3428.

Knudsen, B. and Miyamoto, M. M. (2003). Sequence alignment and pair hidden Markov models using evolutionary history. *Journal of Molecular Biology* **333**, 453–460.

Koonin, E. V. and Galperin, M. Y. (2003). *Sequence – Evolution – Function: Computational Approaches in Comparative Genomics* (Norwell, MA: Kluwer Academic Publishers).

Korber, B., Muldoon, M., Theiler, J. *et al.* (2000). Timing the ancestor of the HIV-1 pandemic strains. *Science* **288**, 1789–1796.

Kozen, D. C. (1999). *Automata and Computability* (New York: Springer).

Krogh, A., Larsson, B., Heijne, G. von, and Sonnhammer, E. L. L. (2001). Predicting transmembrane protein topology with a hidden Markov model: Application to complete genomes. *Journal of Molecular Biology* **305**, 567–580.

Krogh, A., Mian, I. S., and Haussler, D. (1994). A hidden Markov model that finds genes in *E. coli* DNA. *Nucleic Acids Research* **22**, 4768–4778.

Krogh, A. and Mitchison, G. (1995). Maximum entropy weighting of aligned sequences of proteins or DNA. In Rawlings, C., Clark, D., Altman, R., Hunter, L., Lengauer, T., and Wodak, S., eds *Proceedings of the Third International Conference on Intelligent Systems for Molecular Biology* (Menlo Park, CA: AAAI Press), pp. 215–221.

Kullback, S., Kupperman, M., and Ku, H. H. (1962). Tests for contingency tables and Markov chains. *Technometrics* **4**, 573–608.

Kumar, S. and Hedges, S. B. (1998). A molecular timescale for vertebrate evolution. *Nature* **392**, 917–920.

Kumar, S., Tamura, K., and Nei, M. (1993). *Manual for MEGA: Molecular Evolutionary Genetics Analysis Software* (Philadelphia, PA: Pennsylvania State University).

Kumar, S., Tamura, K., and Nei, M. (2004). MEGA3: Integrated software for Molecular Evolutionary Genetics Analysis and sequence alignment. *Briefings in Bioinformatics* **5**, 150–163.

Lake, J. A. (1987). A rate-independent technique for analysis of nucleic acid sequences: Evolutionary parsimony. *Molecular Biology and Evolution* **4**, 167–191.

Larget, B. and Simon, D. L. (1999). Markov chain Monte Carlo algorithms for the Bayesian analysis of phylogenetic trees. *Molecular Biology and Evolution* **16**, 750–759.

Lari, K. and Young, S. J. (1990). The estimation of stochastic context-free grammars using the inside-outside algorithm. *Computer Speech and Language* **4**, 35–56.

Laquer, H. T. (1981). Asymptotic limits for a two-dimensional recursion. *Studies in Applied Mathematics* **64**, 271–277.

Larson, H. J. (1982). *Introduction to Probability Theory and Statistical Inference*, 3rd edn (New York: Wiley).

Lawrence, C. E., Altschul, S. F., Boguski, M. S., Liu, J. S., Neuwald, A. F., and Wootton, J. C. (1993). Detecting subtle sequence signals: A Gibbs sampling strategy for multiple alignment. *Science* **262**, 208–214.

Lewis, P. O. (2001). Phylogenetic systematics turns over a new leaf. *Trends in Ecology and Evolution* **16**, 30–37.

Li, M., Badger, J. H., Chen, X., Kwong, S., Kearney, P., and Zhang, H. (2001). An information-based sequence distance and its application to whole mitochondrial genome phylogeny. *Bioinformatics* **17**, 149–154.

Lipman, D. J., Wilbur, W. J., Smith, T. F., and Waterman, M. S. (1984). On the statistical significance of nucleic acid similarities. *Nucleic Acids Research* **12**, 215–226.

Liu, J. S. (2001). *Monte Carlo Strategies in Scientific Computing* (New York: Springer-Verlag).

Liu, J. S. and Lawrence, C. E. (1999). Bayesian inference on biopolymer models. *Bioinformatics* **15**, 38–52.

Liu, J. S., Neuwald, A. F., and Lawrence, C. E. (1999). Markovian structures in biological sequence alignments. *Journal of American Statistical Association* **94**, 1–15.

Löytynoja, A. and Milinkovitch, C. (2003). A hidden Markov model for progressive multiple alignment. *Bioinformatics* **19**, 1505–1513.

Lukashin, A. V. and Borodovsky, M. (1998). GeneMark.hmm: New solutions for gene finding. *Nucleic Acids Research* **26**, 1107–1115.

Lyngsø, R. B., Pedersen, C. N. S., and Nielsen, H. (1999). Metrics and similarity measures for hidden Markov models. *Proceedings of International Conference in Intelligent Systems for Molecular Biology* (Menlo Park, CA: AAAI Press), pp. 178–186.

MacKay, D. J. C. (2003). *Information Theory, Inference, and Learning Algorithms* (Cambridge: Cambridge University Press).

Maidak, B. L., Cole, J. R., Lilburn, T. G. *et al.* (2000). The RDP (Ribosomal Database Project) continues. *Nucleic Acids Research* **28**, 173–174.

Martí-Renom, M. A., Stuart, A. C., Fiser, A., Sánchez, R., Melo, F., and Šali, A. (2000). Comparative protein structure modeling of genes and genomes. *Annual Review of Biophysics and Biomolecular Structure* **29**, 291–325.

Mathews, D. H., Sabina, J., Zuker, M., and Turner, D. H. (1999). Expanding sequence dependence of thermodynamic parameters improves prediction of RNA secondary structure. *Journal of Molecular Biology* **288**, 911–940.

Mathews, D. H., Disney, M. D., Childs, J. L., Schroeder, S. J., Zuker, M., and Turner, D. H. (2004). Incorporating chemical modification constraints into a dynamic programming algorithm for prediction of RNA secondary structure. *Proceedings of the National Academy of Sciences of the USA* **101**, 7287–7292.

Mau, B., Newton, M. A., and Larget, B. (1999). Bayesian phylogenetic inference via Markov chain Monte Carlo methods. *Biometrics* **55**, 1–12.

Meyer, C. D. (2000). *Matrix Analysis and Applied Linear Algebra* (Philadelphia, PA: Society for Industrial and Applied Mathematics).

Meyer, I. M. and Durbin, R. (2002). Comparative *ab initio* prediction of gene structure using pair HMM. *Bioinformatics* **18**, 1309–1318.

Meyer, I. M. and Durbin, R. (2004). Gene structure conservation aids similarity based gene prediction. *Nucleic Acids Research* **32**, 776–783.

Meyer, P. L. (1970). *Introductory Probability and Statistical Applications*, 2nd edn (Reading, MA: Addison-Wesley).

Moon, J. W. (1970). *Counting Labelled Trees.* Canadian Mathematical Monographs (London and Beccles: William Clowes and Sons Ltd).

Morgenstern, B. (1999). DIALIGN 2: Improvement of the segment-to-segment approach to multiple sequence alignment. *Bioinformatics* **15**, 211–218.

Morgenstern, B., Frech, K., Dress, A., and Werner, T. (1998). DIALIGN: Finding local similarities by multiple sequence alignment. *Bioinformatics* **14**, 290–294.

Mott, R. (1999). Local sequence alignments with monotonic gap penalties. *Bioinformatics* **15**, 455–462.

Mott, R. (2000). Accurate formula for P-values of gapped local sequence and profile alignments. *Journal of Molecular Biology* **300**, 649–659.

Mott, R. and Tribe, R. (1999). Approximate statistics of gapped alignments. *Journal of Computational Biology* **6**, 91–112.

Motwani, R., Ullman, J. D., and Hopcroft, J. E. (2003). *Introduction to Automata Theory, Languages, and Computation*, 2nd edn (Upper Saddle River, N.J.: Pearson Education).

Muse, S. V. and Gaut, B. S. (1994). A likelihood approach for comparing synonymous and nonsynonymous nucleotide substitution rates, with application to the chloroplast genome. *Molecular Biology and Evolution* **11**, 715–724.

Myers, E. W. and Miller, W. (1988). Optimal alignments in linear space. *Computer Applications in the Biosciences* **4**, 11–17.

Nei, M., Chakraborty, R., and Fuerst, P. A. (1976). Infinite allele model with varying mutation rate. *Proceedings of the National Academy of Sciences of the USA* **73**, 4164–4168.

Nei, M., Xu, P., and Glazko, G. (2001). Estimation of divergence times from multiprotein sequences for a few mammalian species and several distantly related organisms. *Proceedings of the National Academy of Sciences of the USA* **98**, 2497–2502.

Neuhauser, C. (1994). A Poisson approximation for sequence comparisons with insertions and deletions. *The Annals of Statistics* **22**, 1603–1629.

Notredame, C., Higgins, D. G., and Heringa J. (2000). T-Coffee: A novel method for fast and accurate multiple sequence alignment. *Journal of Molecular Biology* **302**, 205–217.

Nussinov, R., Pieczenik, G., Griggs, J. R., and Kleitman, D. J. (1978). Algorithms for loop matchings. *SIAM Journal of Applied Mathematics* **35**, 68–82.

Pachter, L., Alexandersson, M., and Cawley, S. (2002). Applications of generalized pair hidden Markov models to alignment and gene finding problems. *Journal of Computational Biology* **9**, 389–399.

Park, J., Teichmann, S. A., Hubbard, T., and Chothia, C. (1997). Intermediate sequences increase the detection of homology between sequences. *Journal of Molecular Biology* **273**, 349–354.

Park, J., Karplus, K., Barrett, C. *et al.* (1998). Sequence comparisons using multiple sequences detect three times as many remote homologues as pairwise methods. *Journal of Molecular Biology* **284**, 1201–1210.

Pearson, W. R. (1995). Comparison of methods for searching protein sequence databases. *Protein Science* **4**, 1145–1160.

Pearson, W. R. (1996). Effective protein sequence comparison. *Methods in Enzymology* **266**, 227–258.

Perriquet, O., Touzet, H., and Dauchet, M. (2003). Finding the common structure shared by two homologous RNAs. *Bioinformatics* **19**, 108–116.

Pevzner, P. A., Borodovsky, M. Yu., and Mironov, A. A. (1989). Linguistics of nucleotide sequences I: The significance of deviations from mean statistical characteristics and prediction of the frequencies of occurrence of words. *Journal of Biomolecular Structure and Dynamics* **5**, 1013–1026.

Posada, D. and Crandall, K. A. (2001). Selecting the best-fit model of nucleotide substitution. *Systematic Biology* **50**, 580–601.

Prüfer, H. (1918). Neuer Beweis eines Satzes über Pemutationen. *Archiv für Mathematik und Physik* **27**, 142–144.

Qi, J., Wang, B., and Hao, B. I. (2004). Whole proteome prokaryote phylogeny without sequence alignment: K-string composition approach. *Journal of Molecular Evolution* **58**, 1–11.

Reese, J. T. and Pearson, W. R. (2002). Empirical determination of effective gap penalties for sequence comparison. *Bioinformatics* **18**, 1500–1507.

Reich, J. G., Drabsch, H., and Däumler, A. (1984). On the statistical assessment of similarities in DNA sequences. *Nucleic Acids Research* **12**, 5529–5543.

Reinert, G., Schbath, S., and Waterman, M. S. (2000a). Probabilistic and statistical properties of words: An overview. *Journal of Computational Biology* **7**, 1–46.

Reinert, K., Stoye, J., and Will, T. (2000). An iterative method for faster sum-of-pairs multiple sequence alignment. *Bioinformatics* **16**, 808–814.

Reza, F. M. (1994). *An Introduction to Information Theory* (New York: Dover Publications).

Rivas, E. and Eddy, S. R. (1999). A dynamic programming algorithm for RNA structure prediction including pseudoknots. *Journal of Molecular Biology* **285**, 2053–2068.

Robin, S. and Schbath, S. (2001). Numerical comparison of several approximations of the word count distribution in random sequences. *Journal of Computational Biology* **8**, 349–359.

Ross, S. M. (1996). *Stochastic Processes*, 2nd edn (New York: John Wiley & Sons, Inc.).

Rost, B. (1999). Twilight zone of protein sequence alignments. *Protein Engineering* **12**, 85–94.

Rychlewski, L., Jaroszewski, L., Li, W., and Godzik, A. (2000). Comparison of sequence profiles. Strategies for structural predictions using sequence information. *Protein Science* **9**, 232–241.

Saitou, N. and Nei, M. (1987). The neighbor-joining method: A new method for reconstructing phylogenetic trees. *Molecular Biology and Evolution* **4**, 406–425.

Salomaa, A., Wood, D., and Yu, S., eds (2001). *A Half-Century of Automata Theory: Celebration and Inspiration* (Singapore: World Scientific).

Salzberg, S. L., Delcher, A. L., Kasif, S., and White, O. (1998). Microbial gene identification using interpolated Markov models. *Nucleic Acids Research* **26**, 544–548.

Sankoff, D. and Cedergren, R. J. (1983). Simultaneous comparison of three or more sequences related by a tree. In Sankoff, D. and Kruskal, J. B., eds, *Time Warps, String Edits, and Macromolecules: The Theory and Practice of Sequence Comparison* (Reading, MA: Addison-Wesley), Chap. 9, pp. 253–264.

Schäffer, A. A., Aravind, L., Madden, T. L. *et al.* (2001). Improving the accuracy of PSI-BLAST protein database searches with composition-based statistics and other refinements. *Nucleic Acids Research* **29**, 2994–3005.

Schbath, S. (2000). An overview on the distribution of word counts in Markov chains. *Journal of Computational Biology* **7**, 193–201.

Schmidler, S. C., Liu, J. S., and Brutlag, D. L. (2000). Bayesian segmentation of protein secondary structure. *Journal of Computational Biology* **7**, 233–248.

Schmidt, H. A., Strimmer, K., Vingron, M., and Haeseler, A. von (2002). TREE-PUZZLE: Maximum-likelihood phylogenetic analysis using quartets and parallel computing. *Bioinformatics* **18**, 502–504.

Schneider, T. D., Stormo, G. D., Gold, L., and Ehrenfeucht, A. (1986). Information content of binding sites on nucleotide sequences. *Journal of Molecular Biology* **188**, 415–431.

Schuler, G. D., Altschul, S. F., and Lipman, D. J. (1991). A workbench for multiple alignment construction and analysis. *Proteins: Structure, Function, and Genetics* **9**, 180–190.

Schwartz, S., Zhang, Z., Frazer, K. A. *et al.* (2000). PipMaker – a web server for aligning two genomic DNA sequences. *Genome Research* **10**, 577–586.

Shannon, C. E. and Weaver, W. (1963). *The Mathematical Theory of Communication* (Urbana-Champaign: University of Illinois Press).

Shimodaira, H. (2002). An approximately unbiased test of phylogenetic tree selection. *Systematic Biology* **51**, 492–508.

Shimodaira, H. and Hasegawa, M. (1999). Multiple comparisons of log-likelihoods with applications to phylogenetic inference. *Molecular Biology and Evolution* **16**, 1114–1116.

Shimodaira, H. and Hasegawa, M. (2001). CONSEL: For assessing the confidence of phylogenetic tree selection. *Bioinformatics* **17**, 1246–1247.

Shindyalov, I. N. and Bourne, P. E. (1998). Protein structure alignment by incremental combinatorial extension (CE) of the optimal path. *Protein Engineering* **11**, 739–747.

Shiryaev, A. N. (1996). *Probability*, 2nd edn (New York: Springer-Verlag).

Siegmund, D. and Yakir, B. (2000). Approximate P-values for local sequence alignments. *The Annals of Statistics* **28**, 657–680.

Siegmund, D. and Yakir, B. (2003). Correction: Approximate P-values for local sequence alignments. *The Annals of Statistics* **31**, 1027–1031.

Simon, M. (1999). *Automata Theory* (Singapore: World Scientific).

Smith, T. F. and Waterman, M. S. (1981). Identification of common molecular subsequences. *Journal of Molecular Biology* **147**, 195–197.

Smith, T. F., Waterman, M. S., and Burks, C. (1985). The statistical distribution of nucleic acid similarities. *Nucleic Acids Research* **13**, 645–656.

Snel, B., Bork, P., and Huynen, M. A. (1999). Genome phylogeny based on gene content. *Nature Genetics* **21**, 108–110.

Snel, B., Huynen, M. A., and Dutilh, B. E. (2005). Genome trees and the nature of genome evolution. *Annual Reviews in Microbiology* **59**, 191–209.

Sokal, R. R. and Michener, C. D. (1958). A statistical method for evaluating systematic relationships. *University of Kansas Scientific Bulletin* **28**, 1409–1438.

Sonnhammer, E. L. L., Eddy, S. R., Birney, E., Bateman, A., and Durbin, R. (1998). Pfam: Multiple sequence alignments and HMM-profiles of protein domains. *Nucleic Acids Research* **26**, 320–322.

Steel, M., Hendy, M. D., and Penny, D. (1998). Reconstructing phylogenies from nucleotide pattern probabilities: A survey and some new results. *Discrete Applied Mathematics* **88**, 367–396.

Strimmer, K. and Haeseler, A. von (1996). Quartet puzzling: A quartet maximum-likelihood method for reconstructing tree topologies. *Molecular Biology and Evolution* **13**, 964–969.

Suzuki, Y., Glazko, G. V., and Nei, M. (2002). Overcredibility of molecular phylogenies obtained by Bayesian phylogenetics. *Proceedings of the National Academy of Sciences of the USA* **99**, 16 138–16 143.

Székely, L. A., Steel, M. A., and Erdös, P. L. (1993). Fourier calculus on evolutionary trees. *Advances in Applied Mathematics* **14**, 200–216.

Tabaska, J. E., Cary, R. B., Gabow, H. N., and Stormo, G. D. (1998). An RNA folding method capable of identifying pseudoknots and base triples. *Bioinformatics* **14**, 691–699.

Tatusov, R. L., Galperin, M. Y., Natale, D.A., and Koonin, E. V. (2000). The COG database: A tool for genome-scale analysis of protein functions and evolution. *Nucleic Acids Research* **28**, 33–36.

Tavaré, S. and Song, B. (1989). Codon preference and primary sequence structure in protein coding regions. *Bulletin of Mathematical Biology* **51**, 95–115.

Tekaia, F., Lazcano, A., and Dujon, B. (1999). The genomic tree as revealed from whole proteome comparisons. *Genome Research* **9**, 550–557.

Thompson, J. D., Higgins, D. G., and Gibson, T. J. (1994a). Improved sensitivity of profile searches through the use of sequence weights and gap excision. *Computer Applications in the Biosciences* **10**, 19–29.

Thompson, J. D., Higgins, D. G., and Gibson, T. J. (1994b). CLUSTAL W: Improving the sensitivity of progressive multiple sequence alignment through sequence weighting, position specific gap penalties and weight matrix choice. *Nucleic Acids Research* **22**, 4673–4680.

Thompson, J. D., Plewniak, F., and Poch, O. (1999a). A comprehensive comparison of multiple sequence alignment programs. *Nucleic Acids Research* **27**, 2682–2690.

Thompson, J. D., Plewniak, F., and Poch, O. (1999b). BAliBASE: A benchmark alignment database for the evaluation of multiple alignment programs. *Bioinformatics* **15**, 87–88.

Thompson, J. D., Plewniak, F., Ripp, R., Thierry, J-C., and Poch, O. (2001). Towards a reliable objective function for multiple sequence alignments. *Journal of Molecular Biology* **314**, 937–951.

Thorne, J. L., Kishino, H., and Felsenstein, J. (1991). An evolutionary model for maximum likelihood alignment of DNA sequences. *Journal of Molecular Evolution* **33**, 114–124.

Thorne, J. L., Kishino, H., and Felsenstein, J. (1992). Inching toward reality: An improved likelihood model of sequence evolution. *Journal of Molecular Evolution* **34**, 3–16.

Tönges, U., Perrey, S. W., Stoye, J., and Dress, A. W. M. (1996). A general method for fast multiple sequence alignment. *Gene* **172**, GC33–GC41.

Tusnády, G. E. and Simon, I. (1998). Principles governing amino acid composition of integral membrane proteins: Application to topology prediction. *Journal of Molecular Biology* **283**, 489–506.

Vingron, M. and Waterman, M. S. (1994). Sequence alignment and penalty choice: Review of concepts, case studies and implications. *Journal of Molecular Biology* **235**, 1–12.

Vinh, L. S. and Haeseler, A. von (2004). IQPNNI: Moving fast through tree space and stopping in time. *Molecular Biology and Evolution* **21**, 1565–1571.

Waterman, M. S. (1995). *Introduction to Computational Biology* (New York: Chapman and Hall).

Waterman, M. S. and Vingron, M. (1994). Rapid and accurate estimates of statistical significance for sequence data base searches. *Proceedings of the National Academy of Sciences of the USA* **91**, 4625–4628.

Webb, B-J. M., Liu, J. S., and Lawrence, C. E. (2002). BALSA: Bayesian algorithm for local sequence alignment. *Nucleic Acids Research* **30**, 1268–1277.

Webber, C. and Barton, G. J. (2001). Estimation of P-values for global alignments of protein sequences. *Bioinformatics* **17**, 1158–1167.

Whelan, S. and Goldman, N. (2001). A general empirical model of protein evolution derived from multiple protein families using a maximum-likelihood approach. *Molecular Biology and Evolution* **18**, 691–699.

Whelan, S., Liò, P., and Goldman, N. (2001). Molecular phylogenetics: State-of-the-art methods for looking into the past. *Trends in Genetics* **17**, 262–272.

Wilbur, W. J. (1985). On the PAM matrix model of protein evolution. *Molecular Biology and Evolution* **2**, 434–447.

Wolf, Y. I., Rogozin, I. B., and Koonin E. V. (2004). Coelomata and not Ecdysozoa: Evidence from genome-wide phylogenetic analysis. *Genome Research* **14**, 29–36.

Wuyts, J., De Rijk, P., Peer, Y. Van de, Winkelmans, T., and De Wachter, R. (2001). The European Large Subunit Ribosomal RNA database. *Nucleic Acids Research* **29**, 175–177.

Yang, Z. (1998). Likelihood ratio tests for detecting positive selection and application to primate lysozyme evolution. *Molecular Biology and Evolution* **15**, 568–573.

Yang, Z. and Bielawski, J. P. (2000). Statistical methods for detecting molecular adaptation. *Tree* **15**, 496–503.

Yang, Z. and Nielsen, R. (2000). Estimating synonymous and nonsynonymous substitution rates under realistic evolutionary models. *Molecular Biology and Evolution* **17**, 32–43.

Yang, Z. and Rannala, B. (1997). Bayesian phylogenetic inference using DNA sequences: A Markov chain Monte Carlo method. *Molecular Biology and Evolution* **14**, 717–724.

Yang, Z., Nielsen, R., Goldman, N., and Pedersen, A-M. K. (2000). Codon-substitution models for heterogeneous selection pressure at amino acid sites. *Genetics* **155**, 431–449.

Younger, D. H. (1967). Recognition and parsing of context-free languages in time n^3. *Information and Control* **10**, 189–208.

Zhu, J., Liu, J. S., and Lawrence, C. E. (1998). Bayesian adaptive sequence alignment algorithms. *Bioinformatics* **14**, 25–39.

Zuker, M. (2000). Calculating nucleic acid secondary structure. *Current Opinion in Structural Biology* **10**, 303–310.

Zuker, M. and Stiegler, P. (1981). Optimal computer folding of large RNA sequences using thermodynamic and auxiliary information. *Nucleic Acids Research* **9**, 133–148.

Zwieb, C., Gorodkin, J., Knudsen, B., Burks, J., and Wower, J. (2003). tmRDB (tmRNA database). *Nucleic Acids Research* **31**, 446–447.

Index